BEGINNER'S GUIDE TO HYDROCOLLOIDS

BEGINNER'S GUIDE TO HYDROCOLLOIDS

Understanding Food Thickening, Emulsification, Stabilizing and Jellification

Beginner's Guide to Hydrocolloids

Understanding Food Thickening, Emulsification, Stabilizing and Jellification

Editor: Edmund J. McCormick, Jr.
Technical Editor: Dr. Mohammad Abdullah Sandhu

TEACHINGPRESS

Published by Teaching Press
Summit, New Jersey 07901

Hardcover: ISBN 978-1-7361391-5-8
Paperback: ISBN 978-1-7361391-4-1
Electronic Edition: ISBN 978-1-7361391-6-5
Library of Congress Control Number: 2023923179

First paperback edition January 2024.
Edited by: Edmund J. McCormick, Jr.
Technical Editor: Dr. Muhammad Abdullah Sandhu

Cover by: Farhan Shahid
Layout and Design by: Farhan Shahid
Art and Infographics: Rai
Photos by Dreamstime.com

Printed in the USA

TEACHINGPRESS

Teaching Press
18 Bank Street / Suite 1
Summit, NJ 07901

DEDICATION

I am especially grateful for the help of Dr. Sandhu whose advice and eagle eye were the critical success factor in creating this guide. I also want to that Gary and Andrew Wine of AEP Colloids for their helpful suggestions. It's also a pleasure by way of thanking the guide's well executed format, designer Farhan Shahid. I think you will agree that its modern layout makes a more complicated subject a bit easier to comprehend. Speaking of comprehending I need to add a special thank you to our infographic artist, Rai. Most of all, I owe my greatest gratitude to my wife who allowed me to work the many hours it took to put this book together without complaint and with unselfish encouragement. Thank you all.

Edmund McCormick, Jr.
Summit, New Jersey

FOREWORD

To be written by Andrew Wine of ADP Colloids

When you step into the world of hydrocolloids – or gums as I have come to call them – there is a fascinating world of possibilities laid out before you on any given day.

We use these building blocks to connect the world, to bring folks closer together through greater access to food, pharmaceutical, and technical applications.

This industry reminds me a lot about rock formations in the American West– erosion presents these different colors and textures of rock. They look different, feel different, behave differently, but are all connected through geological history.

Food science – particularly in hydrocolloids – is like this. You end up with this finished product found on the shelf, but there are layers and stories within it. There are components brought together through the natural world or science, that interact with each other at a molecular level to create the tastes, the textures, and the aromas we enjoy at our kitchen tables.

It is a fascinating corner of the supply chain to be in. In fact, quite a few of the folks we work with do not have food science backgrounds at all. They come from purchasing or sales or management roles elsewhere. I am a product of that too; food science is a family business, and one I came to as a second career. Most learn on the job – and having a resource like this Beginners Guide to Hydrocolloids brings the fundamentals of this world right to you.

This guide will open the world of hydrocolloids and equip you with the tools to elevate your culinary experience. You do not need to be a classically trained chemist to understand gums. With all the information available across academic resources and online, having this curated physical resource to reference is invaluable. It helps to form an agency for someone looking to improve their cooking skills and find creative ways to eat well. It tightens the knowledge gap.

There is a complimenting benefit to this guide in having experts you can lean on to point you towards the right solutions or help supply the products you need to bring a dish to life. That is why AEP Colloids is proud to work with companies like Cape Crystal Brands to broaden the visibility of gums, to work with individuals, and with companies of any size, and industry.

There has never been a day that has gone by where I did not learn something new. That ethos has motivated me to ask questions and unpack the layers of this corner of food science – both at work and at home in the kitchen. This Beginners Guide to Hydrocolloids has taken me on a journey, answering some questions I had not thought to ask. It will certainly challenge you to create outside the box too.

We are glad you're here with us.

Andy Wine

V.P. Operations

AEP COLLOIDS, div. of SARCOM, INC.

HADLEY, NEW YORK

aep COLLOIDS DIVISION OF SARCOM INC.

INTRODUCTION

Welcome to "The Beginner's Guide to Hydrocolloids," a comprehensive and accessible journey into the intriguing world of hydrocolloids. This book is designed to be an essential resource for anyone keen to delve into the realm of these versatile substances, which play a pivotal role in our daily lives, often without our awareness. Whether you are a budding chef, a curious home cook, a DIY enthusiast, or a professional in the food industry, this guide is tailored to demystify the science behind hydrocolloids and illuminate their myriad applications.

Hydrocolloids, encompassing food thickeners, emulsifiers, stabilizers, and jellifiers, are not just modern culinary innovations. Their history is as rich and varied as their uses, dating back to ancient times when they were employed in both food and medicine. This book begins by exploring this historical context, providing a fascinating backdrop to our current understanding and use of hydrocolloids.

As we delve into the various types of hydrocolloids, both natural and synthetic, you will encounter familiar names like agar, alginate, carrageenan, gelatin, pectin, and xanthan gum. Each of these has unique properties and applications, which are thoroughly examined in subsequent chapters. Understanding these properties - such as viscosity, gelation, solubility, hydration, and temperature sensitivity - is crucial for appreciating how hydrocolloids can be effectively utilized in different contexts.

The applications of hydrocolloids are diverse and span multiple industries. In the food industry, they are indispensable for thickening, gelling, stabilizing, and emulsifying a wide range of products. But their utility extends far beyond the culinary realm. In medicine, hydrocolloids are found in wound dressings and drug delivery systems. The cosmetic industry also leverages their unique properties in products like lotions, creams, and shampoos. This book provides a detailed exploration of these applications, offering insights into how hydrocolloids are shaping various industries.

For those with a culinary bent, "The Beginner's Guide to Hydrocolloids" offers a special treat. The book delves into the role of hydrocolloids in modernist cuisine, a field where culinary art meets science. You will discover recipes and techniques for using specific hydrocolloids to enhance texture and mouthfeel in food, opening new horizons in your culinary adventures.

Moreover, the guide addresses the health and nutritional aspects of hydrocolloids. It discusses their digestibility, metabolism, and health benefits, such as their role as dietary fiber and in promoting gut health. The book also sheds light on potential side effects or concerns, providing a balanced view of these substances.

"The Beginner's Guide to Hydrocolloids" is more than just a textbook; it is a journey into the heart of everyday science and its practical applications. It aims to equip you with the knowledge and confidence to experiment with hydrocolloids, whether in the kitchen, the lab, or in industrial applications. By the end of this book, you will not only understand the science behind these fascinating substances but also how to harness their properties effectively in various contexts.

Embark on this journey with us and discover the versatile and exciting world of hydrocolloids. Welcome to "The Beginner's Guide to Hydrocolloids."

Best regards,
Edmund McCormick, Jr.

Table of

CONTENTS

CHAPTER

01

Introduction to Hydrocolloids

PART 1.1

What are Hydrocolloids?

Hydrocolloids are natural and functional ingredients commonly used in the food industry for their stabilizing, thickening, gelling, emulsifying, and texture-modifying properties. They are often derived from plant, animal, and microbial sources and are essential for food formulation across various sectors.

Core Insights

- ☑ Hydrocolloids are natural and functional ingredients.
- ☑ Used in the food industry for stabilizing, thickening, gelling, emulsifying, and texture-modifying properties.
- ☑ These are often derived from plant, animal, and microbial sources.
- ☑ These are essential components of food formulation across various sectors.
- ☑ It is crucial in improving food products' texture, stability, and shelf life.

1.1.1 Understanding Hydrocolloids

Hydrocolloids are polysaccharides and proteins that can form gels or solutions in water. They are widely used in food and other industries due to their unique properties. Hydrocolloids have several functional properties, such as thickening, stabilizing, gelling, emulsifying, and modifying texture. Polymeric molecules with hydrophilic and hydrophobic properties are the basis for understanding hydrocolloids. Their hydrophilic properties allow them to absorb water, whereas their hydrophobic properties help them to interact with other ingredients.

Hydrocolloids' unique properties make them attractive in food formulation, allowing manufacturers to enhance product texture, taste, and consistency. Other than food, hydrocolloids are used in the pharmaceutical, personal care, textile, and other industries due to their functional properties.

1.1.2 Properties of Hydrocolloids

The properties of hydrocolloids depend on their molecular structures. Hydrocolloids like starch, cellulose, and pectin are polysaccharides, whereas proteins like gelatin and casein are protein-based. The functionality of hydrocolloids depends on their solubility, viscosity, gel strength, and interaction with other ingredients.

Hydrocolloids' hydrophilic and hydrophobic properties are determined by their molecular weight, degree of branching, and level of cross-linking. For example, high-molecular-weight hydrocolloids like xanthan gum and carrageenan have a high degree of branching, which results in a high viscosity and gel strength.

1.1.3 Functionality of Hydrocolloids

The functionality of hydrocolloids is related to their ability to interact with water and other ingredients. Hydrocolloids can absorb water to form a stable gel or solution or bind to other ingredients to form a stable emulsion or suspension. Their functionality depends on several factors, such as pH, temperature, concentration, and the presence of other ingredients.

Its functionality is essential for food and other industries, where they are used as thickening agents, stabilizers, gelling agents, emulsifiers, and texture modifiers. For example, guar gum is a thickener in dairy products, whereas pectin is a gelling agent in fruit jams and jellies. Xanthan gum and carrageenan are stabilizers in salad dressings, sauces, and dairy products.

1.1.4 Common Types of Hydrocolloids

Hydrocolloids are widely used in various industries, including food, pharmaceuticals, and personal care. Here are some of the most common types of hydrocolloids:

Table 1: Types of Hydrocolloids

TYPE	EXAMPLES
Carrageenan	Iota, kappa, and lambda carrageenan
Agar	Agar-agar
Pectin	High methoxyl (HM) and low methoxyl (LM) pectin
Guar gum	Guar gum
Xanthan gum	Xanthan gum
Gelatin	Gelatin

Carrageenan is extracted from red seaweed and is commonly used as a gelling agent and stabilizer in dairy products and desserts. Agar, another seaweed extract, is used as a gelling agent in confectionery and microbiology. Pectin derived from fruit is used as a thickener and gelling agent in jams and jellies. Guar gum is a natural thickener in various food products, including sauces and dressings. Xanthan gum is commonly used as a thickener and stabilizer in gluten-free products and salad dressings. Gelatin, derived from animal collagen, is a versatile gelling agent commonly used in gummy candies and desserts.

1.1.5 Hydrocolloids in the Food Industry

Hydrocolloids are vital ingredients in the food industry due to their diverse applications and benefits. They are widely used as stabilizers, thickening agents, gelling agents, emulsifiers, and texture modifiers in various food formulations.

Hydrocolloids as Food Stabilizers

Hydrocolloids are widely used in the food industry as stabilizers due to their unique functionality. They help maintain the desired consistency, prevent ingredient separation, and enhance the stability of food products.

Hydrocolloids form a network of molecules that trap water and other ingredients in a stable structure. This network prevents the movement of particles and helps maintain the desired texture of the product over time.

Food stabilizers are important in the food industry, especially for products that have a long shelf life or are subjected to harsh environments during processing, transportation, and storage. Hydrocolloids, being natural and functional ingredients, are preferred over synthetic alternatives.

Additionally, hydrocolloids can provide other functionalities beyond stabilizing. For example, xanthan gum, a commonly used hydrocolloid, is also a thickening agent and can improve the texture of food products.

Hydrocolloids as Thickening Agents

Hydrocolloids are widely used as thickening agents in the food industry because they increase the viscosity of food products. The thickening effect is achieved by forming a network of long-chain molecules that trap water and other ingredients, increasing thickness and consistency.

They can thicken various food products, including sauces, dressings, and desserts. They can be added in small quantities and still achieve the desired thickness, making them a cost-effective solution for food formulation.

The choice of hydrocolloid used as a thickening agent will depend on the specific application and desired texture. For example, xanthan gum is commonly used to create a smooth and creamy texture in dressings and sauces, while guar gum is preferred to thicken and stabilize dairy products.

Hydrocolloids as thickening agents can also improve the mouthfeel of food products, making them more enjoyable to consume. Additionally, they can help prevent ingredient separation and provide better stability over extended periods, enhancing the shelf life of food products.

Hydrocolloids as Gelling Agents

Hydrocolloids are frequently used as gelling agents in food products. When added to a liquid, hydrocolloids can form a gel structure that provides stability, texture, and functionality to various food formulations.

The gel formation process involves hydrocolloids' hydration, which causes them to swell and form a network of interlocking chains. This network can trap liquid and air to form a stable gel resistant to syneresis (liquid separation) and maintain its shape and texture throughout storage and handling.

"Hydrocolloids can provide unique textures and sensory experiences through their ability to form stable and versatile gels in food products."

Carrageenan, agar, and gelatin are commonly used as gelling agents in the food industry. Carrageenan, derived from seaweed, is often used in dairy products to provide a smooth and creamy texture. Agar, extracted from red algae, is used in desserts and confectionery products to form a firm and cohesive gel. Gelatin, sourced from animal collagen, is commonly used to produce gummy candies and jellied desserts.

Hydrocolloids as Emulsifiers

One of the key functions of hydrocolloids in the food industry is their ability to act as emulsifiers. Emulsions are mixtures of two or more immiscible liquids, such as oil and water, that an emulsifying agent stabilizes. Hydrocolloids, such as xanthan gum, guar gum, and carrageenan, are commonly used as emulsifiers due to their ability to promote emulsion formation and stability.

The emulsification process reduces the interfacial tension between the two liquids by creating small droplets of one liquid dispersed in the other. Hydrocolloids help to stabilize the resulting emulsion by forming a film around the droplets, preventing them from coalescing or separating. This improves the texture and appearance of the final product, as well as extending its shelf life.

Hydrocolloids are useful in preparing salad dressings, mayonnaise, and other sauces requiring stable emulsions. They are also used in producing dairy products, such as ice cream and whipped cream, to provide a smooth and creamy texture.

Hydrocolloids as Texture Modifiers

Hydrocolloids are widely used to enhance the texture of food products. As texture modifiers, hydrocolloids can alter various sensory properties, such as mouthfeel, creaminess, or smoothness. They can also prevent undesirable textural changes during processing or storage, such as ice crystal formation or syneresis.

The effectiveness of hydrocolloids as texture modifiers is due to their ability to interact with water and other ingredients in a food product. These interactions can change the food's viscosity, elasticity, or physical properties.

One example of a commonly used hydrocolloid as a texture modifier is xanthan gum. Xanthan gum is known for increasing viscosity and maintaining stability in various food applications. It is often used in dressings, sauces, and bakery products to provide a smooth and creamy texture.

Another hydrocolloid used as a texture modifier is carrageenan, which is extracted from red seaweed. Carrageenan is particularly effective in dairy products, enhancing texture and providing a pleasant mouthfeel. It is often used in ice cream, yogurt, and other dairy desserts.

Other hydrocolloids used as texture modifiers include guar gum, locust bean gum, and pectin. These hydrocolloids can provide unique textures and enhance the overall quality of various food products.

1.1.6 Hydrocolloids as Natural and Functional Ingredients

Hydrocolloids have recently gained popularity as natural and functional ingredients in the food industry. Derived from natural sources such as seaweed, plant exudates, and microbial fermentation, hydrocolloids offer a range of functional properties to food products.

One of the main benefits of using hydrocolloids as natural ingredients is their ability to enhance the nutritional profile of food products while maintaining their texture and sensory properties. Hydrocolloids can be used to increase the fiber content of foods, reduce fat content, and improve the overall quality of the product.

Hydrocolloids also provide functional benefits to food products, such as improving stability, texture, and mouthfeel. They can act as thickeners, gelling agents, emulsifiers, and stabilizers, contributing to the desired texture and consistency of the final product.

Furthermore, hydrocolloids offer advantages over synthetic alternatives, such as being biodegradable and environmentally friendly. They are also less likely to cause adverse health effects than synthetic food additives.

The use of hydrocolloids as natural and functional ingredients in food products has become increasingly popular due to their potential to enhance the nutritional profile and functionality of foods while maintaining their natural properties.

1.1.7 Hydrocolloids in Other Industries

Hydrocolloids are not limited to the food industry. They have numerous applications outside of the culinary world.

In tablets and capsules, the pharmaceutical industry uses hydrocolloids as binders, disintegrants, and controlled-release agents. They are also used as absorbents and gelling agents in wound dressing manufacturing.

Hydrocolloids also play an essential role in the personal care industry. They can be found in shampoos, lotions, and toothpaste, which are thickeners and stabilizers.

One unexpected use of hydrocolloids is in the textile industry. They are utilized as sizing agents to improve the strength and quality of fibers during the weaving process.

Overall, hydrocolloids are versatile and widely applicable in various industries. Their benefits include improved texture, stability, and product performance, making them an attractive option for many manufacturing processes.

Conclusion

In conclusion, hydrocolloids are vital in various industries, especially food. Hydrocolloids are natural and functional ingredients that provide food products with improved stability, texture, and shelf life. The properties of hydrocolloids, such as their molecular structure, allow them to perform their desired functions as food stabilizers, thickening agents, gelling agents, emulsifiers, and texture modifiers. Their ability to interact with water and other ingredients to create stable gels, emulsions, and desired textures makes them indispensable to food formulation.

Beyond the food industry, hydrocolloids have potential applications in other industries, such as pharmaceuticals, personal care products, and textiles. As the demand for natural and functional ingredients grows, hydrocolloids offer a promising alternative to synthetic alternatives. In summary, the versatility and importance of hydrocolloids in various industries cannot be overstated. Further research and innovation in hydrocolloid applications are required to unlock their full potential and meet the ever-evolving demands of consumers.

FAQ

What are hydrocolloids?

Hydrocolloids are food stabilizers, thickening agents, gelling agents, emulsifiers, and texture modifiers commonly used in the food industry. They are natural ingredients with functional properties that help improve the texture, stability, and overall quality of food products.

How do hydrocolloids work?

Hydrocolloids interact with water and other ingredients to perform their desired functions. Their molecular structure allows them to absorb water and form a gel-like substance, contributing to their thickening, gelling, and stabilizing properties in food formulations.

What are some common types of hydrocolloids?

Some common types of hydrocolloids include carrageenan, agar, pectin, guar gum, xanthan gum, and gelatin. Each hydrocolloid has specific applications in the food industry and offers unique texture, stability, and viscosity functionalities.

How are hydrocolloids used in the food industry?

Hydrocolloids are widely used in the food industry to improve products' texture, stability, and shelf life. They are added to various formulations such as sauces, dressings, dairy products, baked goods, and beverages to enhance their overall quality and appeal.

What is the role of hydrocolloids as food stabilizers?

Hydrocolloids act as food stabilizers by helping maintain the desired consistency, preventing ingredient separation, and enhancing the stability of food products. They contribute to the overall texture and sensory experience, ensuring that food products remain visually appealing and maintain quality over time.

FREQUENTLY ASKED QUESTIONS (FAQS) CONTINUED

How do hydrocolloids function as thickening agents?

Hydrocolloids increase the viscosity of food products, resulting in improved texture and mouthfeel. They contribute to the desired thickness and provide a smooth and creamy texture, enhancing the overall sensory experience of the food product.

What is the function of hydrocolloids as gelling agents?

Hydrocolloids can act as gelling agents, contributing to the formation of stable gels in various food applications. They help create a network structure that provides structure, stability, and texture to products such as jellies, desserts, and confectioneries.

How do hydrocolloids work as emulsifiers?

Hydrocolloids help promote the formation and stabilization of emulsions in food products. They assist in keeping oil and water-based ingredients properly mixed, preventing separation and providing a smooth, homogenous texture.

What is the role of hydrocolloids as texture modifiers?

Hydrocolloids can act as texture modifiers in food products, altering and enhancing the desired texture qualities. They can add creaminess, smoothness, or mouth-coating properties to various food formulations, improving the sensory experience for consumers.

Why are hydrocolloids considered natural and functional ingredients?

Hydrocolloids are derived from natural sources and are considered natural ingredients. They offer functional properties in food formulations, such as improving texture, stability, and mouthfeel. Hydrocolloids provide a natural alternative to synthetic additives, meeting the growing consumer demand for clean labels and naturally derived ingredients.

In what other industries are hydrocolloids used?

Hydrocolloids find applications in industries beyond the food sector. They are used in pharmaceuticals, personal care products, and textiles. Their unique properties contribute to functionalities such as viscosity control, stability enhancement, and texture modification in various non-food applications.

PART 1.2

Historical Use of Hydrocolloids

Hydrocolloids have been an important part of human civilization for centuries. From ancient to modern times, these compounds have played a crucial role in various applications, particularly food. The use of **hydrocolloids** in **food production** and preservation has transformed the industry, ensuring the availability of food to people worldwide.

These natural or synthetic compounds are widely utilized in various applications due to their unique properties, including thickening, gelling, and stabilizing. Hydrocolloids are commonly used in the **food industry** to improve **product texture, taste, and appearance**. They are also used in non-food applications such as cosmetics, pharmaceuticals, and textiles.

Core Insights

- ☑ Hydrocolloids have been used for centuries in various applications, particularly in the food industry.
- ☑ These compounds are widely utilized due to their unique thickening, gelling, and stabilizing properties.
- ☑ Hydrocolloids are commonly used in the food industry to improve product texture, taste, and appearance.
- ☑ They are also used in non-food applications such as cosmetics, pharmaceuticals, and textiles.
- ☑ Their use has transformed various industries, ensuring food availability to people worldwide.

1.2.1 Origins of Hydrocolloids

As we continue our exploration of hydrocolloid history, it is fascinating to learn about the developers and innovators who played a crucial role in the development and initial applications of these essential substances.

Locust Bean Gum

Locust bean gum, also known as carob gum, is derived from the seeds of the carob tree (Ceratonia siliqua). The carob tree is native to the Mediterranean region, and its pods and seeds have been used for thousands of years by various cultures in that area. The ancient Egyptians, Greeks, and Romans utilized carob pods for food and medicine.

While the traditional use of carob pods and seeds has a long history in the Mediterranean region, the commercial extraction and development of locust bean gum as a food additive and stabilizer began in the 20th century. The demand for natural thickeners in the food industry, especially in producing ice cream and other dairy products, led to the exploration and commercialization of locust bean gum.

The exact individual or organization responsible for the initial commercial development of locust bean gum is not widely documented. However, its commercial extraction and application as a standardized food additive resulted from collective efforts by researchers and industries in the 20th century, recognizing its potential as a valuable thickening and stabilizing agent.

Guar Gum

Guar gum is derived from the seeds of the guar plant (Cyamopsis tetragonoloba), which is native to India and Pakistan. Using guar seeds for various purposes, especially as a food ingredient, has been a part of traditional Indian and Pakistani cultures for centuries.

While the traditional use of guar seeds and the knowledge of their properties existed in these regions, the commercial development and extraction process for guar gum as an industrial and food additive began in the 20th century. The increased demand for natural thickeners in various industries, including food, textiles, and oil drilling, led to the exploration and commercialization of guar gum on a larger scale.

It's important to note that the commercial extraction and application of guar gum as a standardized additive was a collective effort involving researchers, industries, and farmers. The individual or organization responsible for its initial commercial development must be more widely documented. Still, the broader development and popularization of guar gum as a commercial product can be attributed to the combined efforts of various stakeholders in the mid-20th century.

Gum Arabic

Gum Arabic, also known as acacia gum, is a natural gum made from the hardened sap of various species of the acacia tree. It has been used for thousands of years in various cultures, primarily in the African Sahel region, which includes parts of Sudan, Senegal, and Somalia.

The use of gum Arabic dates to ancient civilizations. The ancient Egyptians, for instance, used it in the mummification process, as cosmetics, and as an adhesive for papyrus. It was also used in traditional African and Middle Eastern cultures for various purposes, including as a food stabilizer, in medicine, and in arts and crafts.

Given its long history of use, gum Arabic wasn't "developed" in the modern sense by a specific individual or company. Instead, its properties and applications were recognized and utilized by ancient cultures. However, its trade became significant, especially in the Middle Ages, when it was exported to Europe for use in various industries, including the food and art sectors.

Carrageenan

As an example, carrageenan is a natural substance derived from red seaweed, and it has been used for centuries, especially in coastal regions where these seaweeds are abundant. The use of carrageenan as a food thickener can be traced back to Ireland in the 400s AD, where it was known as «Irish moss.» Coastal communities would boil the seaweed to extract the carrageenan and use it as a gelatinous substance in foods.

While the traditional use of carrageenan dates back centuries, its commercial extraction and recognition as a food additive began in the 20th century. The term "carrageenan" comes from the Irish word "carragáin," which means "little rock," likely referring to the rocky habitats where the red seaweed grows.

So, while carrageenan as a substance wasn't "discovered" in the traditional sense, its properties and potential applications were researched and developed in the modern era.

Tara Gum

Tara gum is a natural polysaccharide derived from the seeds of the Tara tree (Caesalpinia spinosa), which is native to South America. The use of Tara seeds for various purposes can be traced back to ancient indigenous cultures in the Andean region.

The commercial development and extraction process for Tara gum as a food additive and thickening agent began in the mid-20th century. While the exact individual or organization responsible for its initial commercial development is not as widely documented as some other gums, the increasing demand for natural thickeners and stabilizers in the food industry during the 20th century led to the exploration and commercialization of various plant-based gums, including Tara gum.

It's worth noting that while the commercial extraction and application of Tara gum as a standardized food additive began in the 20th century, the Tara tree and its seeds have been utilized by local populations in South America for much longer.

Xanthan Gum

Xanthan gum was discovered by a team of researchers at the United States Department of Agriculture (USDA) in the 1960s. Allene Rosalind Jeanes led the team. They were studying the bacterium ***Xanthomonas campestris***, which produces a polysaccharide during fermentation, later named xanthan gum. This polysaccharide was found to have unique properties that made it valuable as a thickening and stabilizing agent in various industries, especially the food industry. As a result of their research, xanthan gum was commercialized and is now widely used in many products.

Pectin

Pectin is a naturally occurring polysaccharide found in the cell walls of plants, especially in fruits. It has been used for centuries in cooking and baking, particularly as a gelling agent for jams and jellies.

The use of pectin in food preparation dates back to ancient times, and its gelling properties were likely discovered through traditional cooking practices. For instance, boiling certain fruits, like apples or quinces, would naturally release pectin and cause the mixture to gel.

The scientific understanding and extraction of pectin began in the 19th century. In 1825, Henri Braconnot, a French chemist, isolated and described pectin from plant tissues, making him one of the first to recognize and study pectin as a distinct substance. However, pectin's commercial production and application as a standardized food additive began in the early 20th century.

Gellan Gum

Gellan gum is a polysaccharide produced by the bacterium Sphingomonas elodea (formerly known as Pseudomonas elodea). The development and commercialization of gellan gum can be attributed to the American multinational corporation CP Kelco (and its predecessor companies), which undertook extensive research on this microbial polysaccharide in the 1970s and 1980s.

CP Kelco introduced Gellan gum to the market under the brands "Gelrite" and "Kelcogel." The gum has since become a valuable additive in food, pharmaceutical, and other industries due to its unique gelling, stabilizing, and thickening properties.

So, while the bacterium naturally produces Gellan gum, CP Kelco spearheaded its development for commercial applications and its introduction to the market.

1.2.2 Early Applications of Hydrocolloids

Historically, hydrocolloids have been used for various applications, particularly in the food industry. In the early days, hydrocolloids were utilized for water retention, thickening, and food gelling.

One of the earliest recorded uses of hydrocolloids was in ancient China, where agar was extracted from seaweed and used as a gelling agent in food. Similarly, the ancient Aztecs used chia seeds as a source of hydrocolloids to thicken their drinks.

In the early modern period, Europeans used various hydrocolloids such as pectin, gelatin, and carrageenan in cooking and food preservation. For instance, pectin was used to produce jellies and jams, while gelatin was utilized to create desserts such as panna cotta and aspic. Conversely, Carrageenan stabilizes and thickens dairy products such as milk and ice cream.

With the advent of the Industrial Revolution in the 19th century, hydrocolloids found new and innovative applications in the food industry. For example, hydrocolloids such as xanthan gum and guar gum were introduced as thickeners and stabilizers in various processed foods, including sauces, dressings, and bakery products.

Today, hydrocolloids remain an integral part of the food industry, with various applications in various products. They are used for everything from improving the texture and mouthfeel of food to increasing shelf life and stability.

1.2.3 Evolution of Hydrocolloid Usage

Hydrocolloids have come a long way since their early applications in the food industry. The evolution of hydrocolloid usage can be traced from the days of simple thickening and gelling agents to the modern era of complex formulations and innovative applications.

Advancements in technology and scientific understanding have played a crucial role in this evolution. With the development of new processing techniques and analytical tools, researchers have better-understood hydrocolloid properties and potential applications.

One major trend in the evolution of hydrocolloid usage has been the shift toward natural and renewable sources. In response to increasing consumer demand for clean-label products, manufacturers have turned to plant-based and polysaccharide-based hydrocolloids, such as carrageenan and guar gum, as alternatives to synthetic or animal-derived ingredients.

Another significant development in the evolution of hydrocolloid usage has been expanding their applications beyond the food industry. Hydrocolloids are now widely utilized in industries such as pharmaceuticals, cosmetics, and textiles, where they play important roles in drug delivery, surface modification, and fiber stabilization, to name a few.

With ongoing research and development, the prospects of hydrocolloids appear promising. New formulations and applications are being explored, and the potential for hydrocolloids to address future challenges in various industries is being actively pursued.

1.2.4 Advancements in Hydrocolloid Research

Hydrocolloids have a rich history of applications in various industries, particularly in the food industry. However, there is still much to be explored in the world of hydrocolloids, and researchers are continuously advancing their use.

One area of active research is the development of new hydrocolloid formulations with unique properties and functionalities. These new formulations can improve food texture, flavor, and nutritional value and enhance the stability of other products.

Researchers are also investigating the potential of hydrocolloids as drug delivery systems. Hydrocolloids can be designed to release drugs over extended periods, providing a more efficient and targeted delivery system than traditional methods.

Advances in hydrocolloid research are also leading to new applications In the cosmetic industry. Hydrocolloids can be used to improve the texture and appearance of cosmetic products and enhance their effectiveness.

Another area of active research is the use of hydrocolloids in 3D printing. Hydrocolloids can be used as a printable material for creating functional food structures, offering exciting possibilities for the future of food manufacturing.

1.2.5 Future Prospects of Hydrocolloids

From their origins as thickening agents in the food industry to their current diverse applications in various industries, the history of hydrocolloids is a testament to their versatility and potential.

The prospects of hydrocolloids are exciting and offer vast possibilities. As the demand for sustainable and environmentally friendly products continues to rise, hydrocolloids may play a critical role in meeting these needs. For instance, natural hydrocolloids such as agar and carrageenan are being explored as alternatives to synthetic plastics in packaging, thanks to their biodegradable properties.

Moreover, advances in hydrocolloid research are paving the way for innovative applications in medicine and biotechnology. Hydrocolloids such as chitosan and alginate are already being used in wound healing and tissue engineering, with the potential for further breakthroughs in regenerative medicine.

With the ever-growing demand for convenience foods, hydrocolloids will continue to play a vital role in the food industry. They offer unique solutions to challenges such as texture modification, stability enhancement, and fat reduction, leading to a better sensory experience for consumers.

Conclusion

In conclusion, we have taken a captivating journey through the history of hydrocolloids and their evolution. From their origins to modern applications, hydrocolloids have transformed various industries, particularly food. We have learned how hydrocolloids were utilized in the past and how they continue to be a crucial component of modern food processing, texture modification, and stability enhancement. As technology and scientific understanding continue to advance, new and innovative applications of hydrocolloids are emerging across a wide range of industries, from pharmaceuticals to cosmetics to textiles. Scientists and researchers continuously push the boundaries of hydrocolloid research, discovering new properties and characteristics of these versatile substances.

Looking to the future, we can expect hydrocolloids to play an increasingly important role in addressing future challenges in various industries. Emerging trends and potential applications offer exciting possibilities for the future, and we eagerly anticipate the continued impact of hydrocolloids. Overall, our exploration into the history of hydrocolloids demonstrates the crucial role that scientific innovation and research play in shaping our world. Whether in the past, present, or future, hydrocolloids remain a fascinating subject of study and an essential component of many industries.

FAQ

What is the historical use of hydrocolloids?

Hydrocolloids have been used throughout history in various applications, particularly in food. They have been utilized for their ability to modify texture, enhance stability, and improve overall quality.

Who were the inventors of hydrocolloids?

The origins of hydrocolloids can be traced back to inventors and innovators who played a crucial role in their development. Some notable figures include Stanley Ros

How were hydrocolloids used in the early days?

In the early days, hydrocolloids were primarily used in the food industry. They were employed for their gelling, thickening, and stabilizing properties, which enabled food production and preservation advancements.

How has the usage of hydrocolloids evolved?

The usage of hydrocolloids has evolved significantly over time due to advancements in technology and scientific understanding. Today, hydrocolloids find applications beyond the food industry, including pharmaceuticals, cosmetics, textiles, etc.

What are the different types of hydrocolloids?

There are various types of hydrocolloids, each with its unique properties and characteristics. Some common types include agar, carrageenan, guar gum, xanthan gum, pectin, and gelatin.

Can you provide examples of hydrocolloids in action?

Certainly! Hydrocolloids are used in a wide range of products and processes. Examples include carrageenan in dairy products, xanthan gum in gluten-free baking, and pectin in jams and jellies.

FREQUENTLY ASKED QUESTIONS (FAQS) CONTINUED

- **How have hydrocolloids revolutionized the food industry?**

  Hydrocolloids have revolutionized the food industry by enabling texture modification, stabilization, and improved overall quality. They have played a crucial role in creating innovative food products, enhancing sensory experiences, and extending shelf life.

- **In which other industries are hydrocolloids used?**

  Hydrocolloids have found applications beyond the food industry. They are utilized in pharmaceuticals, cosmetics, textiles, and other industries, where their unique properties contribute to product formulation and performance.

- **What are the latest advancements in hydrocolloid research?**

  Ongoing research in hydrocolloids focuses on exploring new applications, optimizing production processes, and improving sustainability. Scientists and researchers continuously push the boundaries to unlock the full potential of hydrocolloids.

- **What are the prospects of hydrocolloids?**

  Hydrocolloids offer exciting possibilities for the future. Emerging trends include the development of plant-based alternatives, personalized nutrition solutions, and novel applications in biotechnology and renewable energy.

CHAPTER 02

Types of Hydrocolloids

PART 2.1

Types of Hydrocolloids: Natural vs Synthetic

Hydrocolloids are among the most widely used ingredients in various industries. They are known for improving different products' texture, stability, and quality. Hydrocolloids can be classified into natural and synthetic forms with unique characteristics and applications.

Core Insights:

- ☑ Hydrocolloids are important ingredients in various industries.
- ☑ Natural and synthetic hydrocolloids have different characteristics and applications.

2.1.1 Understanding Hydrocolloids

Hydrocolloids are substances that form gels or solutions when dispersed in water. They can be natural or synthetic and have various applications in various industries, including food, pharmaceuticals, and personal care.

Food-grade hydrocolloids are commonly used in the food industry as thickeners, stabilizers, and emulsifiers. Hydrocolloid gum, such as xanthan gum and guar gum, are some of the most widely used types of hydrocolloids. Hydrocolloid stabilizers, on the other hand, prevent the separation of ingredients in food products and improve their texture.

Hydrocolloids are particularly important in the food industry, enhancing food products' quality, texture, and stability. They are used to improve the mouthfeel of food products, prevent syneresis (the separation of liquid from a gel), and enhance the flavours and aromas of food.

2.1.2 Types of Hydrocolloid Gums

Hydrocolloid gums are a subgroup commonly used as thickeners and stabilizers in food products. Some of the most common hydrocolloid gums used in the food industry include:

Table 2: Types of Hydrocolloids Gums

HYDROCOLLOID GUM	SOURCE	FUNCTION
Xanthan gum	Fermentation of sugars by the bacterium Xanthomonas campestris	Thickener, emulsifier, stabilizer
Guar gum	Seed of the guar plant, Cyamopsis tetragonoloba	Thickener, stabilizer
Locust bean gum	Seed of the carob tree, Ceratonia siliqua	Thickener, stabilizer

Hydrocolloid gums are usually added to food products in small quantities and can significantly improve their texture and stability. They are particularly useful in low-fat or fat-free products, where they can replace some of the texture and mouthfeel lost due to the absence of fat.

Overall, hydrocolloids play a key role in the food industry and many other industries by improving various products' quality, texture, and stability. With an understanding of the different types of hydrocolloids and their functions, manufacturers can create products that meet consumers' preferences for texture, stability, and sensory attributes.

2.1.3 Natural Hydrocolloids

Natural hydrocolloids are polysaccharides derived from plants and animals, with unique physical and chemical properties that make them valuable ingredients in various applications. Their ability to form gels, thicken, stabilize, and emulsify makes them essential in the food, pharmaceuticals, and cosmetics industries.

The benefits of using natural hydrocolloids are manifold. They are renewable, biodegradable, and have a clean label appeal, making them an attractive choice for consumers who value sustainability and natural ingredients. Additionally, natural hydrocolloids offer health benefits, such as improved gut health and reduced blood glucose levels, making them a popular choice in the functional food sector.

Table 3: Natural hydrocolloids, Applications and Benefits

TYPES OF NATURAL HYDROCOLLOIDS	APPLICATIONS	BENEFITS
Agar	Food, pharmaceuticals, microbiology	Thermoreversible gelling agent, vegan alternative to gelatin, high in fiber and minerals
Alginate	Food, pharmaceuticals, personal care	Forms strong, elastic gels, stabilizes emulsions, enhances texture, and is high in fiber and minerals
Carrageenan	Food, pharmaceuticals, personal care	Forms strong, elastic gels, stabilizes emulsions, enhances texture, and may have anti-inflammatory properties
Guar Gum	Food, pharmaceuticals, cosmetics	Thickener, a stabilizer, improves texture and may have cholesterol-lowering properties.
Xanthan Gum	Food, pharmaceuticals, personal care	Thickener, stabilizer, improves texture, may have prebiotic properties

These are just a few examples of the many natural hydrocolloids available. Their diverse characteristics and functionalities enable them to be used in various products, from dairy alternatives to gluten free baked goods to plant-based meat substitutes.

Through ongoing research and innovation, the applications of natural hydrocolloids continue to expand, as does their potential to meet growing consumer demand for sustainable and health-promoting ingredients.

2.1.4 Synthetic Hydrocolloids

Synthetic hydrocolloids are manufactured artificially through chemical processes and are widely used in various industries, including food, pharmaceuticals, and personal care. These versatile compounds have diverse properties, making them effective as thickeners, stabilizers, and emulsifiers.

One of the key advantages of synthetic hydrocolloids is their consistent performance, which is paramount in industries where product quality and stability are critical. They also tend to be more cost-effective than natural hydrocolloids, making them an attractive option for manufacturers.

In the food industry, synthetic hydrocolloids are used extensively to achieve specific texture and sensory attributes in products such as sauces, dressings, and baked goods. They can also enhance food products' shelf life and stability, improving their quality.

The hydrocolloid market for synthetic compounds is expected to grow in the coming years, driven by the increasing demand for convenience foods, processed and packaged goods, and functional ingredients. However, there is also a growing trend towards natural, clean-label ingredients, which may limit the growth potential of synthetic hydrocolloids in some markets.

2.1.5 Characteristics of Hydrocolloids

Hydrocolloids are versatile ingredients that exhibit a range of physical and chemical characteristics. Both natural and synthetic hydrocolloids share similar properties, although there are some differences between the two categories.

Natural hydrocolloids, derived from plant, animal, or microbial sources, have unique properties that make them desirable in various applications. They are often preferred in the food industry due to their clean label appeal and consumer perception of being natural and healthy. Some examples of natural hydrocolloids include carrageenan, xanthan gum, and agar.

On the other hand, synthetic hydrocolloids are derived from chemical processes and designed to mimic natural hydrocolloids' properties. They are often preferred in industrial applications due to their consistent performance and cost-effectiveness. Some examples of synthetic hydrocolloids include carboxymethyl cellulose (CMC), methylcellulose, and polyvinyl alcohol (PVA).

Natural and synthetic hydrocolloids can form gels, thicken, stabilize, and emulsify. They can also modify a product's texture, appearance, and sensory properties. Hydrocolloids typically function by interacting with water molecules, forming a network that provides structural support and stability.

Other important characteristics of hydrocolloids include their solubility, viscosity, and gel strength. The solubility of a hydrocolloid refers to its ability to dissolve in a solvent, such as water or oil. Viscosity refers to a liquid's resistance to flow, which can be influenced by the concentration and type of hydrocolloid used. Gel strength refers to the firmness of a hydrocolloid gel, which can vary depending on the type and amount of hydrocolloid used.

2.1.6 Emerging Trends and Innovations in Hydrocolloids

The hydrocolloid market constantly evolves, driven by consumer demands for clean and sustainable ingredients. As a result, there is a growing trend towards natural hydrocolloids derived from plant- or animal-based sources.

Natural hydrocolloids offer numerous advantages over synthetic alternatives. They are often perceived as healthier and more environmentally friendly, free from chemical additives, and produced using sustainable farming practices. Moreover, the clean-label appeal of natural hydrocolloids has increased its popularity in the food industry, where consumers are looking for products with fewer artificial ingredients.

The demand for natural hydrocolloids drives innovation, with manufacturers exploring novel sources and production methods. For example, chia seeds, seaweed, and flaxseed are emerging as potential sources of natural hydrocolloids, offering unique functional properties and health benefits.

At the same time, developments in processing technologies have enabled the production of more complex hydrocolloid structures, allowing for a wider range of applications. For instance, hydrocolloid-based encapsulation systems are being developed to deliver active ingredients in pharmaceuticals and personal care products.

Synthetic hydrocolloids have a significant presence in the market, particularly in the food industry, offering consistent performance and cost-effectiveness. However, there is a growing demand for natural alternatives, which is driving manufacturers to develop synthetic hydrocolloids that mimic the functionality of natural hydrocolloids.

The hydrocolloid market is expected to grow in the coming years, driven by increasing demand from developing regions and a rising disposable income. However, the industry must navigate regulatory challenges, as there is increasing scrutiny on the safety and sustainability of hydrocolloid production.

As consumer demands for clean and sustainable ingredients continue to grow, the hydrocolloid industry is undergoing significant changes. Natural hydrocolloids are emerging as a popular alternative to synthetic ones, offering a range of functional and health benefits. Innovations in processing technologies and new sources of natural hydrocolloids drive product development and widen the range of applications. However, synthetic hydrocolloids are still widely used, particularly in the food industry, and offer cost-effective solutions. As the hydrocolloid market grows, manufacturers must balance the demand for natural and synthetic hydrocolloids while addressing regulatory challenges and ensuring sustainable production practices.

2.1.7 Outlook for Hydrocolloids

The global hydrocolloid market is expected to grow significantly over the next few years, driven by the increasing demand from various industries, such as food and beverage, pharmaceutical, and personal care. Synthetic hydrocolloids are expected to dominate the market due to their cost-effectiveness and versatility in various applications.

However, the trend towards clean and sustainable ingredients is expected to impact the hydrocolloid market, with a growing preference for natural hydrocolloids. Manufacturers are increasingly investing in research and development to develop new natural hydrocolloid-based products that meet consumer preferences.

Regulatory developments and restrictions on using certain synthetic hydrocolloids in food and pharmaceutical applications could also impact the market share of synthetic hydrocolloids and drive further demand for natural alternatives.

Technological advancements in hydrocolloid production, such as using novel sources and processing techniques, are expected to create new opportunities for the industry. Hydrocolloids with enhanced properties, such as improved stability and texture, will likely become more prevalent in the market.Overall, the hydrocolloid market is poised for significant growth, with opportunities and challenges ahead. The industry is expected to continue innovating and developing new solutions to meet consumers' and manufacturers' evolving needs and preferences.

Conclusion

We explored the distinction between natural and synthetic hydrocolloids and the characteristics of each. Natural hydrocolloids are sourced from plant and animal materials, while synthetic hydrocolloids are derived from chemical processes. We also discussed the different types of hydrocolloids commonly used, such as food-grade hydrocolloids, gums, and stabilizers.

We discussed the benefits of natural hydrocolloids, including their clean label appeal and health benefits, while recognizing synthetic hydrocolloids' advantages, such as their consistent performance and cost-effectiveness.

As the demand for clean and sustainable ingredients continues to grow, we explored the emerging trends and innovations within the hydrocolloid industry, particularly the increasing interest in natural hydrocolloids. Lastly, we considered the outlook for hydrocolloids, recognizing the potential challenges and opportunities in the industry, particularly with synthetic hydrocolloids.

Overall, hydrocolloids are critical ingredients that are vital in numerous industries. Their importance in enhancing product quality, texture, and stability cannot be overstated, and they will continue to play an essential role in the years to come.

FAQ

Q: What is the difference between natural and synthetic hydrocolloids?

Natural hydrocolloids are derived from plants or animals and are minimally processed. On the other hand, synthetic hydrocolloids are chemically synthesized and designed for specific applications.

Q: What are the common applications of hydrocolloids?

Hydrocolloids are widely used in the food industry as thickeners, stabilizers, and texture modifiers. They are also used in pharmaceuticals, personal care, and paints.

Q: What are the emerging trends and innovations in hydrocolloids?

The hydrocolloid industry is experiencing a growing demand for natural hydrocolloids and the development of new products that align with consumer preferences for clean and sustainable ingredients.

Q: What is the future outlook for hydrocolloids?

The future of hydrocolloids is influenced by factors such as market growth, technological advancements, and regulatory developments. Synthetic hydrocolloids may face challenges, while natural hydrocolloids are expected to thrive.

PART 2.2

Common Hydrocolloids: Versatile Additives in the Food Industry

Hydrocolloids are natural polymers, known as polysaccharides, that function as thickening agents, gelling agents, stabilizers, emulsifiers, water-binding agents, and texturizers in various food products. As food additives, they are widely used in the food industry to enhance the texture, stability, and other functional properties of food and beverage products.

This section will explore the common hydrocolloids used in the food industry, their benefits, and practical applications. Hydrocolloids offer various applications in various food and beverage products, from agar agar to pectin.

Core Insights:

- ☑ Hydrocolloids are natural polymers that function as thickening agents, gelling agents, stabilizers, emulsifiers, water-binding agents, and texturizers in various food products.
- ☑ Hydrocolloids are critical in enhancing food and beverage products' texture, stability, and other functional properties.
- ☑ Common hydrocolloids include agar, carrageenan, guar gum, Gellan gum, locust bean gum, alginate, xanthan gum, cellulose, and pectin.
- ☑ Hydrocolloids are natural, functional, and valuable tools for creating high-quality, innovative food products.
- ☑ Their versatility and functionality make them valuable in creating various food formulations.

2.2.1 Agar Agar

Agar agar is a popular hydrocolloid derived from seaweed extract. It is a versatile ingredient in the food industry because it acts as a gelling agent.

As a hydrocolloid, agar agar is widely used for its thickening and stabilizing properties. It is particularly useful as a vegan alternative to gelatin, a common gelling agent derived from animal products.

Agar agar is extracted from various types of seaweed and is available in powder, flake, and bar form. It is known for its superior gelling properties, making it a popular ingredient in producing jellies, jams, and desserts.

Aside from being a vegan-friendly alternative to gelatin, agar agar is a natural thickener for soups and sauces and a stabilizer for emulsions and suspensions. Its ability to form a gel at a low concentration and high melting point makes it a valuable ingredient in various food products.

2.2.2 Carrageenan: A Versatile Hydrocolloid for Improved Texture and Stability

Carrageenan is a hydrocolloid widely used as a thickening agent and stabilizer in the food industry. It is extracted from red seaweed and is a natural ingredient in various food products. Carrageenan is important in improving the texture and stability of dairy products, such as ice cream, yogurt, and cheese.

As a thickening agent, carrageenan helps to create a smooth and creamy texture in dairy products. It prevents syneresis, which is the separation of liquid and solid components, and helps to maintain the desired consistency and appearance of the product. Carrageenan is also known for forming strong gels, making it an ideal ingredient for creating stable dairy-based desserts.

Aside from its use in dairy products, carrageenan is also found in various plant-based alternatives as a natural and vegan-friendly ingredient. It is commonly used in plant-based milk, such as almond and soy milk, to improve texture and create a more milk-like consistency.

Carrageenan's versatility as a hydrocolloid makes it an important ingredient in the food industry. Its functional properties make it a valuable tool for creating high-quality, innovative food formulations.

2.2.3 Guar Gum: A Versatile Hydrocolloid for Gluten-Free Bakery Products

Guar gum is a hydrocolloid widely used as a thickening agent and stabilizer in the food industry. Its natural origin and functionality make it a valuable tool for creating high-quality, innovative food formulations. In the world of gluten-free baking, guar gum plays a crucial role in enhancing the texture and structure of baked goods.

Its high viscosity properties allow it to create a smooth and uniform texture, improving the overall appearance of gluten-free baked goods. Additionally, guar gum is known for improving bakery products' shelf life, reducing the likelihood of going stale and other quality issues.

Guar gum is particularly useful in gluten-free baking because it can replace gluten, a protein found in wheat, rye, and barley that gives baked goods their elasticity and structure. Guar gum acts as a binding agent in gluten-free products, preventing baked goods from becoming crumbly and dry. Its water-binding properties also help to retain moisture, creating a more tender and flavorful product.

Bakery products, such as bread, cakes, and cookies, are just a few examples of guar gum used in gluten-free formulations. Its versatility makes it suitable for other baked goods, such as muffins, scones, and pastries.

Table 4: Benefits of Guar Gum in Bakery

BENEFITS OF GUAR GUM IN BAKERY PRODUCTS
Provides a smooth and uniform texture
Improves shelf life of baked goods
Acts as a binding agent in gluten-free products
Retains moisture, creating a more tender and flavorful product

In conclusion, guar gum is a valuable hydrocolloid in the food industry, particularly in gluten-free bakery products. Its ability to thicken and stabilize baked goods, replace gluten, and improve moisture retention makes it an essential ingredient in many formulations. As the demand for gluten-free products grows, guar gum's versatility and functionality will remain a staple in the gluten-free baking industry.

2.2.4 Gellan Gum: A Versatile Hydrocolloid for Vegan-Friendly Suspensions

Gellan gum is a hydrocolloid that acts as a gelling agent and stabilizer in various food and beverage applications. It is a popular ingredient in vegan-friendly products due to its plant-based origins and functionality.

It is known for its ability to create suspensions, giving liquids a smooth, uniform texture resistant to settling. It is commonly used in dairy-free products such as plant-based milk and yogurt alternatives and vegan-friendly desserts and beverages.

Aside from its stabilizing properties, Gellan gum is also used as a thickening agent in certain applications. It can improve the viscosity and mouthfeel of products such as sauces, dressings, and syrups.

Compared to other hydrocolloids, Gellan gum has a unique property that allows it to withstand high temperatures and acidic conditions, making it suitable for a wide range of applications.

Overall, Gellan gum is a versatile hydrocolloid that offers functional benefits for vegan-friendly suspensions and various other food and beverage products.

2.2.5 Locust Bean Gum: A Versatile Hydrocolloid for Improving Texture and Stability

Locust bean gum is a hydrocolloid widely used as a thickening agent and stabilizer in the food industry. Derived from the endosperm of the seeds of the carob tree, this natural polysaccharide offers various functional benefits that make it a valuable ingredient in ice cream and other dairy-based desserts.

One of the main advantages of locust bean gum is its ability to improve texture and mouthfeel. Adding to ice cream helps prevent ice crystal growth and enhances creaminess, resulting in a smoother and more enjoyable product. It also contributes to the overall stability of the dessert, preventing syneresis and ensuring a longer shelf life.

In addition to its benefits in ice cream and other dairy-based desserts, locust bean gum can be used in various other applications, including sauces, dressings, and bakery products. As a thickener and stabilizer, it helps improve texture and consistency, providing consumers with a more appealing and pleasant experience.

Furthermore, locust bean gum is a natural ingredient that can appeal to consumers who prefer clean labels and are looking for products free from artificial additives. Its functionality and versatility make it a valuable tool for creating high-quality, innovative food formulations that meet the demands of modern consumers.

Locust bean gum is a valuable hydrocolloid that improves texture and stability in various food products. Its natural origins, functionality, and versatility make it a valuable ingredient for food manufacturers looking to create innovative and high-quality products.

2.2.6 Alginate

Alginate is a hydrocolloid commonly used in the food industry for its unique gelling and film-forming properties. This hydrocolloid is derived from brown seaweed and is a popular ingredient in molecular gastronomy. Alginate is used in spherification, where liquids are transformed into gel-like spheres. This method is widely used in high-end restaurants to create visually stunning dishes.

Aside from its use in molecular gastronomy, alginate is also used in the food industry as a gelling and film-forming agent. Its ability to form a thin film on the surface of food products can enhance its texture and stability. Alginate is commonly found in dairy-based products such as yogurt, ice cream, and cream cheese and is also used in fruit-based products such as jams and jellies.

Overall, alginate is a versatile hydrocolloid that offers a range of functional benefits in the food industry. Its unique properties make it valuable for creating innovative food formulations, and its natural origins make it a popular choice for consumers seeking natural and sustainable ingredients.

2.2.7 Xanthan Gum: A Versatile Hydrocolloid for Thickening and Stabilizing

Xanthan gum is a commonly used hydrocolloid in the food industry, particularly in gluten-free products. This natural thickening agent and stabilizer is derived from the fermentation of bacteria that produce polysaccharides. Its unique properties make it ideal for use in various food products.

One of the primary benefits of xanthan gum is its ability to enhance viscosity, which makes it an effective thickening agent. It can be added to sauces and dressings to create a smooth, consistent texture. Xanthan gum can also stabilize emulsions, preventing them from separating over time.

Another advantage of xanthan gum is that it is gluten-free, making it a popular ingredient in gluten-free formulations. It can be used instead of wheat flour as a thickener for soups, gravies, and baked goods.

2.2.8 Cellulose and Pectin

Cellulose and pectin are two commonly used hydrocolloids in the food industry. They act as texture enhancers and provide dietary fiber in various food products. Cellulose can be derived from plant sources, while pectin is commonly found in fruit-based products.

Cellulose is a complex carbohydrate found in the cell walls of plants. It is often used as a bulk-forming agent and provides a low-calorie source of dietary fiber. It is commonly used in food applications to improve texture in bakery, dairy, and meat products. It can also reduce fat content in salad dressings and sauces.

Pectin is a naturally occurring polysaccharide in fruits such as apples, oranges, and berries. It is a commonly used hydrocolloid in the food industry due to its ability to form gels when mixed with sugar and acid in fruit-based products. Pectin is particularly useful in producing jams, jellies, and other fruit spreads. It is also a stabilizer in dairy products, such as yogurt and ice cream, to improve texture and prevent syneresis.

Both cellulose and pectin are safe for consumption and are widely used in the food industry. Both are also important sources of dietary fiber, linked to numerous health benefits such as improved digestion, weight management, and lower cholesterol levels.

Conclusion

In conclusion, hydrocolloids are natural polymers widely used as food additives in the food industry. They provide functional benefits such as texture enhancement, stability, and viscosity control. Hydrocolloids can act as thickening agents, gelling agents, stabilizers, emulsifiers, water-binding agents, and texturizers.

Common hydrocolloids discussed in this article include agar agar, carrageenan, guar gum, Gellan gum, locust bean gum, alginate, xanthan gum, cellulose, and pectin. Each hydrocolloid has specific properties that make it suitable for different applications in the food and beverage industry.

SOURCES OF FOOD **HYDROCOLLOIDS**

(Thickeners, Emulsifiers, Stabilizers, and Jellifiers)

Cellulose Polymers

- Microcrystalline cellulose
- Methylcellulose
- Ethyl cellulose
- Hydroxypropyl methylcellulose
- Hydroxypropyl cellulose
- Carboxymethylcellulose

Microbial and Bacterial Polysaccharides

- Dextran
- Xanthan Gum
- Pullulan Gum
- Gellan Gum

Herbal Secretions

- Karaya gum
- Acacia Gum (Gum Arabic)
- Tragacanth gum
- Combretaceae gum

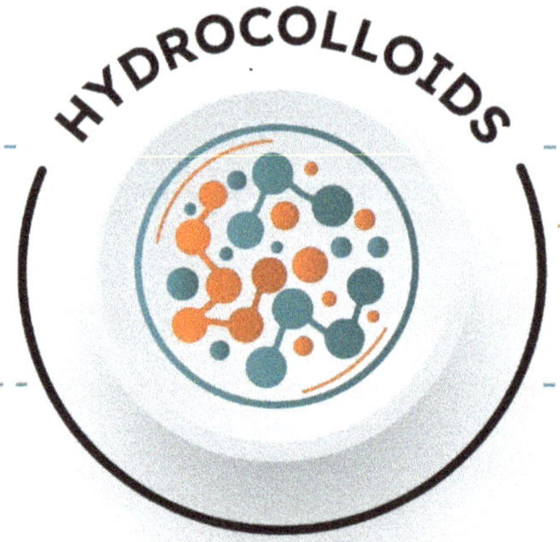

Grains

- Locust Bean Gum
- Tamarind Seed Gum
- Tara Gum

Animal derivatives

- Types of Gelatins
- Types of caseins

Processed Polysaccharides

- Modified starch types
- Amidated pectin
- Propylene glycol alginate (PGA)
- Cellulose derivatives

Pure Herbal

- Agar agar
- Types of alginates
- Types of carrageenan
- Types of pectin
- Types of starch
- Types of Cellulose

FAQ

Q: Is agar agar vegan-friendly?

Agar agar is a vegan alternative to gelatin derived from seaweed extract.

Q: What are the common uses of carrageenan?

Carrageenan is commonly used as a thickening agent and stabilizer in dairy products and plant-based alternatives.

Q: Is guar gum gluten-free?

Yes, guar gum is gluten-free and commonly used in gluten-free bakery products.

Q: What is gellan gum used for?

Gellan gum is a gelling agent and stabilizer, particularly in vegan-friendly products and suspensions.

Q: How is locust bean gum used in ice cream?

Locust bean gum is a thickening agent and stabilizer in ice cream and dairy-based desserts to improve texture and prevent ice crystal growth.

Q: What is alginate used for in molecular gastronomy?

Alginate is commonly used in molecular gastronomy techniques such as spherification, where liquids are turned into gel-like spheres.

Q: Is xanthan gum gluten-free?

Yes, xanthan gum is gluten-free and commonly used as a thickening agent and stabilizer in gluten-free sauces and dressings.

Q: What are cellulose and pectin used for?

Cellulose and pectin are hydrocolloids that act as texture enhancers and dietary fiber in various food products. Cellulose can be derived from plant sources, while pectin is commonly found in fruit-based products.

CHAPTER 03

Properties of Hydrocolloids

PART 3.1

Hydrocolloid Viscosity and Gelation

Hydrocolloids are essential in various industries, contributing to product texture, stability, and functionality. This section will explore the fascinating world of hydrocolloids, focusing on their viscosity and gelation properties.

Understanding the properties of hydrocolloids is crucial for manufacturers to optimize product development. Viscosity and gelation are two significant characteristics that contribute to the functionality of hydrocolloids, allowing them to perform a wide range of applications across various industries.

Core Insights:

- ☑ Hydrocolloids are widely used in various industries.
- ☑ Viscosity and gelation are critical properties of hydrocolloids.
- ☑ Controlling viscosity and gelation is crucial for product development.
- ☑ Hydrocolloids have specific applications in the food and beverage, pharmaceuticals, and cosmetics industries.
- ☑ The field of hydrocolloids is constantly evolving, with new trends and innovations emerging.

3.1.1 The Role of Viscosity in Hydrocolloids

Viscosity is a critical property of hydrocolloids that affects various aspects of their functionality. It refers to a fluid's resistance to flow and is measured in units such as centipoise (cP) or millipascal-seconds (mPa·s). The viscosity of a hydrocolloid solution depends on factors such as concentration, temperature, pH, and other ingredients.

Hydrocolloids with high viscosity contribute to the thickening of products, imparting body and texture. For instance, xanthan gum, a commonly used hydrocolloid, has a high viscosity and is used in food products such as salad dressings and sauces to achieve the desired thickness and uniformity. Viscosity is also important for formulations' stability, preventing ingredients' separation or settling. Additionally, it influences the flow behavior of products through pipelines and equipment during manufacturing processes.

Conversely, hydrocolloids with low viscosity are used as emulsifiers, stabilizers, and suspending agents. They allow for the dispersion of ingredients that would otherwise separate, such as oil and water. Low-viscosity hydrocolloids also exhibit good flow properties, making them suitable for spray drying or atomization applications.

The viscosity of hydrocolloid solutions can be manipulated through varying concentration, pH, and temperature. Furthermore, other ingredients, such as salts, sugars, and acids, can influence viscosity. For example, adding salt to a xanthan gum solution can reduce its viscosity, while adding acid can increase it.

3.1.2 The Role of Gelation in Hydrocolloids

Hydrocolloids are unique in their ability to form gels, making them valuable in various applications. Gelation is the process by which hydrocolloids form a gel structure when mixed with water or another liquid. The gel structure is an essential characteristic of hydrocolloids, as it contributes to their thickening, stabilizing, and emulsifying properties.

The gelation process involves forming a three-dimensional network of molecules within the liquid, resulting in a gel-like structure. Hydrocolloids use different mechanisms to form gels, including hydrogen bonding, electrostatic interactions, and covalent bonding. This allows various gel types to be formed from hydrocolloids, including thermo-reversible, pH-dependent, and ionic gels.

Several factors influence the gelation process, including the hydrocolloid's concentration, the liquid's nature, and other ingredients. The presence of salts, sugars, and other ions can affect the gelation process by altering the strength and stability of the resulting gel.

Gelation is an essential property of hydrocolloids in the food, beverage, and pharmaceutical industries. It forms gels in jams, jellies, and gummy candies. In the pharmaceutical industry, hydrocolloid gels are used in drug delivery systems to control the release of medication.

3.1.3 The Types of Gels Formed by Hydrocolloids

Hydrocolloids can produce different gels, each with unique properties and characteristics. The most common types of hydrocolloid gels include:

- **Thermo-reversible G**els form when hydrocolloids are heated and melt when cooled. They are widely used in food products, such as puddings and custards.
- **pH-Dependent G**els form when the pH of the liquid is adjusted to a specific level. They are used in products such as fruit fillings and yogurt.
- **Ionic G**els form when ions are added to the hydrocolloid solution. They are used in products such as cheese and meat products.

Understanding the gelation properties of hydrocolloids is essential for manufacturers, as it allows for the creation of unique and innovative products. Manufacturers can create products with specific textures, melt-in-the-mouth sensations, and other desirable properties by manipulating the gelation process.

3.1.4 Controlling Viscosity and Gelation

Manufacturers can control the viscosity and gelation of hydrocolloids to achieve specific product characteristics. By manipulating these properties, products' texture, stability, and flow behavior can be adjusted to meet desired standards.

Various techniques are used to control the viscosity of hydrocolloid solutions, such as adjusting temperature, pH levels, and ionic strength. Additionally, the addition of other substances, like salts, acids, or sugars, can also affect viscosity.

Gelation can be controlled by altering the hydrocolloid concentration, temperature, and pH. Other additives, like cross-linking agents or surfactants, can also impact the gelation process.

The ability to control viscosity and gelation gives manufacturers a high level of flexibility in product development, allowing them to create unique and desirable products for consumers in various industries.

3.1.5 Hydrocolloid Applications in the Food and Beverage Industry

Hydrocolloids play an essential role in the food and beverage industry, contributing to various products such as sauces, dressings, desserts, beverages, etc. This section will explore the diverse applications of hydrocolloids in this industry.

Thickening: Hydrocolloids increase the viscosity of food products, providing a desirable texture and consistency. They are commonly used in soups, gravies, and sauces like ketchup and mayonnaise. Thickening also helps prevent sedimentation in beverages, maintaining a homogenous product.

Stabilizing: Hydrocolloids help maintain the uniformity of food products, preventing the separation or settling of ingredients. They are used in dairy products such as ice cream, ensuring a creamy texture and avoiding ice crystal formation. Hydrocolloids also stabilize emulsions, preventing oil and water from separating in salad dressings and sauces.

Emulsifying: Hydrocolloids are used to create stable emulsions, allowing for blending ingredients that would otherwise not mix. They are used in mayonnaise, salad dressings, and chocolate milk.

Creating Unique Textures: Hydrocolloids can create unique textures and shapes, such as gelling fruit fillings in pies or creating foam in whipped cream and mousse. They can also improve mouthfeel, providing a pleasant sensation in the mouth.

3.1.6 Hydrocolloid Applications in Pharmaceuticals and Cosmetics

Hydrocolloids are widely used in the pharmaceutical and cosmetics industries due to their versatile properties. They play a significant role in drug formulation, controlled release systems, and skin care products, among other applications.

Hydrocolloid Applications in Pharmaceuticals

In the pharmaceutical industry, hydrocolloids, inactive ingredients that serve as carriers for active substances, are used as excipients. They help to improve drug stability, solubility, and bioavailability, ensuring optimal drug delivery.

Furthermore, hydrocolloids can formulate controlled release systems designed to release active compounds slowly over a specific period. This technology ensures sustained drug release, reducing dosing frequency and improving patient compliance.

Hydrocolloid Applications in Cosmetics

In cosmetics, hydrocolloids can be used in products such as creams, lotions, and gels. They help to improve product texture, stability, and viscosity.

Hydrocolloids can also be used in skin care products such as facial masks, as they can trap water and help to hydrate the skin. Additionally, they can act as emulsifiers, allowing for the creation of stable oil-in-water or water-in-oil emulsions.

Overall, hydrocolloids are valuable ingredients in pharmaceutical and cosmetics formulations, providing benefits such as improved stability, solubility, and texture.

Conclusion

In conclusion, hydrocolloids are versatile compounds with viscosity and gelation properties, making them valuable resources across various industries. Their ability to manipulate texture, flow behavior, and stability has led to wide-ranging applications in food and beverage, pharmaceuticals, cosmetics, and manufacturing sectors. Controlling viscosity and gelation is essential; manufacturers can achieve these characteristics through various techniques and additives. Hydrocolloids play a significant role in drug formulation, controlled release systems, skin care products, adhesives, coatings, and textiles.

As the field of hydrocolloids continues to evolve, cutting-edge research and development boost their potential further, offering new trends and opportunities. Choosing hydrocolloids over other compounds can lead to unique and improved products in terms of texture, stability, and overall shelf life. In conclusion, hydrocolloids' unique properties and applications make them a must-have compound for manufacturers across different sectors. Understanding and harnessing the full potential of hydrocolloids will continue transforming products in various industries.

FAQ

Q: How are hydrocolloids used?

Hydrocolloids have diverse applications across industries such as food and beverage, pharmaceuticals, cosmetics, etc. They enhance texture, improve stability, and create unique product characteristics.

Q: What is the role of viscosity in hydrocolloids?

Viscosity is a crucial characteristic of hydrocolloids as it affects various products' texture, flow behavior, and stability. It determines the thickness and resistance to flow, impacting the overall quality of the end product.

Q: How does gelation occur in hydrocolloids?

Gelation is the process by which hydrocolloids form a gel structure. It is influenced by factors such as temperature, pH, and concentration. Gelation provides products with the desired texture, stability, and functionality.

Q: How can viscosity and gelation be controlled in hydrocolloids?

Viscosity and gelation in hydrocolloids can be controlled through various techniques and additives. Adjusting parameters such as temperature, pH, and concentration and incorporating stabilizers and other additives allows manufacturers to achieve desired product characteristics.

Q: What are the specific applications of hydrocolloids in the food and beverage industry?

Hydrocolloids play a vital role in the food and beverage industry, where they are used for thickening, stabilizing, emulsifying, and creating unique textures in a wide range of products, including sauces, dressings, dairy products, and desserts.

FREQUENTLY ASKED QUESTIONS (FAQS) CONTINUED

- **Q: How do hydrocolloids contribute to the pharmaceutical and cosmetics industries?**

  Hydrocolloids find applications in pharmaceuticals and cosmetics for their ability to aid in drug formulation, controlled release systems, and the creation of various skincare products. They provide texture, stability, and improved delivery mechanisms.

- **Q: What are the industrial applications of hydrocolloids?**

  In addition to the food, beverage, pharmaceutical, and cosmetics industries, hydrocolloids are used in various industrial applications. They are utilized in adhesives, coatings, textiles, and other manufacturing processes where their properties contribute to improved performance and functionality.

- **Q: What are the future trends and innovations in hydrocolloids?**

  The field of hydrocolloids constantly evolves, with ongoing research and development leading to new trends and innovations. Scientists and manufacturers are exploring novel applications and improving existing hydrocolloid functionality for enhanced performance and sustainability.

- **Q: What is the significance of viscosity and gelation in hydrocolloids?**

  Viscosity and gelation are key properties of hydrocolloids that contribute to their versatile applications. By understanding and controlling these properties, manufacturers can harness the potential of hydrocolloids in various industries, creating products with the desired texture, stability, and functionality.

PART 3.2

Hydrocolloid Solubility and Hydration: Exploring the Key Properties

Hydrocolloids are essential components in various industries, making them valuable materials with numerous applications worldwide. Understanding their properties can help harness their full potential. This article will discuss hydrocolloids' solubility and hydration properties in detail.

We will explore how hydrocolloids dissolve and interact with water and their significance in industries such as food and healthcare. We will also delve into the factors that affect their solubility, such as temperature, pH, and molecular weight. We will also discuss the hydration process of hydrocolloids, including how they absorb and bind water molecules to form gels or viscous solutions.

Core Insights:

- ☑ Hydrocolloids are important materials in various industries.
- ☑ The solubility and hydration properties of hydrocolloids are crucial in understanding their functionality.
- ☑ Factors such as temperature, pH, and molecular weight can affect the solubility of hydrocolloids.
- ☑ Hydration plays a vital role in texture modification and product development.
- ☑ Exploring hydrocolloid solubility and hydration can unlock their potential in food, healthcare, and other industries.

3.2.1 Understanding Hydrocolloid Solubility

Hydrocolloids are widely used in various industries due to their unique properties. One of the essential characteristics of hydrocolloids is their solubility. Solubility refers to the ability of a substance to dissolve in a solvent, typically water. Hydrocolloid solubility plays a significant role in their overall functionality.

The solubility of hydrocolloids depends on several factors, including temperature, pH, and molecular weight. For example, some hydrocolloids are more soluble at higher temperatures, while others are more soluble at lower temperatures. Additionally, the solubility of hydrocolloids may vary depending on the pH of the solution. Hydrocolloids with high molecular weights tend to have lower solubility than those with lower molecular weights.

Various solubility tests can be used to determine the solubility characteristics of hydrocolloids. These tests include the "cloud point" test, which measures the temperature at which a hydrocolloid solution becomes cloudy due to the formation of insoluble aggregates. Another is the "pH sweep" test, which measures solubility at different pH levels to determine the pH range in which a hydrocolloid is most soluble.

Understanding hydrocolloid solubility is crucial for optimal utilization of their properties in various applications. Different solubility characteristics can be tailored to meet specific formulation needs in different industries. For example, hydrocolloids with high solubility are ideal for clear beverages, while those with low solubility are useful for stabilizing emulsions and suspensions.

3.2.2 Hydrocolloid Hydration Process

Hydration is a critical process in the functionality and performance of hydrocolloids. Hydrocolloids are water-soluble biopolymers that can interact with water molecules through hydrogen bonding, electrostatic interaction, and Van der Waals forces.

The hydration process involves the absorption and binding of hydrocolloid molecules of water molecules, resulting in their swelling and dispersion in water. The degree of hydration depends on various factors, including the type of hydrocolloid, the pH and temperature of the solution, and the hydrocolloid concentration.

Hydrocolloids can form gels or viscous solutions when hydrated, depending on the type and concentration of the hydrocolloid and the processing conditions. Gels are formed when the interactions between hydrocolloid molecules result in a three-dimensional network that traps water molecules, forming a solid-like structure. Viscous solutions are formed when hydrocolloid molecules disperse in water, increasing solution viscosity and thickness.

The hydration process of hydrocolloids is crucial in enhancing their functional properties, such as thickening, emulsifying, and stabilizing. For example, the ability of hydrocolloids to form gels enables them to act as thickeners and stabilizers in food products. In contrast, their ability to form viscous solutions is critical in providing a smooth texture and mouthfeel in beverages and other products.

3.2.3 Factors Affecting Hydrocolloid Hydration

The hydration properties of hydrocolloids are affected by various factors, including:

- **pH:** The pH of the solution can affect the ionization and charge density of the hydrocolloid molecules, influencing their interactions with water molecules.
- **Temperature:** The solution's temperature can affect the hydrocolloid's solubility and viscosity and the rate of hydration.
- **Concentration:** The concentration of the hydrocolloid can affect its ability to hydrate and form gels or viscous solutions.

Understanding these factors is essential in controlling and manipulating the hydration properties of hydrocolloids for specific applications.

> "The hydration process of hydrocolloids is crucial in enhancing their functional properties, such as thickening, emulsifying, and stabilizing."

3.2.4 Hydrocolloids in Healthcare

Hydrocolloids have emerged as essential components in several healthcare applications, ranging from wound care products to drug delivery systems.

One of the most common uses of hydrocolloids in healthcare is producing wound dressings. Hydrocolloid dressings are designed to provide a moist environment for optimal wound healing by absorbing exudate, maintaining moisture balance, and supporting autolytic debridement. These dressings are particularly effective for treating chronic wounds, such as pressure ulcers, diabetic foot ulcers, and leg ulcers.

Hydrocolloids are also being used in drug delivery systems to enhance bioavailability and ensure controlled release of drugs. Hydrocolloid-based drug carriers can improve the solubility and stability of poorly soluble drugs, thereby improving their efficacy. They can also protect drugs from degradation and reduce the frequency of dosing.

3.2.5 Hydrocolloids in Biomedical Applications

Hydrocolloids are also gaining attention in biomedical applications due to their unique physicochemical properties. Researchers are exploring the use of hydrocolloids in tissue engineering, drug delivery, and imaging applications. Hydrocolloids, such as alginate, chitosan, and hyaluronic acid, have been shown to support cell growth and tissue regeneration, making them promising candidates for producing biocompatible materials.

3.2.6 The Role of Solubility in Hydrocolloid Functionality

The solubility characteristics of hydrocolloids directly impact their functionality in various industries, including food and healthcare. Solubility is defined as the ability of a substance to dissolve in another substance, and in the case of hydrocolloids, the ability to dissolve and interact with water is crucial.

The solubility of hydrocolloids depends on several factors, including temperature, pH, and molecular weight. Hydrocolloids with low molecular weight and high ionization are more soluble in water. Additionally, the solubility of hydrocolloids can be influenced by the presence of other substances in a formulation.

The solubility of hydrocolloids plays a critical role in their functionality. For instance, in the food industry, the solubility of hydrocolloids affects their ability to thicken, stabilize, and emulsify food products. Hydrocolloids that are more soluble in water are better thickeners, while those with low solubility tend to be better stabilizers.

Furthermore, the solubility characteristics of hydrocolloids can be tailored to meet specific formulation needs. For example, in the healthcare industry, hydrocolloids with controlled solubility are used in drug delivery systems to regulate the release of active ingredients.

3.2.7 Hydration and Texture Modification

The hydration properties of hydrocolloids play a crucial role in the texture modification of food and other products. Hydrocolloids can absorb water and form gels or viscous solutions, which can change the texture and mouthfeel of food. The amount and quality of water absorbed affect the final texture of the product.

Hydration can also control the release of flavors and aromas in food products. For example, hydrocolloids can encapsulate and release flavors over time, providing a gradual and prolonged sensory experience. Hydrocolloids can also create texture gradients, such as a crispy outer layer and a soft inner core in fried foods.

Texture modification using hydrocolloid hydration is not limited to food products. It can also be used in cosmetics and personal care products. Hydrocolloids are used to create gels and creams with desirable textures and stabilize emulsion-based products.

Various techniques can be used to manipulate hydrocolloid hydration to achieve specific textural properties. For example, using different types of hydrocolloids, adjusting the pH, or adding salts can alter the hydration properties and create different textures. Understanding hydrocolloids' hydration properties is crucial for achieving the desired texture and sensory attributes in various products.

3.2.8 Unlocking the Potential of Hydrocolloids

Hydrocolloids are versatile ingredients that offer unique solutions to various complex problems across multiple industries. With their solubility and hydration properties, we can create high-quality products that meet consumers' needs. Hence, it is essential to understand and harness their full potential to improve our daily lives.

Conclusion

In conclusion, understanding the hydrocolloid solubility and hydration properties is vital for maximizing their potential in various industries. The solubility of hydrocolloids is affected by factors such as temperature, pH, and molecular weight, which determine their functionality as stabilizers, thickeners, and emulsifiers in food products. Meanwhile, the hydration process of hydrocolloids influences texture modification and plays a crucial role in wound care and drug delivery systems in the healthcare industry.

Hydrocolloid research continues to advance, leading to emerging trends in their applications in cosmetics, personal care products, paper manufacturing, and other industrial sectors. We can discover innovative solutions in various fields by exploring the scientific principles behind hydrocolloid solubility and hydration.

FAQ

Q: What factors affect the solubility of hydrocolloids?

The solubility of hydrocolloids is influenced by factors such as temperature, pH, and molecular weight. Higher temperatures and lower pH levels generally increase solubility, while larger molecular weight hydrocolloids may have decreased solubility.

Q: How do hydrocolloids hydrate and bind water molecules?

Hydrocolloids absorb water and form gels or viscous solutions through hydration. Due to their molecular structure and interactions, they can bind water molecules, contributing to their functional properties in various industries.

Q: What are some examples of hydrocolloids used in the food industry?

Hydrocolloids are widely used in the food industry as stabilizers, thickeners, emulsifiers, and texture modifiers. Common examples include carrageenan, xanthan gum, guar gum, and pectin.

Q: How are hydrocolloids utilized in wound care products and healthcare?

Hydrocolloids are commonly used in wound care products, such as hydrocolloid dressings, to promote healing and maintain a moist environment for optimal wound management. They are also utilized in drug delivery systems and have applications in biomedical research.

Q: How does solubility affect the functionality of hydrocolloids?

Solubility is closely linked to the functional properties of hydrocolloids, such as their thickening, gelling, and stabilizing abilities. The solubility characteristics of hydrocolloids can be tailored to meet specific formulation needs in different industries.

FREQUENTLY ASKED QUESTIONS (FAQS) CONTINUED

Q: How do the hydration properties of hydrocolloids impact texture modification?

The hydration properties of hydrocolloids play a crucial role in the texture modification of food and other products. Manipulating hydration can lead to desired changes in texture, mouthfeel, and sensory attributes.

Q: What recent advancements have been made in hydrocolloid research?

Hydrocolloid research continually evolves, leading to advancements in understanding their solubility, hydration, and functionality. Recent developments include new insights into their applications and potential future innovations.

Q: Where else are hydrocolloids used besides food and healthcare?

Hydrocolloids find industrial applications in diverse sectors, such as cosmetics, personal care products, and paper manufacturing. Their unique properties make them valuable in these industries.

Q: Why is it important to understand hydrocolloid solubility and hydration?

Understanding hydrocolloid solubility and hydration is crucial for harnessing their full potential in various industries. These properties directly impact functionality, texture modification, and product development.

PART 3.3

Hydrocolloid Temperature Sensitivity

Have you ever wondered why some materials behave differently when exposed to varying temperatures? Hydrocolloids are one such material that exhibits unique behavior in response to temperature changes. This section will explore the fascinating world of hydrocolloid and its temperature sensitivity.

Core Insights:

- ☑ Hydrocolloids exhibit unique behavior when exposed to different temperatures.
- ☑ Understanding hydrocolloid composition and structure is crucial to comprehending their temperature sensitivity.
- ☑ Hydrocolloids have a wide range of industrial applications, with their temperature sensitivity being particularly interesting in the food, pharmaceutical, and cosmetics industries.

3.3.1 Understanding Hydrocolloid

Hydrocolloids are substances that form gels or viscous dispersions when dispersed in water. They are widely used in the food, pharmaceutical, and cosmetic industries for their unique properties. Hydrocolloids can be derived from animal, plant, or synthetic sources, and their chemical composition and physical behavior can vary widely.

Chemical Composition

Hydrocolloids consist of long, branched chains of molecules, usually polysaccharides or proteins, which can attract and bind water molecules. These chains have hydrophilic (water-loving) and hydrophobic (water-repelling) regions, which enable them to form gel-like structures. Hydrocolloids' molecular weight and structure determine their solubility, viscosity, and gelation properties.

3.3.2 The Chemistry of Hydrocolloid

Hydrocolloids are long-chain polymers that disperse or dissolve in water to form a viscous or gel-like solution. Their unique properties are a result of their chemical composition and molecular structure. The chemical formula of hydrocolloids varies depending on the type of polymer, but they generally contain repeating units of hydrophilic and hydrophobic monomers.

The hydrophilic monomers are attracted to water and form hydrogen bonds with the water molecules, while the hydrophobic monomers are repelled by water and tend to aggregate together. This combination of hydrophilic and hydrophobic properties forms a three-dimensional network that can trap water and other molecules.

The chemical composition of hydrocolloids and the structure of their molecular network determine their functionality and behavior. For example, some hydrocolloids are highly soluble in cold water but form a gel when heated. At the same time, others remain stable at high temperatures but break down in acidic or alkaline environments.

3.3.3 Applications of Hydrocolloids in Personal Care Products

Hydrocolloids are commonly used in personal care products such as shampoos, conditioners, and lotions. They help to improve the texture and viscosity of these products while providing moisturization and protecting the skin from irritation. Some commonly used hydrocolloids in personal care products include carboxymethyl cellulose (CMC), hydroxypropyl guar (HPG), and carrageenan.

In summary, hydrocolloids have a wide range of applications across many industries. Their unique properties, such as thickening, stabilizing, and gelling, make them a valuable resource in formulating various products.

3.3.4 Understanding Temperature Sensitivity

Temperature sensitivity is a fascinating concept that refers to changes in the behavior of materials when exposed to different temperatures. In the context of hydrocolloids, temperature sensitivity is of immense importance. Understanding hydrocolloids' temperature sensitivity helps us optimize their performance and unlock their full potential in various industrial applications.

The temperature sensitivity of hydrocolloids is affected by several factors, such as their chemical composition, structure, and concentration. The behavior of hydrocolloids may change with temperature, affecting their solubility, viscosity, gelation, and stability in different ways.

Hydrocolloids can exhibit various temperature sensitivity behaviors, such as thermal reversibility or irreversibility. A thermally reversible hydrocolloid can change its behavior when heated or cooled, while a thermally irreversible hydrocolloid permanently alters its behavior after being subjected to a particular temperature.

The temperature range that affects the behavior of hydrocolloids can also vary widely. Some hydrocolloids may be sensitive to small temperature fluctuations, while others may require extreme temperature changes to exhibit temperature-sensitive behavior.

Overall, understanding temperature sensitivity is crucial in optimizing the performance of hydrocolloids in different industrial applications. In the following sections, we will explore the effects of temperature on hydrocolloids and techniques used to manage their temperature sensitivity.

3.3.5 Effects of Temperature on Hydrocolloids

Temperature plays a critical role in the properties and behavior of hydrocolloids. Understanding the effects of temperature on hydrocolloids is essential in optimizing their functionality and performance in various applications.

Viscosity: The viscosity of hydrocolloids is highly temperature dependent. As the temperature increases, the viscosity decreases, and vice versa. This is due to the changes in the molecular structure of hydrocolloids caused by temperature fluctuations. For instance, carrageenan, a popular hydrocolloid used in the food industry, exhibits a gel-like consistency at low temperatures, but it becomes more fluid at higher temperatures.

Gelation: The ability of hydrocolloids to form gels is also affected by temperature. Some hydrocolloids require heat to form gels, while others require cooling. For instance, gelatin gels are formed by cooling a heated solution, while agar gels require heating up to a specific temperature for gelation.

Solubility: Temperature has a direct impact on the solubility of hydrocolloids. For example, pectin, a hydrocolloid commonly used in producing jams and jellies, requires high temperatures and low pH levels to dissolve completely.

Stability: The stability of hydrocolloids is also affected by temperature changes. Hydrocolloids can degrade or denature when exposed to high temperatures, leading to loss of functionality. For instance, the stability of xanthan gum, a popular thickening agent, decreases significantly above a temperature of 88°C.

Overall, temperature sensitivity is a critical aspect of hydrocolloids and should be considered when formulating products that require their use. Knowing how temperature affects the properties and behavior of hydrocolloids enables researchers to develop new and innovative solutions to meet the needs of different industries.

3.3.6 Hydrocolloid Temperature Sensitivity Treatment

Managing temperature sensitivity in hydrocolloids is critical for many industrial applications. Fortunately, several techniques and processing methods are available to optimize their performance under specific temperature conditions.

One approach is to modify the chemical structure of hydrocolloids to enhance their thermal stability. For example, cross-linking can increase the hydrocolloid's resistance to temperature changes, resulting in improved functional properties.

Another method combines different types of hydrocolloids to create a synergistic effect that reduces temperature sensitivity. This approach is particularly useful in food products, where a combination of hydrocolloids can provide the desired texture and stability while minimizing sensitivity to temperature changes.

Encapsulation is another technique used to protect hydrocolloids from temperature fluctuations. Encapsulating hydrocolloids in a protective matrix can reduce their temperature sensitivity, improving their performance and stability in various applications.

Finally, processing methods such as spray-drying and freeze-drying can produce hydrocolloids with improved temperature stability. These processes can help to reduce the water content in hydrocolloids, minimizing their sensitivity to temperature changes and enhancing their functional properties.

By employing these techniques and processing methods, we can manage and manipulate the temperature sensitivity of hydrocolloids, unlocking their full potential in various industrial applications.

3.3.7 Industrial Applications of Temperature-Sensitive Hydrocolloids

Hydrocolloids that exhibit temperature-sensitive behavior have various applications in various industries. Here are some specific examples:

Table 5: Hydrocolloids Temperature-Sensitivity

INDUSTRY	APPLICATION
Food	Hydrocolloids are used to create products such as pudding, yogurt, and ice cream, where their temperature sensitivity helps maintain a stable texture and consistency.
Pharmaceuticals	Hydrocolloids are used in controlled-release formulations, where temperature sensitivity plays a critical role in the rate and duration of drug release.
Cosmetics	Hydrocolloids create temperature-sensitive gels for personal care products, such as hair styling gels and face masks.

These applications are just the tip of the iceberg, as hydrocolloids continue finding innovative uses in various fields.

FOOD

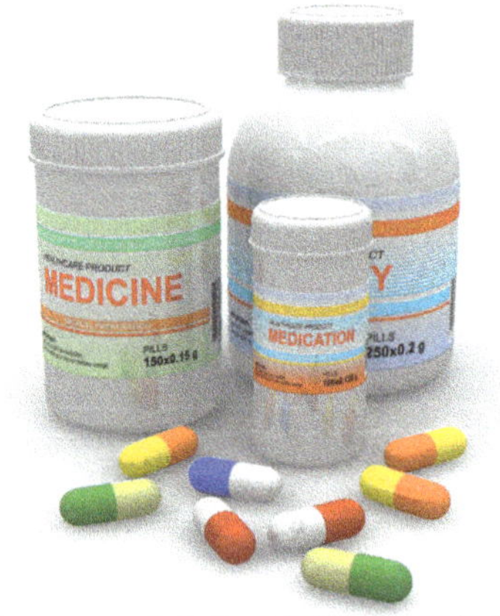

PHARMACEUTICALS

COSMETICS

3.3.8 Future Trends and Research in Hydrocolloid Temperature Sensitivity

As the demand for hydrocolloids continues to grow in various industries, researchers are focusing on exploring new applications and innovating existing formulations to enhance their performance. Here are some future trends and research areas in hydrocolloid temperature sensitivity:

- **Nanotechnology:** Researchers are exploring the use of nanotechnology to enhance the properties of hydrocolloids, such as their solubility, stability, and temperature sensitivity.
- **Bio-based hydrocolloids:** With the increasing demand for sustainable and eco-friendly products, researchers are developing hydrocolloids from renewable sources such as seaweed, plant-based materials, and microbial fermentation.
- **Polymer blends:** Researchers are investigating using polymer blends to improve the temperature sensitivity of hydrocolloids. Combining different polymers can create hydrocolloids with enhanced properties, such as improved gelation and viscosity at specific temperature ranges.
- **Application-specific formulations:** With advancements in processing technologies, researchers are developing hydrocolloid formulations tailored to specific applications and temperature conditions, such as high-temperature processing and freeze-thaw stability.

3.3.9 The Future of Hydrocolloid Temperature Sensitivity

Research in hydrocolloid temperature sensitivity is continuously evolving, with ongoing efforts to optimize and expand the use of hydrocolloids in industrial applications. The development of new types of hydrocolloids with specific temperature-sensitive properties and improved performance is an area of active research.

Furthermore, emerging trends in natural and sustainable ingredients drive the development of new hydrocolloid-based formulations that offer environmentally friendly solutions in various industries. The future of hydrocolloid temperature sensitivity is bright, with exciting potential for innovative applications and discoveries.

In summary, hydrocolloid temperature sensitivity is a fascinating field of study that offers significant potential for various industrial applications. By continuing to advance our understanding of hydrocolloid behavior under different temperature conditions, we can unlock their full potential and improve the performance of various products and processes.

Conclusion

The study of hydrocolloid temperature sensitivity has revealed fascinating properties that can be harnessed for various industrial applications. By understanding how hydrocolloids behave under different temperature conditions, we can optimize their use in food, pharmaceuticals, and personal care products.

Hydrocolloids' chemical composition and molecular structure play a vital role in their temperature sensitivity. Different types of hydrocolloids exhibit varying degrees of sensitivity to temperature changes, which can be manipulated to achieve specific effects.

Temperature can significantly impact the properties of hydrocolloids, affecting their viscosity, gelation, solubility, and stability. However, through careful formulation, processing techniques, and innovative solutions, we can manage and optimize the temperature sensitivity of hydrocolloids to improve their performance in various applications.

FAQ

Q: How are hydrocolloids used in food?

Hydrocolloids are widely used in the food industry for various purposes, such as thickening, stabilizing, emulsifying, and gelling. They help improve food products' texture, consistency, and shelf life.

Q: What are some examples of hydrocolloids in food?

A: Some common examples of hydrocolloids used in food include agar, pectin, carrageenan, xanthan gum, and guar gum. Each has specific functions and applications in different food products.

Q: How does temperature affect hydrocolloids?

Temperature can significantly impact the behavior of hydrocolloids. It can affect their viscosity, gelation, solubility, and stability. Understanding temperature sensitivity is important when working with hydrocolloids.

Q: How can the temperature sensitivity of hydrocolloids be managed?

The temperature sensitivity of hydrocolloids can be managed through various techniques, formulations, and processing methods. These include adjusting pH, adding salts, utilizing specific heating or cooling processes, or combining different hydrocolloids to achieve desired results.

Q: What are some industrial applications of temperature-sensitive hydrocolloids?

Temperature-sensitive hydrocolloids find applications in food, pharmaceuticals, cosmetics, and biotechnology industries. They are used for controlled release systems, drug delivery, encapsulation, and as smart materials with temperature-triggered responses.

Q: What are the future trends and research in hydrocolloid temperature sensitivity?

Ongoing research focuses on developing new temperature-responsive hydrocolloids, exploring their potential in advanced applications, and optimizing their performance under specific conditions. The field continues to evolve, offering exciting prospects for future developments.

CHAPTER 04

Applications of Hydrocolloids

PART 4.1

Applications of Hydrocolloids in the Food Industry: Thickening, Gelling, Stabilizing, Emulsifying

Hydrocolloids are widely used in the food industry for various applications. They are primarily thickening and gelling agents in soups, gravies, salad dressings, sauces, and toppings. Hydrocolloids like pectin, agar, carrageenan, and gelatin are extensively used in jams, jellies, marmalades, and low-sugar/calorie gels. They also can stabilize emulsions and improve the texture of food products.

Core Insights:

- ☑ Hydrocolloids are versatile ingredients used in the food industry for thickening, gelling, stabilizing, and emulsifying.
- ☑ Common hydrocolloids used include pectin, agar, carrageenan, and gelatin.
- ☑ They are used in various food products such as soups, gravies, salad dressings, sauces, jams, jellies, and marmalades.
- ☑ Hydrocolloids are crucial in improving food products' texture, stability, and overall quality.
- ☑ Ongoing research and advancements continue to expand our understanding of hydrocolloids and their potential applications in the food industry.

4.1.1 Introduction to Hydrocolloids and their Properties

Hydrocolloids are a diverse group of compounds that play a crucial role in the food industry. These compounds, which include polysaccharides and proteins, can form viscous dispersions and gels when dispersed in water. They are characterized by their hydrophilic nature, meaning they have a high affinity for water.

The properties of hydrocolloids make them versatile ingredients in food formulations. One of their key properties is their impact on the rheology of food systems, including flow behavior (viscosity) and mechanical solid property (texture). Hydrocolloids like starch, xanthan gum, guar gum, and gelatin are commonly used in the food industry for their thickening and gelling properties.

Moreover, hydrocolloids can stabilize emulsions and improve the overall texture of food products. They play a crucial role in preventing oil and water separation in emulsions, such as mayonnaise and salad dressings. Hydrocolloids are known for their water-binding properties, which help retain moisture in food products and contribute to their desired texture.

Properties of Hydrocolloids

Table 6: Properties of Hydrocolloids

PROPERTY	EXAMPLES OF HYDROCOLLOIDS
Thickening	Starch, xanthan gum, guar gum
Gelling	Alginate, pectin, carrageenan
Stabilizing	Gums (guar gum, locust bean gum)
Emulsifying	Lecithin, gum Arabic, modified starches
Water Binding	Guar gum, xanthan gum, carrageenan

Overall, hydrocolloids are essential ingredients in the food industry due to their wide range of properties and functionalities. Their ability to modify food products' texture, stability, and sensory experience makes them invaluable tools for food formulators and manufacturers.

4.1.2 Thickening Agents in the Food Industry

Thickening agents play a crucial role in food formulations, helping to increase the viscosity and improve the texture of various food products. In the food industry, hydrocolloids are widely used as effective thickening agents due to their ability to modify food systems' flow behavior and thickness. Different hydrocolloids exhibit varying thickening properties, offering a wide range of options for food formulators.

Starch, xanthan gum, guar gum, and carboxymethyl cellulose (CMC) are commonly utilized hydrocolloids for thickening in the food industry. Starch, derived from various sources such as corn, rice, or wheat, is a versatile thickening agent that provides a smooth, creamy texture to soups, sauces, and gravies. Xanthan gum, a microbial polysaccharide, offers excellent thickening and stabilizing properties, making it suitable for hot and cold applications. Guar gum, extracted from the guar bean, is known for its high viscosity at low concentrations, making it ideal for thickening a wide range of food products. Lastly, carboxymethyl cellulose (CMC), derived from cellulose, is often used in processed foods and beverages to enhance the viscosity and impart a smooth mouthfeel.

Table 7: Hydrocolloid Thickening Properties

HYDROCOLLOID	THICKENING PROPERTIES	APPLICATIONS
Starch	Smooth, creamy texture	Soups, sauces, gravies
Xanthan Gum	Excellent thickening and stabilizing	Dressings, sauces, bakery fillings
Guar Gum	High viscosity at low concentrations	Dairy products, desserts, processed meats
Carboxymethyl Cellulose (CMC)	Enhanced viscosity, smooth mouthfeel	Processed foods, beverages

When formulating food products, choosing the right thickening agent is essential based on the desired viscosity, texture, and application. The selection of hydrocolloids depends on factors such as processing conditions, pH, temperature, and compatibility with other ingredients. Food manufacturers carefully consider these factors to ensure that the desired texture and consistency are achieved while maintaining the overall quality and stability of the product.

4.1.3 Gelling Agents in the Food Industry

Hydrocolloids possess the unique ability to form gels, making them valuable gelling agents in the food industry. These gels are created by cross-linking polymer chains, resulting in a three-dimensional network that entraps water and forms a rigid structure. Gelling agents are commonly used in various food products, including puddings, jellies, dairy desserts, and bakery items, to enhance their textural properties and improve the overall sensory experience.

Several hydrocolloids are commonly used as gelling agents in the food industry; among the most popular choices are alginate, pectin, carrageenan, gelatin, Gellan, and agar. These gelling agents have distinct textural properties, allowing food manufacturers to achieve specific gel characteristics in their products. By selecting the appropriate gelling agent, formulators can ensure the desired texture and consistency of the final food product.

For example, agar is often used in desserts like custards and fruit jellies due to its firm and brittle gel texture. On the other hand, pectin is commonly employed in jams and spreads for its ability to create a soft and spreadable gel. By carefully selecting the gelling agent, food manufacturers can meet consumer preferences and create superior mouthfeel and texture products.

The versatility of hydrocolloid gels

Hydrocolloid gels offer a wide range of possibilities in the food industry. These gels can be adjusted to provide different textures, from soft and delicate to firm and resilient. In addition to their textural properties, hydrocolloid gels can contribute to food products' stability, appearance, and overall quality. They can help retain moisture, control syneresis (the separation of liquid from a gel), and prevent the migration of water or oil in composite food systems.

Furthermore, hydrocolloid gels can encapsulate flavors, colors, or active ingredients, protecting them from degradation, oxidation, or premature release. This encapsulation ability is particularly useful in the development of functional foods and in improving the shelf life of various food products.

Table 8: Gelling Agents and their Common Applications

GELLING AGENT	COMMON APPLICATIONS
Agar	Custards, fruit jellies, confectionery
Pectin	Jams, spreads, fruit fillings
Carrageenan	Dairy desserts, chocolate milk, ice cream
Gelatin	Gummy candies, marshmallows, desserts
Gellan	Sauces, dressings, bakery products

4.1.4 Stabilizing Agents in the Food Industry

Hydrocolloids play a crucial role in stabilizing food products and improving their stability. They help prevent phase separation or the breakdown of emulsions in products like mayonnaise, salad dressings, and sauces. The unique properties of hydrocolloids, such as their ability to absorb and retain water, allow them to form a stable network that helps maintain the texture and overall quality of the food products.

Guar gum is a commonly used stabilizing agent in the food industry. This hydrocolloid has excellent water-binding properties, making it ideal for preventing moisture loss and maintaining the desired texture in various food products. Guar gum forms a gel-like structure that helps stabilize emulsions and prevent oil and water separation. It also enhances the mouthfeel of food products, providing a smooth and creamy texture.

The Role of Hydrocolloids in Food Stability

Hydrocolloids, such as guar gum, play a vital role in maintaining the stability of food products. Hydrocolloids ensure that food products maintain their desired texture and appearance throughout their shelf life by stabilizing emulsions and preventing phase separation.

In addition to guar gum, locust bean gum is another popular stabilizing agent used in the food industry. Its high viscosity and adhesive properties help improve the stability of various food formulations. Locust bean gum acts as a thickener and stabilizer, preventing sedimentation and maintaining uniformity in ice creams, sauces, and beverages.

Overall, hydrocolloids have proven valuable stabilizing agents in the food industry. By preventing phase separation, preserving texture, and ensuring product consistency, these hydrocolloids play a key role in enhancing the quality and shelf life of a wide range of food products.

Table 9: Hydrocolloids Stabilizing Properties

HYDROCOLLOID	STABILIZING PROPERTIES	APPLICATIONS
Guar Gum	Excellent water-binding properties prevent phase separation and enhance texture.	Mayonnaise, salad dressings, sauces
Locust Bean Gum	High viscosity, adhesive properties, prevents sedimentation	Ice creams, sauces, beverages

4.1.5 Emulsifying Agents in the Food Industry

Hydrocolloids play a crucial role in the food industry as emulsifying agents, helping to stabilize emulsions and prevent the separation of oil and water. Emulsions like oil-in-water and water-in-oil are commonly found in products like mayonnaise, salad dressings, and ice creams. Hydrocolloids create a stable interface between the oil and water phases, ensuring a homogeneous mixture and improved texture in these food products.

Commonly used hydrocolloids as emulsifying agents include lecithin, gum Arabic, and some modified starches. These hydrocolloids can form a protective layer around the dispersed oil droplets, preventing them from coalescing and separating from the water phase. This enhances emulsion-based food products' stability, appearance, and overall quality.

By using hydrocolloids as emulsifying agents, food manufacturers can create products with a smooth and creamy texture, where the oil and water phases are well integrated. This not only improves the sensory experience for consumers but also extends the shelf life of the products by preventing phase separation and maintaining the desired consistency over time.

> "Hydrocolloids create a stable interface between the oil and water phases, ensuring a homogeneous mixture and improved texture in emulsion-based food products."

4.1.6 Water Binding Properties of Hydrocolloids

One of the key properties of hydrocolloids is their excellent water-binding capabilities. Hydrocolloids can retain moisture in food products, ensuring they stay fresh and maintain their desired texture. This is particularly important in moisture-sensitive applications such as bakery products and processed meats.

Hydrocolloids like guar gum, xanthan gum, and carrageenan are known for their exceptional water-binding properties. They can absorb and hold onto water molecules, preventing them from evaporating or migrating from the food product. As a result, the food product remains moist and doesn't become dehydrated.

This water-binding property also contributes to the improved texture of the food product. By holding onto moisture, hydrocolloids help create a desirable mouthfeel that is moist, smooth, and enjoyable to eat. They prevent the food from becoming too dry or crumbly, enhancing its overall sensory experience.

Table 10: Water Binding Properties

HYDROCOLLOID	WATER BINDING CAPACITY
Guar Gum	High
Xanthan Gum	High
Carrageenan	Medium

It is worth noting that different hydrocolloids have varying water-binding capacities. Some hydrocolloids, like guar gum and xanthan gum, have high water binding capacities, making them suitable for applications where moisture retention is crucial. Others, like carrageenan, have a medium water-binding capacity but still contribute positively to the moisture and texture of food products.

The water-binding properties of hydrocolloids make them valuable ingredients in the food industry. They ensure that food products have an optimal moisture content, maintain their desired texture, and provide an enjoyable eating experience for consumers.

4.1.7 Hydrocolloids and Food Ingredient Functionality

Hydrocolloids are not limited to their role as thickening, gelling, stabilizing, and emulsifying agents in the food industry. They possess a wide range of functional properties that contribute to the overall quality of food products. One of their key functionalities is controlling the crystal growth of ice and sugar, which is particularly important in frozen desserts and confectionery items. By interacting with water molecules, hydrocolloids can help create a smooth and creamy texture, enhancing the sensory experience for consumers.

Another significant functionality of hydrocolloids is their ability to improve the mouthfeel of food products. They can provide a desirable viscosity and contribute to a pleasant texture, making the eating experience more enjoyable.

Whether it's a creamy sauce, a velvety pudding, or a smooth ice cream, hydrocolloids play a crucial role in achieving the desired mouthfeel that consumers expect.

In addition to their impact on texture and mouthfeel, hydrocolloids can also enhance food products' overall sensory experience. They can improve the flavor release and perception, adding depth and complexity to the taste profile. Hydrocolloids can also help retain moisture in food products, preventing dehydration and maintaining the desired freshness and chewiness in bakery items and processed meats.

Functional Properties of Hydrocolloids

- Control crystal growth of ice and sugar
- Improve mouthfeel and texture
- Enhance flavor release and perception
- Maintain moisture and freshness

> "Hydrocolloids are versatile ingredients beyond their traditional roles in the food industry. Their functional properties are crucial in achieving various food products' desired quality, texture, and sensory experience."

Table 11: Food Functionality of Hydrocolloids

FOOD FUNCTIONALITY	HYDROCOLLOID
Thickening	Starch, xanthan gum, guar gum, carboxymethyl cellulose (CMC)
Gelling	Alginate, pectin, carrageenan, gelatin, Gellan, agar
Stabilizing	Guar gum, locust bean gum, xanthan gum
Emulsifying	Lecithin, gum Arabic, modified starches

As research and advancements continue in the field of hydrocolloids, we can expect even more applications and functional properties to be discovered. These versatile ingredients will continue to play a crucial role in food formulations, enabling manufacturers to create innovative and high-quality products that meet the ever-changing demands of consumers.

4.1.8 Hydrocolloids in Food Industry Applications

Hydrocolloids, with their versatile properties, find numerous applications in the food industry. These functional ingredients are used in a wide range of food formulations, contributing to various products' texture, stability, and overall quality. Hydrocolloids are added to soups, gravies, salad dressings, sauces, toppings, ice creams, jams, jellies, bakery products, confectionery items, and more to enhance their sensory experience and improve their shelf life.

One major application of hydrocolloids is as thickening agents. They help increase the viscosity of food products, such as soups, gravies, and sauces, improving their mouthfeel and consistency. Hydrocolloids like starch, xanthan gum, guar gum, and carboxymethyl cellulose (CMC) are commonly used in the food industry for their thickening properties. These hydrocolloids provide different viscosities at varying concentrations, allowing formulators to achieve the desired texture in their products.

In addition to thickening, hydrocolloids also serve as gelling agents, creating a three-dimensional network that traps water and forms a rigid structure. This property is utilized in various food products like puddings, jellies, dairy desserts, and bakery items. Alginate, pectin, carrageenan, gelatin, Gellan, and agar are commonly used hydrocolloids for gelling in the food industry. These gels not only enhance the texture but also contribute to the overall sensory experience of the products.

Furthermore, hydrocolloids act as stabilizing agents, preventing phase separation or breakdown of emulsions in products like mayonnaise, salad dressings, and sauces. Guar gum and locust bean gum are commonly used stabilizers that help maintain the texture and quality of the food products, extending their shelf life. Additionally, hydrocolloids can function as emulsifying agents, stabilizing oil-in-water and water-in-oil emulsions in various food products, ensuring a homogeneous mixture.

GLOBAL HYDROCOLLOIDS MARKET

Market Segmentation and Scope

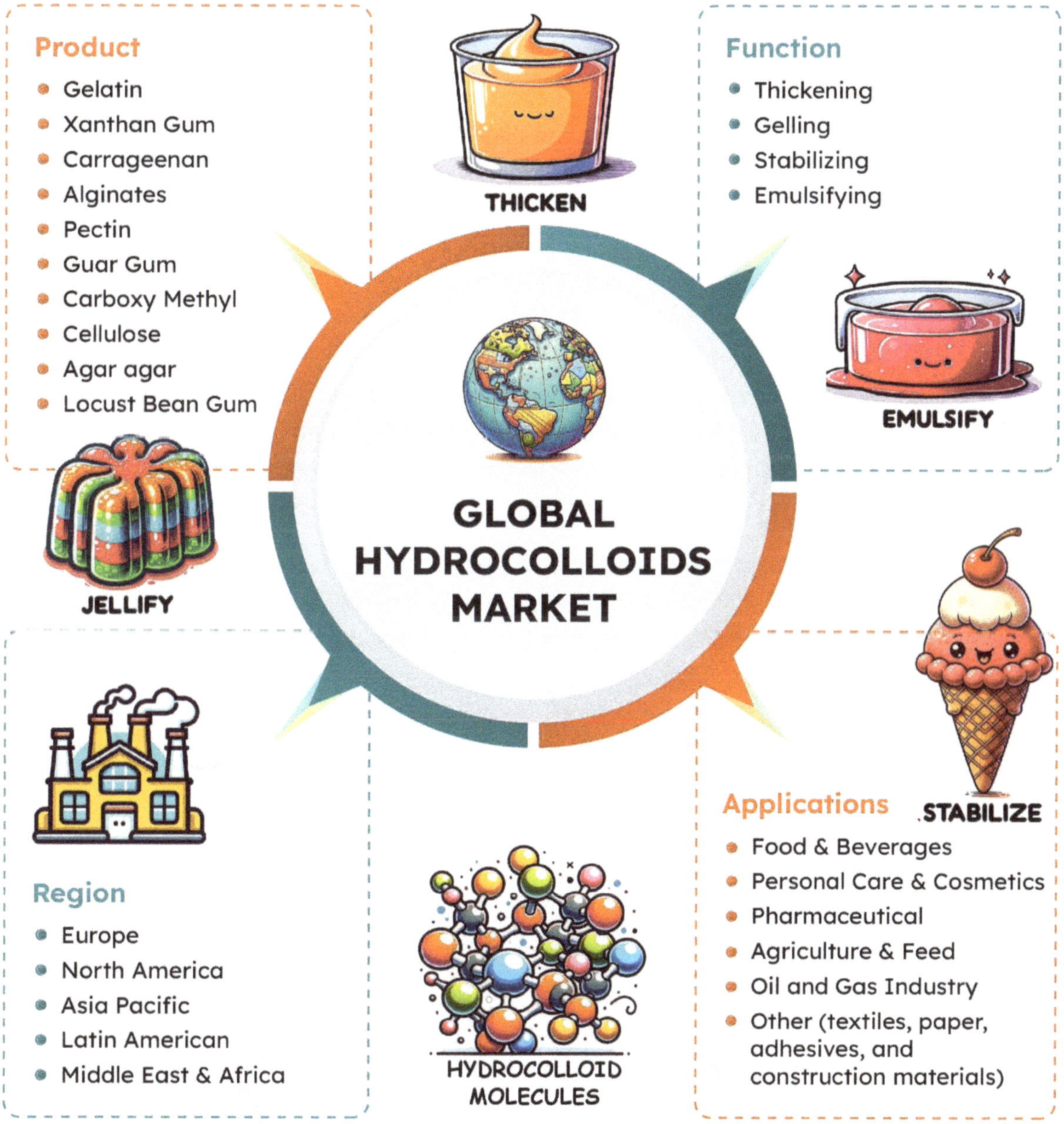

Conclusion

Hydrocolloids are indispensable in the food industry, offering a wide range of applications and properties that enhance the texture, stability, and overall quality of food products. With their thickening, gelling, stabilizing, and emulsifying properties, hydrocolloids find extensive use in soups, gravies, sauces, dressings, bakery items, and more. These versatile ingredients contribute to the desirable characteristics of food formulations, creating an enjoyable sensory experience for consumers.

Continued research and hydrocolloid advancements have deepened our understanding of their properties and potential applications. Scientists and researchers are exploring hydrocolloids' structure, functionality, and interactions with other food ingredients, aiming to unlock new possibilities for their use in various food products. These ongoing studies drive innovation and open doors to optimizing the performance of hydrocolloids in food formulations.

As the demand for functional food ingredients grows, hydrocolloids will continue to play a crucial role in the food industry. Their multifunctional nature and ability to improve texture, stability, and shelf life make them valuable tools for food formulators and manufacturers. With ongoing research, hydrocolloids have the potential to offer even more exciting applications and contribute to the development of innovative food products in the future.

FAQ

What are the primary applications of hydrocolloids in the food industry?

Hydrocolloids primarily thicken and gell in soups, gravies, salad dressings, sauces, and toppings. They are also used to stabilize emulsions, improve texture, enhance the mouthfeel of food products, and improve overall sensory experience.

What are some examples of hydrocolloids used in food products?

Some common examples of hydrocolloids used in food products are pectin, agar, carrageenan, gelatin, starch, xanthan gum, guar gum, and carboxymethyl cellulose (CMC).

How do hydrocolloids thicken food products?

Hydrocolloids increase the viscosity of food products by modifying the flow behavior and mechanical solid properties. Different hydrocolloids have different thickening properties, some giving low viscosities at high concentrations, while others give high viscosities at low concentrations.

What are gels, and how are hydrocolloids used as gelling agents?

Gels are formed by cross-linking the polymer chains of hydrocolloids, creating a three-dimensional network that traps water and forms a rigid structure. Hydrocolloids like alginate, pectin, carrageenan, gelatin, Gellan, and agar are commonly used as gelling agents in food products like puddings, jellies, dairy desserts, and bakery items.

How do hydrocolloids stabilize food products?

Hydrocolloids help prevent phase separation or the breakdown of emulsions in food products like mayonnaise, salad dressings, and sauces. They create a stable interface between oil and water phases, ensuring a homogeneous mixture and maintaining the texture and overall quality of the food products.

FREQUENTLY ASKED QUESTIONS (FAQS) CONTINUED

- **Can hydrocolloids act as emulsifying agents?**

  Yes, hydrocolloids can act as emulsifying agents and help stabilize emulsions in food products. They create a stable interface between oil and water, preventing the separation of the two phases. Common hydrocolloids used as emulsifying agents include lecithin, gum Arabic, and some modified starches.

- **What are the water-binding properties of hydrocolloids?**

  Hydrocolloids have excellent water-binding properties, meaning they can retain moisture in food products. This property is beneficial in preventing dehydration and maintaining the desired texture of the products, especially in bakery products and processed meats.

- **Besides their thickening, gelling, stabilizing, and emulsifying properties, do hydrocolloids have any other functionalities in food?**

  Yes, hydrocolloids have other functionalities in food products. They can control the crystal growth of ice and sugar, improve the mouthfeel of food products, and enhance the overall sensory experience.

- **Where are hydrocolloids commonly used in the food industry?**

  Hydrocolloids are found in many food products, including soups, gravies, salad dressings, sauces, toppings, ice creams, jams, jellies, bakery products, confectionery items, and many more.

- **What advancements and research are being done on hydrocolloids in the food industry?**

  Scientists and researchers are constantly studying hydrocolloid properties, functionality, and potential applications in the food industry. They are focused on the structure of hydrocolloids, their interactions with other food ingredients, and their impact on food product quality.

PART 4.2

Applications of Hydrocolloids in the Medical Field: Wound Dressings, Drug Delivery

Hydrocolloids are versatile water-binding agents that have numerous applications in the medical industry. Their unique properties make them particularly useful in wound dressings and drug delivery systems, offering significant benefits for patient care and treatment outcomes.

Core Insights:

- ☑ Hydrocolloids are crucial in wound dressings, creating a moist environment that promotes optimal wound healing.
- ☑ They have shown promise in drug delivery systems, enhancing the therapeutic efficacy of medications through controlled release.
- ☑ Hydrocolloid dressings effectively manage chronic wounds, pressure ulcers, and diabetic foot ulcers.
- ☑ They can protect drugs from degradation, maintain stability, and increase bioavailability in drug delivery.
- ☑ Ongoing research and innovation in hydrocolloid-based medical applications have the potential to revolutionize wound care and medical treatment.

4.2.1 The Role of Hydrocolloids in Wound Dressings

Hydrocolloids play a significant role in wound dressings, offering numerous benefits for wound care and management. These versatile water-binding agents are widely used in medical settings due to their exceptional moisture management properties. One of the key advantages of hydrocolloid dressings is their ability to absorb wound exudate effectively while creating a moist environment that promotes optimal wound healing.

Hydrocolloid dressings form a gel-like barrier over the wound, which helps protect it from external contaminants and reduces the risk of infection. The moist environment created by the dressings supports the natural healing process, facilitating new tissue growth and reducing the appearance of scars. Hydrocolloids also relieve pain by providing a cushioning effect and reducing friction between the wound and external surfaces.

These dressings have shown remarkable effectiveness in managing chronic wounds, such as pressure ulcers and diabetic foot ulcers. The combination of moisture management, pain relief, and wound protection makes hydrocolloid dressings ideal for healthcare professionals looking to optimize wound care outcomes. By harnessing the power of hydrocolloids, medical professionals can provide patients with the best possible care for successful wound healing.

Table 12: Hydrocolloids Dressing for Wound Care

HYDROCOLLOID DRESSING	BENEFITS
Brand A	Excellent exudate absorption
Brand B	Long-lasting adhesion for extended wear time
Brand C	Conformable and easy to apply on irregular wound sites
Brand D	Provides a waterproof barrier for showering and bathing

When choosing a hydrocolloid dressing, healthcare professionals should consider factors such as exudate absorption, adhesion duration, ease of application, and the ability to provide a waterproof barrier. Selecting the most appropriate hydrocolloid dressing for each patient's needs can help optimize wound care and improve overall patient outcomes.

4.2.2 Enhancing Drug Delivery with Hydrocolloids

Hydrocolloids have revolutionized drug delivery systems, offering numerous advantages in enhancing therapeutic efficacy. These versatile water-binding agents can encapsulate drugs and facilitate controlled release, ensuring targeted and sustained drug delivery. Hydrocolloids shield the drug molecules from degradation by forming a protective barrier around the drug molecules, maintaining their stability and bioavailability.

The controlled-release properties of hydrocolloids play a crucial role in optimizing drug efficacy. Through the precise modulation of drug release rates, hydrocolloid-based drug delivery systems can achieve a more consistent and prolonged therapeutic effect. This controlled release mechanism not only improves treatment outcomes but also minimizes potential adverse effects by reducing the frequency of drug administration.

> "Hydrocolloids have emerged as a promising tool for improving drug delivery systems."

In addition to their controlled release capabilities, hydrocolloids can also enhance the bioavailability of drugs. Hydrocolloids improve their absorption and distribution in the body by improving the solubility and dissolution rate of poorly water-soluble drugs. This increased bioavailability translates to greater therapeutic efficacy, allowing lower drug doses while maintaining the desired pharmacological effect.

Advantages of Hydrocolloid-based Drug Delivery Systems

Hydrocolloid-based drug delivery systems offer several advantages over conventional approaches. These include:

- Improved drug stability and protection against degradation
- Enhanced drug bioavailability through improved solubility and dissolution
- Prolonged and controlled drug release for optimal therapeutic effect
- Potential for reduced dosing frequency and improved patient compliance
- Possibility of targeted drug delivery to specific sites in the body

As researchers continue to explore the potential of hydrocolloids in drug delivery, developing advanced hydrocolloid-based systems holds promise for the future of pharmaceutical formulations.

Table 13: Advantages of Hydrocolloids-based Drug Delivery

ADVANTAGES OF HYDROCOLLOID-BASED DRUG DELIVERY SYSTEMS
Improved drug stability and protection against degradation
Enhanced drug bioavailability through improved solubility and dissolution
Prolonged and controlled drug release for optimal therapeutic effect
Potential for reduced dosing frequency and improved patient compliance
Possibility of targeted drug delivery to specific sites in the body

4.2.3 Types of Hydrocolloid Wound Dressings

Hydrocolloid dressings have proven to be highly effective in managing chronic wounds. These dressings are designed to create a moist environment that promotes optimal healing. Hydrocolloid dressings are available in various types, sizes, and shapes, making them versatile for different body parts.

One common type of hydrocolloid dressing is the thin film dressing. This transparent dressing allows for easy wound monitoring without frequent dressing changes. It provides a barrier against external contaminants while still allowing oxygen to reach the wound for faster healing.

Another type is the foam dressing, which is particularly suitable for wounds with heavy exudate. The foam material absorbs the excess moisture while maintaining a moist wound environment. Foam dressings also offer cushioning and protection, making them ideal for pressure ulcers or areas prone to friction.

A variant of hydrocolloid dressings is the hydroactive dressing, which combines hydrocolloid technology with an absorbent layer. This type of dressing is highly absorbent and can handle moderate to high exudate levels. The absorbent layer draws excess moisture away from the wound, while the hydrocolloid layer maintains a moist environment for optimal healing.

Table 14: Hydrocolloid Features and Benefits

TYPE OF HYDROCOLLOID DRESSING	FEATURES	BENEFITS
Thin Film Dressings	Transparent, allows wound monitoring, oxygen permeable	Minimizes dressing changes, promotes faster healing
Foam Dressings	Highly absorbent, cushioning, protects against friction	Ideal for heavy exudate, pressure ulcers, and areas prone to friction
Hydro active Dressings	Combines hydrocolloid technology with absorbent layer	Handles moderate to high exudate levels and maintains a moist wound environment

Choosing the appropriate type of hydrocolloid dressing is important based on the individual's wound characteristics and exudate level. Consulting with a healthcare professional can help determine the best dressing option for effectively managing chronic wounds.

4.2.4 Hydrocolloids in Burn Wound Management

Hydrocolloid dressings have emerged as valuable tools in the management of burn wounds. These dressings create a moist environment that promotes optimal healing and protects the wound from external factors. Additionally, hydrocolloids have been found to facilitate epithelialization, reduce pain, and minimize scarring, making them particularly effective in burn wound care.

When applied to burn wounds, hydrocolloid dressings form a gel-like barrier that helps maintain a balanced moisture environment. This barrier not only prevents excessive drying of the wound but also aids in removing necrotic tissue and promotes the formation of healthy granulation tissue. The ability of hydrocolloid dressings to provide a moist environment is crucial in burn wound management, as it supports the natural healing process and improves overall outcomes.

Table 15: Benefits of Hydrocolloid Dressings in Burn Wound

BENEFITS OF HYDROCOLLOID DRESSINGS IN BURN WOUND MANAGEMENT
Facilitate epithelialization
Reduce pain and discomfort.
Minimize scarring
Aid in debridement of necrotic tissue
Promote the formation of healthy granulation tissue.

The moist environment created by hydrocolloid dressings also helps to alleviate pain and discomfort associated with burn wounds. By providing a protective barrier, these dressings reduce friction and minimize exposure to external stimuli, allowing the patient to experience enhanced comfort during the healing process.

In addition to their wound-healing benefits, hydrocolloid dressings have been shown to contribute to scar reduction. The ability to create a moist environment, coupled with the protective barrier provided by the dressings, helps to prevent excessive scar formation. This can lead to improved aesthetic outcomes and a better quality of life for burn patients.

4.2.5 Hydrocolloids for Advanced Wound Care

Advanced wound care requires specialized treatments and interventions to promote healing and prevent complications. Hydrocolloids are highly beneficial In this regard, as they offer a range of advantages for managing complex wounds. Here, we will explore how hydrocolloids aid in debridement, promote granulation tissue formation, and help manage excessive exudate.

Debridement

Debridement is an essential step in wound care, as it involves the removal of dead or infected tissue to facilitate healing. Hydrocolloid dressings provide a gentle and effective debridement method. They adhere to the wound bed and create a moist environment, which helps soften and detach necrotic tissue. This process minimizes the need for more invasive debridement techniques, reducing patient discomfort and accelerating healing.

Granulation Tissue Formation

Granulation tissue formation is crucial for wound healing, providing a foundation for new skin growth. Hydrocolloid dressings promote granulation tissue formation by maintaining a moist environment that supports cell proliferation. The gel-like barrier created by hydrocolloids protects the wound bed, prevents drying, and optimizes conditions for healthy tissue growth. This aids in forming a strong and resilient wound closure, minimizing the risk of complications.

Exudate Management

Excessive exudate can delay wound healing and increase the risk of infection. Hydrocolloid dressings effectively manage exudate by absorbing it, forming a gel-like mass that locks in moisture. This helps maintain an optimal moisture balance within the wound, preventing maceration of the surrounding healthy tissue. By controlling exudate levels, hydrocolloid dressings promote a moist environment that supports healing and reduces the risk of complications.

Table 16: Advantages of Hydrocolloids in Advances Wound Care

ADVANTAGES OF HYDROCOLLOIDS FOR ADVANCED WOUND CARE
Effective debridement
Promotion of granulation tissue formation
Exudate management

4.2.6 Future Trends and Innovations in Hydrocolloid-based Medical Applications

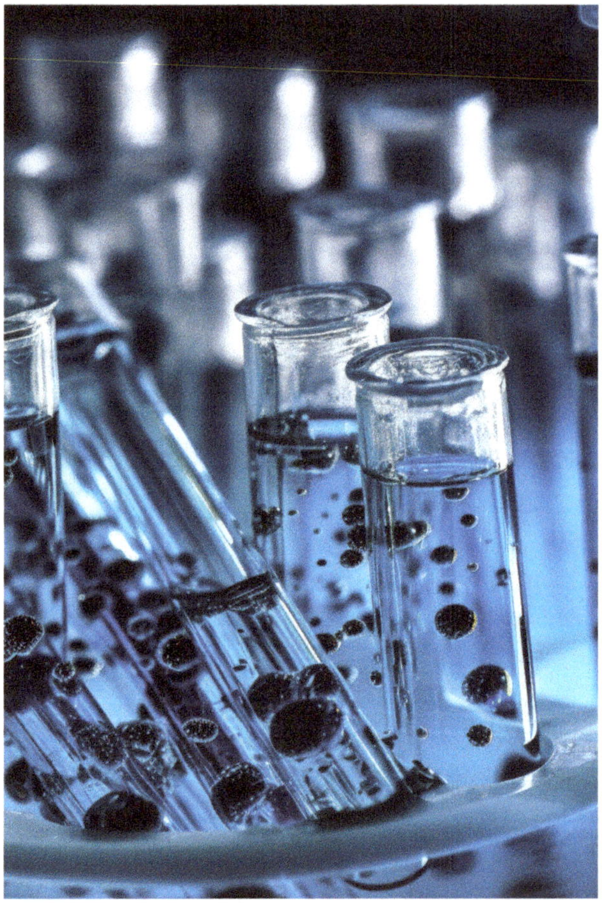

Ongoing research and development in hydrocolloid-based medical applications pave the way for exciting future trends and innovations. Scientists and healthcare professionals are focused on maximizing the function and effectiveness of hydrocolloids in wound healing and exploring new materials for improved wound dressings.

One of the key areas of interest is the development of advanced materials that can enhance the properties of hydrocolloids. These materials aim to optimize wound healing outcomes by providing better moisture management, pain relief, and overall wound care. Researchers are exploring novel techniques to incorporate advanced materials, such as nanofibers and nanoparticles, into hydrocolloid dressings to enhance their functionality and performance.

Another area of innovation is the integration of smart technology into hydrocolloid-based dressings. This approach involves using sensors and electronic components to monitor wound healing progress and deliver targeted therapies. Smart dressings have the potential to revolutionize wound care by providing real-time feedback, enabling personalized treatment plans, and improving patient outcomes.

Table 17: Future Trends and Innovations in Hydrocolloid-based Medical Applications

TREND/INNOVATION	DESCRIPTION
Advanced Materials	Exploration of materials like nanofibers and nanoparticles to enhance the properties of hydrocolloid dressings for improved wound healing.
Smart Technology Integration	Incorporation of sensors and electronic components into hydrocolloid dressings to monitor wound healing progress and provide targeted therapies.
Biocompatible Hydrocolloids	Developing biocompatible hydrocolloids that seamlessly integrate with the body's natural healing processes, minimizing adverse reactions.
Drug Delivery Enhancements	Investigation of novel drug delivery systems utilizing hydrocolloids to improve therapeutic efficacy and targeted drug release.

The future of hydrocolloid-based medical applications holds immense potential for wound care and medical treatment advancements. With ongoing research and innovation, we can expect to see more sophisticated dressings, advanced materials, and personalized approaches to wound healing. These developments will undoubtedly contribute to improved patient outcomes, reduced healing time, and enhanced quality of life for individuals needing medical care.

4.2.7 Polymer Materials for Dressing Applications

Polymer materials have gained significant attention in developing dressing applications in the medical field. These versatile materials offer a range of properties that make them suitable for various wound care needs. One important aspect of polymer materials is their biocompatibility, which ensures they do not cause harmful reactions when in contact with the human body.

Biocompatibility is crucial in dressing applications as it allows for safe and effective wound healing. Polymer-based dressings can mimic the wound's natural environment, providing optimal conditions for tissue regeneration and recovery. These materials can also be tailored to have specific properties, such as high moisture absorption or retention, essential for maintaining a moist wound environment.

Another advantage of polymer materials in dressing applications is their biodegradability. Biodegradable polymers break down over time and are naturally eliminated from the body, reducing the need for frequent dressing changes. This simplifies wound care and minimizes the risk of infection and discomfort for the patient.

Conclusion

Hydrocolloids are versatile water-binding agents with valuable applications in the medical field, particularly in wound dressings and drug delivery. These hydrophilic substances play a significant role in wound care by creating a moist environment that promotes optimal wound healing. They have also shown promise in enhancing drug delivery systems, increasing therapeutic efficacy by controlled release, and improving bioavailability.

The future of hydrocolloid-based medical applications looks promising, with ongoing research and development focused on exploring new materials and dressings. These innovations aim to maximize the function and effectiveness of hydrocolloids in wound healing and advance the field of medical treatment. From managing chronic wounds to facilitating burn wound healing, hydrocolloid dressings offer a range of benefits, including moisture management, pain relief, and scar reduction.

In summary, hydrocolloids are key players in the medical industry, and their use in wound dressings and drug delivery is set to continue growing. With the potential for future advancements and the exploration of advanced materials, the field of hydrocolloid-based medical applications holds exciting opportunities for improving wound care and medical treatment.

FAQ

What are hydrocolloids used for in the medical field?

Hydrocolloids are used in the medical field's wound dressings and drug delivery systems.

How do hydrocolloids promote wound healing?

Hydrocolloid dressings create a moist environment that promotes optimal wound healing by absorbing wound exudate and forming a gel-like barrier over the wound.

What are the benefits of hydrocolloid dressings in managing chronic wounds?

Hydrocolloid dressings effectively manage chronic wounds, pressure ulcers, and diabetic foot ulcers by absorbing exudate and creating a moist environment that promotes healing.

How do hydrocolloids enhance drug delivery?

Hydrocolloids can encapsulate drugs and release them controlled, increasing therapeutic efficacy and protecting drugs from degradation.

What types of wounds can hydrocolloid dressings be used for?

Hydrocolloid dressings are versatile and can be used for various types of wounds, including chronic, pressure, and diabetic foot ulcers.

How do hydrocolloid dressings benefit burn wound management?

Hydrocolloid dressings create a moist environment that promotes optimal healing, facilitates epithelialization, reduces pain, and minimizes scarring in burn wound management.

What are the benefits of hydrocolloids in advanced wound care?

Hydrocolloids aid in debridement, promote granulation tissue formation, and help manage excessive exudate in advanced wound care.

FREQUENTLY ASKED QUESTIONS (FAQS) CONTINUED

What are the future trends and innovations in hydrocolloid-based medical applications?

Ongoing research is focused on developing advanced materials and dressings to maximize the function and effectiveness of hydrocolloids in wound healing and explore new materials for improved wound dressings.

What are polymer materials used for dressing applications?

Polymer materials offer high biocompatibility and biodegradability and can potentially enhance wound healing outcomes in dressing applications.

PART 4.3

Applications of Hydrocolloids in the Cosmetic Industry: Lotions, Creams,

Hydrocolloids play a crucial role in the cosmetics industry, particularly in formulating lotions, creams, and shampoos. These versatile substances, known for their unique properties and abilities, are responsible for enhancing product stability, improving texture, and providing various skin benefits. Using hydrocolloids, cosmetic manufacturers can create high-quality products that meet consumers' demands.

Core Insights:

- ☑ Hydrocolloids are widely used in cosmetics to enhance stability, texture, and skin benefits.
- ☑ They have unique properties such as viscosity, stability, and emulsifying capabilities.
- ☑ Hydrocolloids are commonly found in lotions, creams, and shampoos.
- ☑ They contribute to product stability by preventing phase separation and ensuring ingredient distribution.
- ☑ Hydrocolloids enhance texture, providing a smooth and luxurious feel to cosmetic products.

4.3.1 Hydrocolloids for Product Stability

Hydrocolloids are invaluable for improving product stability in the cosmetics industry. These substances form a stable network within the product, preventing phase separation and ensuring the uniform distribution of ingredients. This property is particularly essential in emulsions, where hydrocolloids act as emulsifiers, stabilizing the mixture and preventing the separation of oil and water phases.

In addition to emulsion stabilization, hydrocolloids also play a crucial role in preventing settling or clumping of particles. This ensures that the product maintains its desired consistency and texture over time. By incorporating hydrocolloids into cosmetic formulations, manufacturers can create products that withstand various environmental factors, such as temperature changes and physical stress, while maintaining their integrity.

To further illustrate the importance of hydrocolloids for product stability, consider the following table:

Table 18: Importance of Hydrocolloids in Product Stability

PRODUCT TYPE	HYDROCOLLOID APPLICATION	BENEFITS
Lotions	Stabilizes emulsions, prevents ingredient separation	Ensures consistent texture and improved shelf life
Creams	Prevents settling and clumping of particles	Creates a smooth, luxurious texture
Shampoos	Stabilizes foam and prevents phase separation	Enhances overall product performance and user experience

Using hydrocolloids in their formulations, cosmetic manufacturers can create products that deliver superior stability, ensuring consistent performance and customer satisfaction.

4.3.2 Hydrocolloids for Texture Enhancement

Hydrocolloids are invaluable ingredients for enhancing the texture of cosmetic products. These versatile substances contribute to the smooth and silky feel that consumers desire, making product application and absorption effortless. Additionally, hydrocolloids play a crucial role in controlling and modifying the viscosity of formulations, allowing formulators to achieve the desired thickness or thinness for their products.

Different types of hydrocolloids offer unique textures and consistencies, enabling formulators to create products with specific textures. Hydrocolloids can be tailored to meet the desired feel and user experience, whether a lightweight lotion or rich cream. By incorporating hydrocolloids, cosmetic manufacturers can ensure their products provide the luxurious, velvety textures consumers crave.

Furthermore, hydrocolloids contribute to the stability of the product texture over time. They help prevent the settling or clumping of particles, ensuring that the consistency remains consistent throughout the product's lifetime. This stability not only enhances user satisfaction but also increases the product's shelf life.

Table 19: Hydrocolloids and Their Texture-Enhancing Properties

HYDROCOLLOID	TEXTURE ENHANCING PROPERTIES
Guar Gum	Provides a smooth and creamy texture
Xanthan Gum	Imparts a thick and dense texture
Cellulose Gum	Creates a lightweight and smooth texture
Carrageenan	Offers a gel-like texture with excellent suspension properties

As seen in the table above, different hydrocolloids exhibit varying texture-enhancing properties. Guar gum, for example, provides a smooth and creamy texture, ideal for lotions and creams. On the other hand, Xanthan gum imparts a thick and viscous texture, perfect for products requiring a more substantial consistency. Cellulose gum creates a lightweight and smooth texture, making it suitable for serums and lightweight moisturizers. Carrageenan offers a gel-like texture with excellent suspension properties, making it an excellent choice for face masks and body washes.

With the diverse range of hydrocolloids available, formulators can create cosmetic products with textures that meet consumers' unique needs and preferences. Hydrocolloids are essential ingredients for achieving the luxurious textures that consumers love, from lotions and creams to serums and body washes.

4.3.3 Hydrocolloids for Skin Benefits

Hydrocolloids offer numerous benefits to the skin, making them valuable ingredients in cosmetic formulations. One of the key advantages of hydrocolloids is their ability to provide hydration and moisturization to the skin. By attracting and retaining moisture, hydrocolloids help to alleviate dryness and keep the skin supple and smooth.

In addition to hydration, hydrocolloids can also help reduce inflammation and redness, making them suitable for sensitive or irritated skin. Their soothing properties can provide relief and promote a healthier complexion. Hydrocolloids can also improve skin texture, leaving it soft and revitalized.

Furthermore, hydrocolloids can reduce the appearance of fine lines and wrinkles. By attracting moisture to the skin, they help plump the skin cells, resulting in a smoother and more youthful appearance. The long-lasting hydration provided by hydrocolloids can give the skin a radiant and glowing look.

> "Hydrocolloids are a game-changer in skincare formulations. Their ability to hydrate, soothe, and improve skin texture makes them invaluable for creating products that deliver visible results."
>
> - Skincare Formulator.

> "The moisturizing and anti-aging benefits of hydrocolloids have made them popular among consumers. It's no wonder why skincare brands incorporate these ingredients into their formulations."
>
> - Beauty Industry Expert.

Overall, hydrocolloids offer a range of skin benefits, making them an essential component in cosmetic products. Hydrocolloids contribute to healthier, more vibrant skin, from hydration and moisturization to soothing and anti-aging effects.

Table 20: Benefits of Hydrocolloids for Skin

BENEFITS OF HYDROCOLLOIDS FOR SKIN	
Hydration and moisturization	Reduced dryness and improved skin texture
Soothing and anti-inflammatory properties	Calms sensitive or irritated skin
Reduction of fine lines and wrinkles	Improved skin elasticity and youthful appearance

4.3.4 Hydrocolloids in Lotions

Hydrocolloids are frequently used in lotions to improve stability, texture, and hydration. These versatile ingredients are crucial in creating high-quality lotions that deliver optimal performance and user satisfaction.

One of the key benefits of hydrocolloids in lotions is their ability to stabilize emulsions. By forming a stable network within the product, hydrocolloids prevent phase separation and ensure a uniform distribution of ingredients. This results in a lotion that remains consistent over time and provides a smooth and lightweight consistency.

In addition to stability, hydrocolloids in lotions also contribute to hydration. They create a protective barrier on the skin, preventing moisture loss and ensuring long-lasting hydration. This helps to keep the skin moisturized and nourished, leaving it feeling soft and smooth.

Furthermore, hydrocolloids enhance the texture of lotions, providing a smoother and more luxurious feel. They contribute to the overall sensory experience of using the lotion, making it more appealing to customers. Whether it's a lightweight, fast-absorbing lotion or a rich, creamy formulation, hydrocolloids help to create the desired texture and consistency.

Table 21: Comparison of Hydrocolloids in Lotions

HYDROCOLLOID	TEXTURE BENEFIT	HYDRATION BENEFIT	STABILITY BENEFIT
Sodium Carboxymethyl Cellulose	Creates a smooth and lightweight texture	Forms a protective barrier for long-lasting hydration	Stabilizes emulsions and prevents phase separation
Xanthan Gum	Imparts a silky, luxurious feel	Helps retain moisture in the skin	Enhances stability and prevents settling of particles
Gellan Gum	Provides a gel-like texture	Soothes and moisturizes the skin	Improves stability and prevents clumping

Using hydrocolloids in lotions improves the product's overall quality and enhances the user experience. These multifunctional ingredients contribute to stability, hydration, and texture, making lotions more effective in nourishing and promoting healthy skin.

4.3.5 Hydrocolloids in Creams

Hydrocolloids are indispensable ingredients for the desired consistency, texture, and emulsion stability in cream formulations. These versatile substances play a crucial role in creating creams with a rich and luxurious texture that provides intense moisturization and nourishment for the skin.

Hydrocolloids act as emulsion stabilizers, preventing creams' separation of oil and water phases. This ensures the formulation remains uniform over time and maintains its desired texture and appearance. Additionally, hydrocolloids enable the controlled release of active ingredients, enhancing creams' efficacy and overall skin benefits.

Furthermore, hydrocolloids in creams create a protective barrier on the skin, helping to prevent moisture loss and ensuring long-lasting hydration. This barrier also contributes to a smoother and more luxurious skin feel, enhancing the overall user experience of the cream.

Table 22: Benefits of Hydrocolloids in Creams

BENEFITS OF HYDROCOLLOIDS IN CREAMS	EXAMPLES OF HYDROCOLLOIDS
Emulsion stabilization	Guar gum
Controlled release of active ingredients	Xanthan gum
Moisture retention	Cellulose gum
Enhanced skin feel	Carrageenan

By incorporating hydrocolloids into cream formulations, cosmetic manufacturers can create products that provide effective moisturization and offer a luxurious sensory experience for the consumer.

4.3.6 Hydrocolloids in Shampoos

Regarding shampoo formulations, hydrocolloids enhance texture, improve foam stability, and provide conditioning properties. These versatile ingredients offer a range of benefits that contribute to a satisfying and effective hair-cleansing experience.

One of the key advantages of hydrocolloids in shampoos is their ability to enhance texture. They provide a thickening effect, creating a rich lather that makes hair washing a pleasurable experience. The increased viscosity of the shampoo formulation allows for better coverage and distribution of the product on the scalp and hair strands.

In addition to texture enhancement, hydrocolloids also offer conditioning properties. They help reduce hair tangling and make combing easier, resulting in smoother and more manageable hair. Hydrocolloids contribute to a soft and silky feel, nourishing and revitalizing the hair.

> "Hydrocolloids in shampoos create a rich lather and provide conditioning properties, resulting in smoother and more manageable hair."

Key Benefits of Hydrocolloids in Shampoos:

- **Texture enhancement:** Hydrocolloids thicken the shampoo formulation, resulting in a rich lather and improved coverage on the scalp and hair.
- **Conditioning properties:** Hydrocolloids reduce tangling and make combing easier, providing a smoother and more manageable hair texture.
- **Enhanced user experience:** The soft and silky feel imparted by hydrocolloids enhances the overall shampooing experience, nourishing and revitalizing the hair.

Table 23: Hydrocolloids in Shampo and Benefits

HYDROCOLLOIDS IN SHAMPOOS	BENEFITS
Texture enhancement	Creates a rich lather and improves coverage on the scalp and hair
Conditioning properties	Reduces tangling and makes combing easier for smoother and more manageable hair
Enhanced user experience	Imparts a soft and silky feel, leaving the hair feeling nourished and revitalized

4.3.7 Natural Hydrocolloids in Personal Care Products

As consumer demand for "all-natural" personal care products continues to rise, manufacturers are incorporating natural hydrocolloids into their formulations. These natural ingredients offer various functions and benefits, making them an attractive choice for creating high-quality and environmentally friendly products.

Cellulose gum, derived from plant cell walls, is a commonly used natural hydrocolloid in personal care products. It acts as a thickener, providing viscosity and enhancing the texture of formulations such as lotions and creams. Additionally, cellulose gum contributes to emulsions' stability and helps create a smooth and luxurious feel on the skin.

Carrageenan, extracted from red seaweed, is another natural hydrocolloid that finds its application in personal care products. It functions as a gelling agent and stabilizer, allowing for the creation of gel-based products. Carrageenan also provides moisturization and soothing effects, making it ideal for skincare products such as masks and serums.

Gellan gum, derived from bacterial fermentation, is a natural hydrocolloid with exceptional gelling properties. It can create firm gels at low concentrations and uniquely texture personal care products. Gellan gum is commonly used in haircare products, such as styling gels, to provide hold and control without leaving a sticky residue.

> "Natural hydrocolloids are a sustainable alternative to synthetic ingredients, aligning with the growing consumer preference for eco-friendly and organic personal care products."
>
> \- Industry Expert.

Lastly, pectin, derived from fruits, is a natural hydrocolloid with adhesive and stabilizing properties. It is often used in haircare products, such as hair masks and conditioners, to improve the texture and manageability of hair. Pectin also provides hydration and nourishment, leaving the hair soft, smooth, and more manageable.

Table 24: Natural Hydrocolloids Functions and Applications

NATURAL HYDROCOLLOID	FUNCTION	APPLICATION
Cellulose gum	Thickening, stability	Lotions, creams
Carrageenan	Gelling, stabilizing	Skincare products
Gellan gum	Gelling	Haircare products
Pectin	Adhesive, stabilizing	Haircare products

Manufacturers can meet consumer demands for sustainable and eco-friendly options by utilizing natural hydrocolloids in personal care products. These ingredients offer functional advantages, enhance product performance, and provide a natural and organic appeal. With increasing emphasis on natural formulations, incorporating natural hydrocolloids in personal care products is a smart choice to meet the evolving needs of consumers.

Conclusion

Hydrocolloids are indispensable in the cosmetic industry, offering numerous applications and benefits. From lotions and creams to shampoos, these versatile ingredients enhance product stability, improve texture, and provide valuable skin benefits. Hydrocolloids' ability to form gels, thicken solutions, and stabilize emulsions ensures that cosmetic formulations maintain their desired consistency and uniformity over time.

The cosmetic industry relies on hydrocolloids to create products with a smooth, luxurious texture that is easy to apply and absorb. These ingredients provide a satisfying user experience by providing long-lasting hydration, reducing fine lines and wrinkles, and soothing the skin. Their incorporation in lotions, creams, and shampoos promotes intense moisturization and nourishment, leaving the skin and hair feeling soft, silky, and conditioned.

As the demand for natural and organic personal care products grows, natural hydrocolloids have gained popularity among manufacturers. Ingredients such as cellulose gum, carrageenan, Gellan gum, and pectin offer similar functionalities to their synthetic counterparts, allowing formulators to create high-quality products that meet consumer preferences. Hydrocolloids are at the forefront of innovation in the cosmetic industry, delivering optimal performance and satisfying the evolving needs of consumers.

FAQ

How do hydrocolloids improve product stability?

Hydrocolloids improve product stability by forming a stable network within the product, preventing phase separation and ensuring the uniform distribution of ingredients. They also act as emulsifiers, stabilizing emulsions and preventing the separation of oil and water phases. In addition, hydrocolloids prevent settling or clumping of particles, ensuring the product maintains its desired consistency and texture.

How do hydrocolloids enhance the texture of cosmetic products?

Hydrocolloids contribute to the smooth, silky feel of cosmetic products such as lotions and creams, allowing easy application and absorption. They also help control and modify the viscosity of formulations, imparting the desired thickness or thinness. Different types of hydrocolloids provide unique textures and consistencies, allowing formulators to achieve specific product textures.

What skin benefits do hydrocolloids provide?

Hydrocolloids offer numerous benefits to the skin, including hydration, moisturization, and soothing effects. They can help reduce inflammation and redness, improve skin texture, and provide long-lasting hydration. By attracting and retaining moisture, hydrocolloids can also reduce the appearance of fine lines and wrinkles and leave the skin feeling soft and smooth.

How are hydrocolloids used in lotions?

Hydrocolloids are used in lotions to improve stability, texture, and hydration. They help stabilize emulsions, providing the lotion with a smooth and lightweight consistency. Hydrocolloids also create a protective barrier on the skin, preventing moisture loss and ensuring long-lasting hydration. Additionally, they contribute to a smoother and more luxurious skin feel, enhancing the overall user experience.

FREQUENTLY ASKED QUESTIONS (FAQS) CONTINUED

How do hydrocolloids contribute to creams?

Hydrocolloids are essential to achieve consistency, texture, and emulsion stability in cream formulations. They contribute to creams' rich, luxurious texture, promoting intense moisturization and nourishment for the skin. Hydrocolloids also help prevent oil and water separation in creams, ensuring the formulation remains uniform over time. Moreover, they allow for the controlled release of active ingredients, improving efficacy and skin benefits.

What role do hydrocolloids play in shampoos?

Hydrocolloids enhance the texture of shampoos, improve foam stability, and provide conditioning properties. They offer a thickening effect, creating a rich lather and promoting a satisfying cleaning experience. Hydrocolloids also help reduce the tangling of hair and make combing easier. Additionally, they contribute to a soft, silky feel, making the shampoo formulation more appealing to customers and enhancing the overall user experience.

Are there natural hydrocolloids used in personal care products?

Many personal care products incorporate natural hydrocolloids in response to consumer demand for "all-natural" products. Natural hydrocolloids such as cellulose gum, carrageenan, Gellan gum, and pectin offer various functions in personal care products, including thickening, stabilizing, and forming gels and films. These natural hydrocolloids provide similar benefits to their synthetic counterparts and are favored by manufacturers aiming to create natural and organic personal care products.

What are the applications of hydrocolloids in the cosmetic industry?

Hydrocolloids have various applications in the cosmetic industry, including in lotions, creams, and shampoos. They are crucial in enhancing product stability, texture, and skin benefits. Hydrocolloids are valuable ingredients that help improve cosmetic products' overall quality and performance.

What are the benefits of using hydrocolloids in cosmetics?

Hydrocolloids offer a wide range of benefits in cosmetics, including improved product stability, enhanced texture, and skin benefits such as hydration and moisturization. They contribute to the overall performance and user satisfaction of cosmetic products.

CHAPTER 05

Hydrocolloids in Cooking and Baking

PART 5.1

Hydrocolloids in Cooking and Baking: Role in Modernist Cuisine

Hydrocolloids are essential ingredients in modernist cuisine, revolutionizing our cooking and baking. These versatile substances, known for their role as food thickeners, gelling agents, stabilizers, emulsifiers, viscosity modifiers, and texturizing agents, have found their place in the world of culinary arts.

Whether you're a professional chef or a passionate home cook, understanding the power and potential of hydrocolloids can take your culinary creations to the next level. Hydrocolloids offer a new realm of possibilities for the modern gastronomist, from achieving the perfect texture to enhancing flavors and extending shelf life.

This article will explore the world of hydrocolloids, their types, benefits, mixing techniques, vegan alternatives, and their impact on modernist cuisine. Join us as we unlock the secrets of these natural and plant-based hydrocolloids and discover how they have transformed the art of cooking and baking.

Core Insights:

- ☑ Hydrocolloids play a crucial role in modernist cuisine, serving as food thickeners, gelling agents, stabilizers, and more.
- ☑ They offer benefits beyond texture, including improved stability, moisture retention, and enhanced flavors and aromas.
- ☑ Understanding the different types of hydrocolloids, their properties, and their applications is key to mastering their use in the kitchen.
- ☑ Vegan alternatives to gelatin, such as agar, carrageenan, and xanthan gum, allow for plant-based culinary delights.
- ☑ Hydrocolloids have revolutionized modernist cuisine, pushed the boundaries of culinary creativity, and enabled chefs to create unique textures and flavors.

5.1.1 Hydrocolloid Types: An Overview of Culinary Magic

Gelatin, agar, carrageenan, and xanthan gum are some of the most popular hydrocolloids in the culinary world. Each of these hydrocolloids has unique properties suitable for specific culinary applications.

Gelatin is a protein-based hydrocolloid commonly used for gelling and thickening. It is derived from animal collagen and is well-known for its use in desserts like panna cotta and jellies.

Agar is a polysaccharide-based hydrocolloid derived from seaweed. It has excellent gelling properties and is often used in Asian cuisine to create firm gels in dishes like agar, agar jelly, and Kanten.

Carrageenan, also derived from seaweed, is a versatile hydrocolloid used as a thickener and stabilizer in various food products. It is commonly found in dairy alternatives, such as plant-based milk and ice cream, to improve texture and mouthfeel.

Xanthan gum is a versatile hydrocolloid produced by fermentation. It is often used as a thickening, stabilizing, and emulsifying agent in various culinary applications. Xanthan gum can be found in sauces, dressings, and gluten-free baked goods to enhance texture and improve consistency.

Table 25: Hydrocolloids and their culinary applications

HYDROCOLLOID	CULINARY APPLICATIONS
Gelatin	Desserts, gels, and thickening
Agar	Asian cuisine, jellies, and firm gels
Carrageenan	Dairy alternatives, thickening, and stabilizing
Xanthan gum	Sauces, dressings, and gluten-free baking

5.1.2 Unlocking Hydrocolloid's Potential: Benefits in Food Preparation

Hydrocolloids offer numerous benefits in food preparation, making them a valuable ingredient in modern cooking and baking. One of the primary advantages of hydrocolloids is their ability to modify the texture of liquid and semi-solid foods. Whether you want to create a smooth and creamy sauce or a light and airy foam, hydrocolloids can help you achieve the desired consistency. In addition to texture modification, hydrocolloids contribute to food products' stability and shelf life. They can prevent separation and microbial growth, ensuring your dishes remain fresh and visually appealing for longer periods. Moreover, hydrocolloids have been found to enhance the flavors and aromas of prepared foods, elevating the overall dining experience for consumers.

Another noteworthy benefit of using hydrocolloids in food preparation is their ability to improve nutrient retention. By binding with water and forming gels, hydrocolloids can help retain essential nutrients in the food matrix, making them more bioavailable to the body. This nutritional advantage adds to the overall health value of dishes prepared with hydrocolloids.

Enhancing Texture, Stability, and Flavor

To summarize, hydrocolloids bring a host of benefits to food preparation. They allow texture modification, ensuring your sauces, mousses, and foams have the desired consistency. Hydrocolloids also contribute to food products' stability and shelf life, preventing separation and spoilage. Additionally, they enhance flavors and aromas and improve the retention of nutrients in prepared dishes. By harnessing the potential of hydrocolloids, chefs and home cooks alike can elevate their culinary creations to new heights.

Table 26: Benefits of Hydrocolloids in Food Preparation

BENEFITS OF HYDROCOLLOIDS IN FOOD PREPARATION
Texture modification
Improved stability and shelf life
Enhanced flavors and aromas
Nutrient retention

5.1.3 Mastering the Art of Mixing and Incorporating Hydrocolloids

Using hydrocolloids in cooking and baking requires a deep understanding of their properties and behavior. These versatile ingredients have different thickening properties and resistance to heat and cold, which can greatly impact their effectiveness in recipes. Cheers can achieve perfect textures and consistencies in their culinary creations by mastering the art of mixing and incorporating hydrocolloids.

When working with hydrocolloids, mixing them with other dry ingredients before adding them to liquids is important. This ensures they are evenly distributed throughout the mixture, preventing clumping or uneven texture. Additionally, the hydrocolloid quantity should be carefully measured to achieve the desired consistency in the final dish.

Hydrocolloids can behave differently depending on the recipe and cooking method. Some hydrocolloids may require a longer hydration time or specific temperature conditions to activate their thickening or gelling properties fully. Understanding these nuances and adjusting the cooking process is key to achieving optimal results.

Experimentation and practice are essential when working with hydrocolloids. By exploring different recipes and techniques, chefs can better understand how hydrocolloids behave in different culinary applications. With time and experience, they can master the art of mixing and incorporating hydrocolloids, unlocking the full potential of these incredible ingredients.

Tips for Mixing and Incorporating Hydrocolloids:

- Mix hydrocolloids with other dry ingredients before adding them to liquids.
- Carefully measure the quantity of hydrocolloids for the desired consistency.
- Consider the specific cooking method and recipe requirements for each hydrocolloid.
- Experiment and practice with different recipes and techniques to gain expertise.

> "Mastering the art of mixing and incorporating hydrocolloids is crucial to achieving perfect textures and consistencies in culinary creations."

Table 27: Hydrocolloid Properties for Temperature

HYDROCOLLOID	THICKENING PROPERTIES	RESISTANCE TO HEAT	RESISTANCE TO COLD
Gelatin	Forms gels and thickens	Melts at high temperatures	Solidifies when chilled
Agar	Forms firm gels	Can withstand high heat	Remains stable when chilled
Carrageenan	Thickens and stabilizes	Stable at high temperatures	Stable when chilled
Xanthan Gum	Thickens stabilizes, and emulsifies	Stable at high temperatures	Stable when chilled

5.1.4 Exploring Vegan Options: Hydrocolloid Alternatives to Gelatin

Gelatin is a commonly used ingredient in cooking and baking, but it may not suit those following a vegan lifestyle. Luckily, several hydrocolloid alternatives to gelatin can provide the same thickening and gelling properties. These vegan options include agar, carrageenan, and xanthan gum.

Agar is derived from seaweed and is known for its strong gelling properties. It can create a firm gel perfect for desserts, jellies, and custards. Agar is easy to work with and is a great alternative to gelatin for vegan recipes.

Carrageenan is another hydrocolloid derived from seaweed. It is a versatile thickener and stabilizer, making it an excellent substitute for gelatin in various applications. Carrageenan can be used in vegan yogurts, ice creams, and plant-based cheeses.

Xanthan gum is a popular vegan alternative to gelatin due to its versatility. It is a fermented product that is a thickener, stabilizer, and emulsifier. Xanthan gum can be used in various recipes, including vegan sauces, dressings, and gluten-free baked goods.

Table 28: Comparison of Vegan Hydrocolloid Alternatives to Gelatin

HYDROCOLLOID	SOURCE	PROPERTIES	APPLICATIONS
Agar	Seaweed	Gelling	Desserts, jellies, custards
Carrageenan	Seaweed	Thickening, stabilizing	Vegan yogurts, ice creams, plant-based cheeses
Xanthan gum	Fermentation	Thickening, stabilizing, emulsifying	Sauces, dressings, gluten-free baked goods

With these vegan hydrocolloid alternatives, you can create delicious and visually appealing dishes without animal-based gelatin. Whether you're a vegan cook experimenting with new recipes or adding more plant-based options to your culinary repertoire, these hydrocolloids provide the perfect gelatin substitute.

5.1.5 Modernist Cuisine: Pushing Boundaries with Hydrocolloids

Modernist cuisine has revolutionized the culinary world, pushed the boundaries of traditional cooking techniques, and allowed chefs to create unique and innovative dishes. Hydrocolloids are at the heart of this culinary revolution, crucial in manipulating textures, creating stunning presentations, and exploring exciting flavor combinations.

Hydrocolloids offer an array of possibilities for culinary experimentation. Chefs can use them to transform liquids into stable gels, create foams, and even build intricate structures. For example, xanthan gum is ideal for stabilizing emulsions and creating creamy textures, while agar-agar can produce gels with varying textures, from soft and delicate to firm and resilient.

With hydrocolloids, chefs can push the boundaries of texture, presentation, and flavor. They can create dishes with unexpected textures, such as liquid spheres that burst in the mouth or "noodles" made from fruit puree. Hydrocolloids allow for the creation of visually stunning plates, with colorful gels, foams, and suspensions adding depth and intrigue. By combining different hydrocolloids and experimenting with flavor pairings, chefs can unlock new taste sensations that surprise and delight diners.

Unlocking Creativity with Hydrocolloids

The use of hydrocolloids in modernist cuisine is not just about novelty; it's about unlocking creativity and expanding the possibilities of what can be achieved in the culinary world. By understanding the properties of different hydrocolloids and how they interact with other ingredients, chefs can create dishes that taste amazing and provide an exciting sensory experience.

As modernist cuisine evolves, chefs constantly explore new ways to utilize hydrocolloids. Hydrocolloids offer a new palette of culinary tools and techniques, whether using gels, foams, or suspensions. By embracing these ingredients, chefs can elevate their dishes to new heights and provide diners with unforgettable culinary experiences.

Table 29: Hydrocolloids Main Applications

HYDROCOLLOID	MAIN APPLICATIONS
Gellan gum	Gels, suspensions, and foams
Agar-agar	Gels with varying textures
Methylcellulose	Thickening, foaming, and emulsifying
Sodium Alginate	Spherification and gelling

These are just a few examples of the hydrocolloids used in modernist cuisine, each with unique properties and applications. With the right knowledge and creativity, chefs can harness the power of hydrocolloids to create dishes that push the boundaries of what is possible in the culinary world.

5.1.6 Hydrocolloid Culinary Adventures: Real-life Examples

Real-life examples of hydrocolloid culinary adventures can be found in cooking classes and demonstrations. These interactive sessions offer a hands-on experience for aspiring chefs and home cooks to learn innovative techniques and modern applications of hydrocolloids. From creating unique textures to adding a unique spin to classic recipes, hydrocolloids open a world of possibilities in the kitchen.

One popular recipe that showcases the creative potential of hydrocolloids is Nutella Mousse. A smooth and fluffy mousse can be created without relying on traditional cooking or baking methods by incorporating hydrocolloids like carrageenan or xanthan gum. The hydrocolloids help stabilize the mixture, resulting in a light and creamy texture that perfectly complements Nutella's rich chocolate hazelnut flavors. Another impressive dish that highlights the versatility of hydrocolloids is Potato and Foie Gras Mousse. A velvety and savory mousse can be achieved by utilizing hydrocolloids such as gelatin or agar-agar. The hydrocolloids help bind the ingredients together, giving the mousse a smooth and luxurious consistency. This unique spin on a classic combination of ingredients is sure to impress any dinner party guest.

These real-life examples demonstrate how hydrocolloids can elevate dishes and create culinary wonders. Whether in a cooking class or experimenting in your kitchen, incorporating hydrocolloids opens endless possibilities for innovative and delicious creations.

Table 30: Nutella Mousse Recipe

INGREDIENTS	QUANTITY
Nutella	1 cup
Heavy cream	1 cup
Sugar	1/4 cup
Gelatin or xanthan gum	1 teaspoon
Water	1/4 cup

> "Nutella Mousse is a perfect example of how hydrocolloids can transform a classic dessert into something extraordinary. By using hydrocolloids like gelatin or xanthan gum, you can achieve a light and airy mousse without the need for excessive whipping or chilling. It's a crowd-pleasing dessert that will leave your guests impressed."

5.1.7 Introduction to Hydrocolloids: Understanding the Science Behind the Magic

Hydrocolloids have revolutionized cooking and baking, allowing chefs to create a wide range of textures and consistencies without compromising flavor. While they were once perceived as additives used by industrial food producers, more consumers recognize their potential as ingredients in creating better and more interesting food. Hydrocolloids are large molecules interacting with water and forming gels or thickening liquids. Derived from natural sources such as seaweed, seeds, and tree sap, hydrocolloids are classified as either linear or branched. Understanding the science behind hydrocolloids is key to unlocking their magic in the kitchen.

Hydrocolloids offer unique properties that make them suitable for specific culinary applications. By interacting with water, they can modify liquid and semi-solid foods' texture, viscosity, and stability. They can improve food products' stability and shelf life, preventing separation and microbial growth. Hydrocolloids can also enhance the texture of dishes, creating a smooth sauce or a light foam. Additionally, hydrocolloids can improve the retention and bioavailability of nutrients in food. It's important to use hydrocolloids appropriately and in moderation to achieve the desired culinary results.

Hydrocolloids have emerged as a game-changer in modernist cuisine, allowing chefs to manipulate textures, create unique presentations, and experiment with flavor combinations. Different hydrocolloids have specific applications, such as using xanthan gum for stable emulsions or agar-agar for gelling with various textures. These ingredients enable chefs to create textures and flavors that were once unimaginable, pushing the boundaries of culinary creativity. Hydrocolloids have become an essential ingredient in modern gastronomy with their versatility and ability to create innovative dishes.

"Hydrocolloids have transformed the culinary world, offering endless possibilities for chefs to create unique textures, enhance flavors, and elevate their cooking to new heights."

5.1.8 Gel Flow Properties: Exploring Textures and Characteristics

Hydrocolloid gels offer a wide range of textures and characteristics influenced by their unique gel flow properties. One important property is shear thinning, where the gel becomes thinner as sheared or stirred vigorously. This property allows for easy pouring or spreading of the gel, making it ideal for applications such as sauces or dressings. Additionally, shear-thinning gels can provide a smooth and creamy mouthfeel, enhancing the overall sensory experience of a dish.

The elasticity of a hydrocolloid gel is another key characteristic. Some gels exhibit high elasticity, which gives them a resilient and springy texture. These gels can withstand deformation and retain their shape when subjected to external forces, making them suitable for confectionery or baked goods applications. On the other hand, gels with low elasticity tend to be more brittle, easily breaking or crumbling when handled. The choice of gel elasticity depends on the desired texture and mouthfeel in a culinary creation.

Cohesiveness is also an important factor to consider when exploring hydrocolloid gels. Cohesive gels have a strong internal structure, allowing them to retain their shape and prevent syneresis or bleeding of liquids. They are often used in applications where a gel needs to hold its forms, such as fillings or mousses. On the other hand, gels with lower cohesiveness may exhibit a more fluid or gelatinous texture, making them suitable for applications such as custards or panna cottas.

Table 31: Gel Texture and Flow Properties

GEL TEXTURE	GEL FLOW PROPERTIES
Hard	Low **shear thinning**, high elasticity, low **cohesiveness**
Soft	Medium shear thinning, medium elasticity, medium cohesiveness
Brittle	Low shear thinning, low elasticity, low cohesiveness
Elastic	High shear thinning, high elasticity, medium cohesiveness
Gummy	Medium shear thinning, medium elasticity, high cohesiveness
Chewy	High shear thinning, medium-high elasticity, medium cohesiveness

Understanding the gel flow properties of hydrocolloids allows chefs to choose the right ones for their culinary creations. By manipulating these properties, chefs can create textures that range from soft and creamy to firm and chewy, enhancing the overall sensory experience of a dish. Whether a delicate mousse or a hearty gel-filled dessert, hydrocolloid gels offer endless possibilities for culinary innovation.

5.1.9 Hydrocolloid Sources and Structure: From Nature to the Kitchen

Hydrocolloids play a crucial role in modern culinary techniques, and their diverse range of applications can be attributed to their unique sources and structures. These natural ingredients, derived from seaweed, seeds, roots, and tree sap, provide the foundation for the magic of hydrocolloids in the kitchen.

Some hydrocolloids, like agar and carrageenan, are cellulose-derived and sourced from seaweed. These natural compounds undergo processing to extract the desired hydrocolloid properties. Other hydrocolloids, such as xanthan gum, are produced through fermentation, transforming natural ingredients into modified compounds suitable for culinary applications.

The structure of hydrocolloids is equally important in determining their behavior and functionality. Hydrocolloids can exhibit different solubility properties, gelling abilities, and textures depending on their molecular weight and composition. The presence of side groups on their main sugar chain further influences their behavior, providing unique attributes that chefs can leverage in their creative cooking endeavors.

> "Hydrocolloid sources and structures offer intriguing possibilities for culinary innovation. Understanding the roots and composition of these natural ingredients empowers chefs to unlock their full potential in creating fascinating textures and delightful flavors."

Table 32: Hydrocolloids - Sources and Structures

HYDROCOLLOID	SOURCE	STRUCTURE
Agar	Seaweed	Polysaccharides with unique gelling properties
Carrageenan	Seaweed	Polysaccharides are used as a thickener and stabilizer
Xanthan gum	Fermentation of natural ingredients	Polysaccharides with thickening, stabilizing, and emulsifying properties
Locust bean gum	Seeds of the carob tree	Polysaccharides are used as a thickener and stabilizer
Pectin	Fruits and vegetables	Polysaccharide responsible for gelling in jams and jellies

Understanding hydrocolloid sources and structures empowers chefs to harness their capabilities and create innovative dishes. By incorporating hydrocolloids derived from natural sources or modified compounds, culinary creativity can flourish, allowing for the exploration of diverse textures, consistencies, and flavors.

Conclusion

Hydrocolloids have revolutionized modern cooking and baking, offering endless culinary possibilities. Their versatility and unique properties have opened a new world of creative cooking and innovative dishes. Whether you are a professional chef or a home cook, embracing hydrocolloids can elevate your culinary creations to new heights.

With hydrocolloids, you can explore a wide range of textures and consistencies, from smooth and creamy to firm and gel-like. They can enhance flavors, stabilize mixtures, and even create surprising presentations. By understanding the science behind hydrocolloids and mastering the art of mixing and incorporating them, you can achieve perfect textures and consistencies in your culinary creations.

There is a wide variety of hydrocolloids, from natural hydrocolloids like agar and carrageenan to plant-based options like xanthan gum. Each hydrocolloid has unique properties and applications, allowing you to experiment and push the boundaries of traditional cooking techniques.

So why not embrace the world of hydrocolloids and unlock your culinary creativity? Hydrocolloids can transform ordinary dishes into extraordinary culinary masterpieces and amaze your guests with innovative textures and flavors. Explore the culinary possibilities, unleash your creativity, and let hydrocolloids take your cooking to the next level.

FAQ

What are some popular hydrocolloids used in cooking?

Gelatin, agar, carrageenan, and xanthan gum are some of the most popular hydrocolloids in the culinary world.

How should hydrocolloids be used in recipes?

Hydrocolloids should be mixed with other dry ingredients before adding them to liquids for even distribution, and the quantity used should be carefully measured.

What are some vegan alternatives to gelatin?

Agar, carrageenan, and xanthan gum are popular vegan alternatives to gelatin.

How have hydrocolloids impacted modernist cuisine?

Hydrocolloids have allowed chefs to manipulate textures, create unique presentations, and experiment with flavor combinations.

Where can examples of hydrocolloid culinary adventures be found?

Real-life examples can be found in cooking classes and demonstrations showcasing innovative hydrocolloid techniques.

What are hydrocolloids derived from?

Hydrocolloids can be derived from natural sources such as seaweed, seeds, roots, and tree sap.

What are some properties that affect gel textures?

Gel textures can vary in hardness, elasticity, cohesiveness, and chewiness, among other characteristics.

FREQUENTLY ASKED QUESTIONS (FAQS) CONTINUED

- **How are hydrocolloids structured and sourced?**

  Hydrocolloids can be linear or branched and derived from natural sources or modified compounds.

- **What can hydrocolloids offer in terms of culinary possibilities?**

  Hydrocolloids provide chefs with endless possibilities to create unique textures, enhance flavors, and push the boundaries of traditional culinary techniques.

PART 5.2

Hydrocolloids in Cooking and Baking: Recipes & Techniques Using Specific Hydrocolloids

Hydrocolloids are versatile ingredients that play a crucial role in the food industry for improving texture and stability in cooking and baking. These natural hydrocolloids offer endless possibilities when it comes to creating unique dishes. In this article, we will explore the applications of hydrocolloids in cooking and baking, provide recipes and techniques, and discuss the benefits and properties of different types of hydrocolloids.

Core Insights:

- ☑ Hydrocolloids are used in cooking and baking to enhance texture and stability.
- ☑ There are various types of hydrocolloids, such as agar-agar, carrageenan, and xanthan gum.
- ☑ Hydrocolloids can create edible films, foams, and stable emulsions.
- ☑ They are widely used in the food industry as thickening agents, gelling agents, stabilizers, and emulsifiers.
- ☑ Hydrocolloids have revolutionized modern cuisine, allowing chefs to experiment with new techniques and textures.

5.2.1 Hydrocolloids in Cooking and Baking

Hydrocolloids are crucial in cooking and baking, offering many benefits and applications. Both professional chefs and home cooks use these versatile ingredients to enhance texture, create unique presentations, and experiment with new culinary techniques. Whether you're looking to create stable foams, thicken sauces, or explore modern cuisine, hydrocolloids provide endless possibilities for elevating your dishes.

In cooking, hydrocolloids create edible films, foams, and stable emulsions. Chefs can use hydrocolloids like xanthan gum and agar-agar to add a touch of creativity to their recipes. These ingredients can create visually stunning presentations, creamy textures in sauces and dressings, and stable foams that enhance the dining experience. With hydrocolloids, the possibilities are truly endless.

Regarding baking, hydrocolloids are essential for achieving the desired texture and stability in various products. They can be used as thickening agents in batters and fillings and stabilizers in doughs and creams. Incorporating hydrocolloids into your baking recipes ensures that your cakes, cookies, and pastries have the perfect texture and mouthfeel.

Hydrocolloid Recipes:

- ⇨ **Stable Foams:** Create a stable foam with hydrocolloids like lecithin and xanthan gum for savory or sweet applications. Use these foams as finishing touches for your dishes or as delightful toppings for desserts.
- ⇨ **Thickened Sauces:** Experiment with different hydrocolloids to achieve the desired consistency in your sauces. Whether making a creamy béchamel or a tangy barbecue sauce, hydrocolloids can help you achieve the perfect texture.
- ⇨ **Vegan Alternatives:** If you want to create vegan-friendly recipes, hydrocolloids can achieve creamy textures without the need for dairy. Try using aquafaba, the liquid from canned chickpeas, as a hydrocolloid-rich ingredient in recipes like vegan chocolate mousse.

Hydrocolloids have revolutionized the culinary world, allowing chefs and home cooks alike to push the boundaries of creativity in the kitchen. From stable foams to thickened sauces, these ingredients offer endless possibilities for enhancing texture and creating unique dishes. Incorporate hydrocolloids into your cooking and baking repertoire and explore the exciting world of modern cuisine.

Table 33: Hydrocolloid Recipe's

HYDROCOLLOID RECIPE	DESCRIPTION
Stable Foams	Create airy and visually stunning foams using hydrocolloids like xanthan gum and lecithin. These foams can be used as a finishing touch for savory dishes or a delightful dessert topping.
Thickened Sauces	Experiment with different hydrocolloids to achieve the perfect consistency in your sauces. Hydrocolloids can help you create thick and creamy sauces that elevate the flavors of your dishes.
Vegan Chocolate Mousse	Try a vegan-friendly alternative to traditional chocolate mousse using aquafaba, the liquid from canned chickpeas. Aquafaba acts as a hydrocolloid-rich ingredient, providing a creamy base for the mousse.

5.2.2 Gelling Agents in Food

Gelling agents are essential ingredients used in the food industry to create gels in various dishes and desserts. Hydrocolloids play a crucial role in this process, providing the necessary texture and stability to the final product. Some popular examples of gelling agents include gelatin, pectin, and agar-agar, each with its unique properties and applications.

Gelatin, derived from animal collagen, is commonly used in jellies, custards, and other desserts. Its ability to form a gelatinous structure gives these dishes their desired consistency and texture.

Pectin, a natural hydrocolloid found in fruits and vegetables, creates gels in jams, fruit fillings, and other sweet preparations. It provides stability and thickening properties to these products.

Agar-agar, derived from seaweed, is a versatile gelling agent used in various dishes, including jellies, custards, and soups. It has a strong gelling power and is often used in vegan and vegetarian recipes as a plant-based alternative to gelatin.

Table 34: Gelling Agents, Sources and Applications

GELLING AGENT	SOURCE	APPLICATIONS
Gelatin	Animal collagen	Jellies, custards, desserts
Pectin	Fruits and vegetables	Jams, fruit fillings
Agar-agar	Seaweed	Jellies, custards, soups

Gelling agents are crucial in the food industry, allowing chefs and manufacturers to create visually appealing and structurally stable products. By understanding the properties and applications of hydrocolloid gelling agents, chefs can experiment with different textures, flavors, and presentations, offering unique dining experiences to their customers.

5.2.3 The Role of Hydrocolloids in Modern Cuisine

Hydrocolloids have revolutionized modern cuisine, offering chefs and food enthusiasts a wide range of possibilities for creativity and innovation. With their unique properties, hydrocolloids have become essential ingredients in molecular gastronomy, allowing chefs to push the boundaries of traditional cooking and create visually stunning dishes.

One of the key roles of hydrocolloids in modern cuisine is their ability to create unique textures. Chefs use hydrocolloids like agar-agar and xanthan gum to create stable foams, creamy sauces, and emulsions. These hydrocolloids provide the perfect balance of viscosity and stability, resulting in dishes that taste delicious and have a visually appealing presentation.

Molecular gastronomy, which embraces the use of hydrocolloids, explores the science behind food and uses innovative techniques to create new flavors and textures. By carefully manipulating hydrocolloids, chefs can create light and airy foams, crispy textures, and delicate gels that tantalize the senses. Applying hydrocolloids in modern cuisine has opened a new world of culinary experiences.

> "Hydrocolloids have transformed the culinary landscape, allowing chefs to create dishes that are not only delicious but also visually stunning. These ingredients offer endless possibilities for experimentation and innovation in the kitchen."
>
> - Chef James Smith.

Hydrocolloids play a vital role in enhancing modern cuisine, whether it's the velvety smoothness of a sauce, the unexpected burst of flavor from foam, or the delicate gel that holds a dish together. The creative use of hydrocolloids has elevated cooking to an art form, enabling chefs to craft dishes that surprise and delight their guests.

Table 35: Hydrocolloid Applications

HYDROCOLLOID	APPLICATION
Agar-agar	Creation of stable foams, gels, and edible films
Sodium Alginate	Spherification of liquids, creating flavorful "caviar"
Lecithin	Creation of airy foams and stable emulsions
Xanthan gum	Thickening agent for sauces and dressings, creation of stable foams

As modern cuisine continues to evolve, the role of hydrocolloids will undoubtedly expand. Chefs, home cooks, and culinary enthusiasts alike can explore the possibilities of hydrocolloids to create dishes that are not only visually impressive but also delightfully delicious.

5.2.4 Hydrocolloids in Modern Cuisine Recipe

A popular use of hydrocolloids in modern cuisine is creating unique textures and presentations in dishes. Chefs use hydrocolloids to create foams, like savory foams for finishing dishes or sweet foams for dessert toppings, by stabilizing the foam with ingredients like lecithin and xanthan gum. This technique creates airy textures that enhance the overall dining experience. Chefs can experiment with different hydrocolloids and flavors to create unique foam recipes.

Hydrocolloid Foam Recipe

Here is a recipe for a delicious and visually stunning hydrocolloid foam:

Ingredients:

- 500ml liquid of your choice (fruit juice, vegetable broth, etc.)
- 5g hydrocolloid of your choice (agar-agar, carrageenan, etc.)
- 2g lecithin
- 2g xanthan gum
- Optional: flavorings and seasonings

Instructions:

- In a saucepan, bring the liquid to a simmer.
- Add the hydrocolloid and stir until fully dissolved.
- Remove from heat and let the mixture cool to room temperature.
- Using an immersion blender or a whisk, blend the mixture while gradually adding the lecithin and xanthan gum until fully incorporated.
- Transfer the mixture to a whipping siphon or a stand mixer with a whisk attachment.
- Whip the mixture until it reaches a light and fluffy consistency.
- Season and flavor the foam as desired.
- Serve the foam as a topping or accompaniment to your dish.

Remember, the possibilities are endless with hydrocolloids. Don't be afraid to experiment with different flavors, colors, and textures to create stunning hydrocolloid foam recipes.

Table 36: Comparison of Hydrocolloid Foam Recipes

HYDROCOLLOID	LIQUID	TEXTURE	FLAVOR
Agar-Agar	Vegetable broth	Firm, stable foam	Savory, umami
Carrageenan	Coconut milk	Smooth, creamy foam	Sweet, tropical
Xanthan Gum	Fruit juice	Airy, light foam	Tart, refreshing

5.2.5 Technique: Creating Stable Foams with Hydrocolloids

Creating stable foams with hydrocolloids is a technique that can add a touch of elegance and creativity to culinary presentations. By incorporating the right hydrocolloids and utilizing proper techniques, chefs can achieve optimal stability and texture foams. The key to creating stable foams is selecting hydrocolloids with the necessary thickening and stabilizing properties.

Hydrocolloid Selection

Not all hydrocolloids are suitable for creating stable foams. Hydrocolloids with high viscosity and gelling properties, such as xanthan gum and agar-agar, are commonly used for this technique. These hydrocolloids provide the necessary structure and stability to the foam, allowing it to hold its shape and texture. It is important to experiment with different hydrocolloids to understand their unique properties and how they interact with other ingredients.

Technique

The technique for creating stable foams with hydrocolloids involves several steps. First, the hydrocolloid of choice is dissolved in a liquid, such as water or juice, to hydrate the particles. Once fully hydrated, the mixture is blended using a blender or hand mixer until a foam is formed. The foam can then be used to top dishes or incorporated as a component in various recipes.

> "Creating stable foams with hydrocolloids requires stabilizing ingredients like lecithin and xanthan gum. These hydrocolloids help maintain the texture and appearance of the foam."

It is important to note that the stability of the foam can be influenced by factors such as temperature and pH. Proper control of these variables is crucial to achieving desired results. Additionally, the foam should be used soon after preparation, as it may gradually lose its stability and collapse over time.

Table 37: Hydrocolloid Properties and Applications

HYDROCOLLOID	PROPERTIES	APPLICATIONS
Xanthan gum	High viscosity, stabilizing	Top foams, creamy dressings
Agar-agar	Strong gelling, stabilizing	Sweet and savory foams, jellies
Lecithin	Emulsifying, stabilizing	Aerated creams, foams

Experimentation and practice are key to mastering the technique of creating stable foams with hydrocolloids. Chefs can explore different combinations of hydrocolloids and liquids and incorporate flavors and aromatics to create unique and exciting foam creations.

5.2.6 Recipe: Vegan Chocolate Mousse with Aquafaba

Looking for a delicious vegan dessert that will satisfy your chocolate cravings? Try this amazing vegan chocolate mousse recipe made with aquafaba, the liquid from canned chickpeas. Aquafaba is a natural hydrocolloid-rich ingredient that provides the perfect base for achieving a creamy texture without using dairy products.

Aquafaba is a natural hydrocolloid-rich ingredient that provides the perfect base for achieving a creamy texture without using dairy products.

To make this vegan chocolate mousse, you will need the following:

Ingredients:

- 1 can of chickpeas (liquid reserved)
- 200g dairy-free dark chocolate, melted
- 1 teaspoon vanilla extract
- 2 tablespoons powdered sugar

Instructions:

- Drain the liquid from the chickpeas (aquafaba) can into a mixing bowl.
- Using an electric mixer, whisk the aquafaba quickly until it forms stiff peaks.
- Gently fold in the melted dark chocolate, vanilla extract, and powdered sugar until well combined.
- Transfer the mixture into individual serving glasses or bowls and refrigerate for at least 2 hours to set.
- Optional: Top with fresh berries or shaved chocolate for an extra touch of indulgence.

> "This vegan chocolate mousse is rich, creamy, and full of intense chocolate flavor. It's the perfect dessert for anyone looking for a delicious dairy-free alternative."

Enjoy this decadent vegan chocolate mousse as a delightful ending to any meal. It's a simple and satisfying dessert that will impress vegans and non-vegans alike.

Table 38: Vegan Chocolate Mousse Ingredients

INGREDIENTS	INSTRUCTIONS
1 can of chickpeas (liquid reserved)	Drain the liquid from the can of chickpeas (aquafaba) into a mixing bowl.
200g dairy-free dark chocolate, melted	Using an electric mixer, whisk the aquafaba quickly until it forms stiff peaks.
1 teaspoon vanilla extract	Gently fold in the melted dark chocolate, vanilla extract, and powdered sugar until well combined.
2 tablespoons powdered sugar	Transfer the mixture into individual serving glasses or bowls and refrigerate for at least 2 hours to set.

5.2.7 Technique: Using Hydrocolloids for Thickening Sauces

Hydrocolloids are invaluable tools for achieving the desired consistency when thickening sauces. These versatile ingredients can absorb water and contribute to the texture and stability of sauces, resulting in a smooth and rich mouthfeel. By understanding the properties of various hydrocolloids and their thickening properties, chefs and home cooks can elevate their sauce-making game to new heights.

When selecting the right hydrocolloid for thickening sauces, it's essential to consider factors such as temperature and pH. Different hydrocolloids have different thickening properties, with some being more effective at higher temperatures or acidic environments. For example, xanthan gum is known for creating stable, thick textures in sauces, even in low concentrations. On the other hand, guar gum is highly effective in thickening cold sauces and dressings.

To ensure optimal results when using hydrocolloids for thickening sauces, it's important to properly disperse them in the liquid before applying heat. This allows the hydrocolloids to hydrate and fully activate their thickening properties. It's also crucial to avoid overcooking the sauce, as prolonged exposure to high heat can diminish the thickening effect of hydrocolloids.

Table 39: Hydrocolloid Thickening Properties and Applications

HYDROCOLLOID	THICKENING PROPERTIES	BEST APPLICATION
Xanthan Gum	Creates stable, thick textures in sauces	Hot sauces, gravies
Guar Gum	Highly effective in thickening cold sauces and dressings	Cold sauces, dressings
Agar-Agar	Forms a gel-like texture, ideal for thickening fruit sauces	Fruit sauces, coulis

By harnessing the power of hydrocolloids, chefs and home cooks can achieve the perfect consistency in their sauces, elevating the overall dining experience. Whether creating a velvety gravy, a thick and creamy pasta sauce, or a luscious fruit coulis, hydrocolloids are an essential tool in the culinary arsenal for achieving the desired texture and stability in sauces.

Conclusion

Hydrocolloids have revolutionized the world of cooking and baking, offering many benefits and possibilities. These versatile ingredients have become essential tools for chefs and home cooks alike, allowing them to enhance texture, stability, and mouthfeel in their culinary creations.

With hydrocolloids, you can explore new techniques and experiment with unique recipes. From creating stable foams to thickening sauces, these ingredients open up a world of culinary possibilities. Whether you're a professional chef or a passionate home cook, incorporating hydrocolloids into your recipes will elevate your dishes.

Cooking with hydrocolloids enhances flavor and presentation and provides an opportunity to explore vegan alternatives. By utilizing hydrocolloids like aquafaba, you can create delicious plant-based dishes without compromising on taste or texture.

So, embrace the potential of hydrocolloids in your kitchen. With their ability to improve texture, stability, and mouthfeel, these ingredients will elevate your cooking and baking to a new level.

FAQ

- **What are some examples of hydrocolloids?**

  Some common hydrocolloids include agar-agar, carrageenan, and xanthan gum. They can be found in jelly, pudding, ice cream, sauces, and salad dressings.

- **How are hydrocolloids used in cooking and baking?**

  Hydrocolloids are used in cooking and baking to enhance texture and create unique presentations. They can create edible films, foams, and stable emulsions. Chefs and home cooks can use hydrocolloids to experiment with new techniques and textures.

- **What are the benefits of using hydrocolloids?**

  By incorporating hydrocolloids into recipes, chefs and home cooks can improve various food products' texture, stability, and mouthfeel. Hydrocolloids also offer endless possibilities for enhancing cooking and baking by creating unique dishes and presentations.

PART 5.3

Hydrocolloids in Cooking and Baking: How Hydrocolloids Add Texture to Food

Hydrocolloids may not be a term most people come across daily, but they play an essential role in the food industry. They are used as food additives to improve the texture, stability, and sensory experience of various food preparations in cooking and baking.

Hydrocolloids are thickening agents that bind liquids and stabilize mixtures. In this section, we will explore the versatility of hydrocolloids and how they add texture to food. We will also discuss their role as food additives and thickening agents.

Core Insights:

- ☑ Hydrocolloids are versatile ingredients used in cooking and baking
- ☑ They contribute to the texture of food
- ☑ Hydrocolloids act as food additives and thickening agents
- ☑ They help improve the stability and sensory experience of various food preparations
- ☑ Hydrocolloids are essential for creating the desired consistency and stability in sauces, dressings, and other culinary creations

5.3.1 Understanding Hydrocolloids

Hydrocolloids are food additives used to modify the texture of food. They are derived from various animal, vegetable, and synthetic sources. Common hydrocolloids include gelling agents, stabilizers, emulsifiers, and water-binding agents.

Gelling Agents

Gelling agents are hydrocolloids used to create gels. They are commonly used in desserts, jellies, and pie fillings. Some of the most common gelling agents include:

Table 40: Gelling Agents, Source, Properties and Uses

GELLING AGENT	SOURCE	PROPERTIES	COMMON USES
Agar agar	Seaweed	Firm and brittle gels	Cheesecake, ice cream, custards
Carrageenan	Seaweed	Semi-firm and elastic gels	Chocolate milk, pudding, tofu
Pectin	Fruits	Soft and elastic gels	Jams, jellies, fruit preserves

Stabilizers

Stabilizers are hydrocolloids used to improve the stability and consistency of food. They are commonly used in sauces, dressings, and ice cream. Some of the most common stabilizers include:

- Xanthan gum
- Guar gum
- Carboxymethyl cellulose (CMC)

Emulsifiers

Emulsifiers are hydrocolloids that blend two immiscible substances, such as oil and water, into stable mixtures. They are commonly used in salad dressings, mayonnaise, and margarine. Some of the most common emulsifiers include:

- Lecithin
- Monoglycerides and diglycerides
- Polyglycerol esters

Water-Binding Agents

Water-binding agents are hydrocolloids that retain moisture in food, improving texture and shelf life. They are commonly used in baked goods, such as bread and cakes. Some of the most common water-binding agents include:

- Sodium Alginate
- Methylcellulose
- Hydroxypropyl methylcellulose (HPMC)

Understanding the different types of hydrocolloids and their unique properties can help you choose the perfect ingredient to achieve the desired texture in your culinary creations.

5.3.2 Comparing Natural and Synthetic Hydrocolloids

While both natural and synthetic hydrocolloids have unique properties and uses, there are some key differences between them. One significant difference is their source; natural hydrocolloids are derived from plants and animals, while synthetic hydrocolloids are produced in a laboratory.

Table 41: Comparison of Natural and Synthetic Hydrocolloids

PROPERTIES	NATURAL HYDROCOLLOIDS	SYNTHETIC HYDROCOLLOIDS
Cost	Relatively expensive	Less expensive
Availability	It may not be easy to source	Readily available
Functionality	It may have limited functionality and stability	Generally more versatile and stable

Ultimately, the choice between natural and synthetic hydrocolloids comes from personal preference and the specific application.

5.3.3 Thickening Agents and Stabilizers

Hydrocolloids are widely used as thickening agents and stabilizers in cooking and baking. These ingredients are essential for consistency and stability in various culinary preparations, from sauces and dressings to baked goods and desserts.

Thickening agents, as their name suggests, are used to increase the viscosity of a mixture. Hydrocolloids with thickening properties include carrageenan, xanthan gum, and guar gum. These agents absorb water and swelling, creating a gel-like texture that thickens the mixture.

Stabilizers, however, prevent ingredients from separating or breaking down. Hydrocolloids used as stabilizers include carrageenan, agar agar, and pectin. These agents work by interacting with other ingredients in the mixture, creating a stable network that helps maintain the structure of the preparation.

Tips for Using Hydrocolloids as Thickening Agents and Stabilizers

- Follow the recommended dosage carefully, as too much hydrocolloid can result in an overly thick or gummy texture.
- Hydrate the hydrocolloid properly before adding it to the mixture, as this will ensure even distribution and prevent clumping.
- Take into account the pH and temperature of the preparation, as these factors can affect the performance of the hydrocolloid.
- Consider the type of hydrocolloid and its properties when selecting it for a recipe, as different agents have different thickening and stabilizing abilities.

> "Hydrocolloids are like magic ingredients in my kitchen. They allow me to create the perfect texture in my sauces, soups, and baked goods every time."
>
> **– Chef Maria Rodriguez**

5.3.4 Gelling Agents

Hydrocolloids are essential in creating gels and are popular in desserts, jellies, jams, and pie fillings. Gelling agents bind with water to form a three-dimensional network of molecules that traps liquid, creating a stable and structured gel. Different hydrocolloids offer varying gel textures and firmness.

> Hydrocolloids act as gelling agents to create stable and structured gels, perfect for desserts, jellies, jams, and pie fillings.

Agar and carrageenan are popular natural gelling agents. Agar has a strong gel set, while carrageenan is softer. Pectin is another popular natural gelling agent used in making jams and jellies.

Among synthetic hydrocolloids, gelatin is the most used gelling agent. However, there is a growing demand for vegetarian and vegan alternatives, and hydrocolloids like agar and carrageenan are becoming more widely used. Xanthan gum and guar gum, primarily used as thickening agents, can also produce gels in high concentrations.

Table 42: Comparing the Gelling Properties of Hydrocolloids

GELLING AGENT	GEL TEXTURE	GEL STRENGTH
Agar	Firm and brittle	High
Carrageenan	Soft and elastic	Low to medium
Pectin	Soft and spreadable	Low to medium
Gelatin	Soft and elastic	Low to medium
Xanthan gum	Soft and elastic	Low to medium
Guar gum	Soft and elastic	Low to medium

When choosing a gelling agent, consider the desired texture and firmness of the finished product. For example, firmer gels are ideal for pie fillings, while softer gels work better for jellies and jams.

Using the appropriate hydrocolloid type and concentration is crucial when using gelling agents. Excessive amounts can cause a rubbery or unpleasant texture, while insufficient amounts can result in a weak or runny gel. Experimenting with different types and ratios of hydrocolloids is vital to achieving your gels' desired texture and consistency.

5.3.5 Emulsifiers: How Hydrocolloids Make Oil and Water Play Nice

Emulsions are common in the culinary world, from creamy salad dressings to velvety sauces. However, blending oil and water is no easy feat, as these two substances repel each other due to their different polarities. This is where emulsifiers, including hydrocolloids, come in handy.

Hydrocolloids can act as emulsifiers by stabilizing emulsions and preventing them from separating. They do this by coating the oil droplets and water molecules, making them more compatible and forming a stable mixture.

The Science of Emulsions

Emulsions are mixtures of two immiscible liquids, such as oil and water, that an emulsifying agent holds together. The oil and water will quickly separate without an emulsifier due to their differing densities and molecular structures.

Emulsifiers reduce the surface tension between the two liquids, allowing them to mix more uniformly. This is achieved by having one end of the emulsifier molecule bond with the water molecules while the other end bonds with the oil molecules, forming a bridge between them and keeping them suspended.

Hydrocolloids, in particular, are excellent emulsifiers due to their large molecular size and water-attracting properties. They can form a protective layer around oil droplets, preventing them from coalescing and separating from the water phase.

Table 43: Common Hydrocolloid Emulsifiers

HYDROCOLLOID	SOURCE	EMULSIFYING PROPERTIES
Guar Gum	Seed of guar plant	Stabilizes oil-in-water emulsions
Xanthan Gum	Fermented corn sugar	Stabilizes oil-in-water and water-in-oil emulsions
Carrageenan	Red Seaweed	Stabilizes dairy and vegan creamers, milk substitutes, and other liquid emulsions

Guar gum is a common emulsifying agent in salad dressings, mayonnaise, and ice cream. Xanthan gum is often used in gluten-free baking to mimic the texture of wheat flour, while carrageenan is used to stabilize vegan milk substitutes and creamers.

Tips for Using Hydrocolloids as Emulsifiers:

- Always hydrate hydrocolloids in cold water before adding them to the oil and water mixture.
- Use the proper dosage of hydrocolloid for the amount of oil and water in the recipe.
- Whisk the emulsion vigorously while adding the hydrocolloid to ensure even distribution.
- Avoid overheating the emulsion during cooking; the hydrocolloid may break down and lose its emulsifying properties.

5.3.6 Water-Binding Agents

Hydrocolloids with water-binding properties are essential to creating moist and tender baked goods. They work by absorbing and retaining water, which helps to reduce staling and extends the shelf life of baked goods.

One example of a natural water-binding agent is pectin. It's commonly used in jams and jellies because it forms a gel and holds water. Another natural water-binding agent is carrageenan, often used in dairy products like ice cream, to prevent ice crystal formation and improve texture.

On the synthetic side, xanthan gum is a popular water-binding agent in gluten-free baking. It helps to mimic the elasticity and structure of gluten, which is absent in gluten-free flour. Guar gum is another synthetic option, commonly used in sauces and dressings, to improve texture and stability.

Table 44: Hydrocolloid Sources and Applications

HYDROCOLLOID	SOURCE	APPLICATIONS
Pectin	Extracted from fruits like apples and citrus	Jams, jellies, fruit fillings
Carrageenan	Derived from seaweed	Dairy products, desserts
Xanthan gum	Fermented from corn sugar	Gluten-free baking
Guar gum	Extracted from guar beans	Sauces, dressings, ice cream

When using water-binding agents, following the recommended dosage and hydration instructions is crucial to achieve the desired texture and consistency. Adding too much water-binding agent can result in a gummy or slimy texture, while using too little may not provide the desired moisture retention.

Overall, water-binding agents are essential tools for achieving the perfect texture and extending the shelf life of baked goods. Whether natural or synthetic, these agents offer unique properties that can enhance the quality of your culinary creations.

5.3.7 Benefits of Using Hydrocolloids

Hydrocolloids are incredibly versatile ingredients that offer numerous benefits when incorporated into cooking and baking recipes. Here are some of the reasons why you should consider using hydrocolloids in your culinary creations:

Enhance Texture and Stability

Hydrocolloids can be used as thickening agents and stabilizers, helping to create the desired consistency and stability in a wide range of culinary preparations. They can prevent sauces and dressings from separating, ensure that ice creams and sorbets stay smooth and creamy, and maintain the shape of molded desserts and confections.

Improve Mouthfeel

Hydrocolloids also play a crucial role in improving the mouthfeel of food preparations. They can provide a silky-smooth texture to sauces, dressings, and custards, create a dense and chewy texture in baked goods, and add a satisfying crunch to fried foods.

Reduce Syneresis

Syneresis, or the loss of liquid from gels and other food preparations, can be a common problem when cooking and baking. Hydrocolloids can help reduce syneresis, keeping your dishes fresher for longer and providing a more appealing visual appearance.

Improve Nutritional Profile

Some hydrocolloids, such as soluble fiber, can help improve the nutritional profile of dishes by adding dietary fiber and reducing fat and sugar content.

Gluten-Free and Vegan-Friendly

Many hydrocolloids are gluten-free and can be used as substitutes for gluten-containing thickeners like flour and cornstarch. They are also vegan-friendly, making them a great plant-based cooking and baking option.

Overall, hydrocolloids offer a range of benefits that can help take your culinary creations to the next level. Whether you are a professional chef or a home cook, experimenting with hydrocolloids can help you achieve the desired texture, stability, and overall sensory experience in your dishes.

5.3.8 Applications of Hydrocolloids in Cooking and Baking

Hydrocolloids are utilized in a wide range of cooking and baking applications. These versatile ingredients are particularly useful for creating specific textures and improving product quality. Below are some practical examples of how hydrocolloids can be incorporated into your culinary creations:

Soups and Sauces

Hydrocolloids like xanthan gum and carrageenan can thicken and stabilize soups and sauces, ensuring a smooth and consistent texture. These hydrocolloids are particularly beneficial when making low-fat or low-sugar options, as they help maintain the desired thickness and mouthfeel.

Bakery Products

A wide range of hydrocolloids can be used in bakery products to improve texture, enhance flavor release, and extend shelf life. For example, guar gum and carboxymethyl cellulose (CMC) can be added to bread dough to improve its elasticity and prevent going stale. Meanwhile, Gellan gum can be used in cake glazes to create a glossy finish and prevent moisture loss.

Dairy Products

Hydrocolloids like pectin and carrageenan can be used in dairy products like yogurt, ice cream, and cheese to enhance texture and stability. Pectin can thicken and stabilize yogurt, while carrageenan can be added to ice cream to prevent ice crystal formation and improve creaminess.

Meat Products

Hydrocolloids can also improve the texture and juiciness of meat products like sausages and deli meats. For example, carrageenan can help bind meat proteins and improve slice ability, while xanthan gum can be added to meat fillings to improve viscosity and texture.

Beverages

Hydrocolloids are widely used in the beverage industry to improve texture, stability, and sensory experience. For example, carrageenan can be added to chocolate milk to prevent sedimentation and improve mouthfeel, while xanthan gum can be added to smoothies to improve viscosity and prevent separation.

These are just a few examples of how hydrocolloids can be used in cooking and baking. By experimenting with different types and dosages, you can discover the unique benefits and applications of hydrocolloids in your culinary creations.

5.3.9 Tips for Using Hydrocolloids

Hydrocolloids can be tricky to work with, but they can take your culinary creations to the next level with the right techniques. Here are some tips for using hydrocolloids effectively:

Understand Dosages

Hydrocolloids are typically used in small amounts, and too much or too little can affect the texture and consistency of your dish. Be sure to follow the recommended dosages carefully. As a rule, start with smaller amounts and gradually increase as needed.

Use the Right Hydration Method

Hydrocolloids require hydration before mixing them with a liquid to activate their gelling or thickening properties. Hydrocolloids have different hydration methods, so follow the instructions carefully to achieve the desired texture.

Consider Synergistic Effects

Hydrocolloids can have synergistic effects when used in combination with other ingredients. For example, combining xanthan gum and guar gum can produce a more elastic gel than either ingredient alone. Experiment with different combinations to achieve the desired texture and consistency.

Be Patient

Hydrocolloids can take time to activate and thicken fully, so be patient and allow the mixture to set for the recommended amount of time. Avoid the temptation to adjust the recipe prematurely, as this can result in unexpected outcomes.

Avoid Overheating

Hydrocolloids can break down if exposed to high temperatures, so avoid overheating the mixture. If using a thickener such as gelatin or agar agar, it's best to dissolve it in warm liquid before adding it to a hot mixture to avoid clumping.

Keep in Mind the Final Product's Texture

The final product's texture can be influenced by the type and amount of hydrocolloid used. Keep in mind the desired texture when selecting and using a hydrocolloid. For example, agar agar can create a firm and brittle texture, while xanthan gum can create a smooth and creamy texture.

By following these tips for using hydrocolloids, you can elevate the texture and overall quality of your culinary creations. Happy experimenting!

Conclusion

Hydrocolloids are versatile and essential ingredients used in cooking and baking to achieve the desired texture, consistency, and stability in various dishes. Each type offers unique properties and benefits to the culinary world, from natural hydrocolloids like agar agar, carrageenan, and pectin to synthetic hydrocolloids like xanthan gum and guar gum.

Taking Advantage of Hydrocolloids

You can take advantage of their numerous benefits in your culinary creations by understanding how hydrocolloids function as gelling agents, stabilizers, emulsifiers, and water-binding agents. They enhance texture and stability, and hydrocolloids can also improve mouthfeel and increase the shelf life of baked goods.

Applications of Hydrocolloids

The applications of hydrocolloids in cooking and baking are endless. They can be used in soups, sauces, dressings, ice cream, bread, and pastries. By incorporating hydrocolloids effectively, you can elevate your dishes' texture and sensory experience to the next level.

Tips for Success

When using hydrocolloids, following the dosage guidelines and hydration methods is important to avoid mistakes that can ruin your recipe. Always experiment with small amounts before incorporating hydrocolloids into larger batches. With practice and patience, you can master the art of working with hydrocolloids and achieve consistent, successful results.

In conclusion, hydrocolloids are essential in any chef's or home cook's arsenal. By taking advantage of their unique properties and benefits, you can elevate your dishes' texture, stability, and overall sensory experience. So, don't be afraid to experiment and explore the world of hydrocolloids in your culinary adventures.

FAQ

How do hydrocolloids contribute to the texture of food?

Hydrocolloids provide texture to food by acting as thickening agents, stabilizers, gelling agents, and emulsifiers. They help create desired consistencies, improve mouthfeel, and enhance the sensory experience.

What are some examples of natural hydrocolloids?

Natural hydrocolloids include agar agar, carrageenan, and pectin. These substances are derived from plant-based sources and offer unique properties for enhancing food texture.

Are there synthetic hydrocolloids available?

Yes, synthetic hydrocolloids like xanthan gum and guar gum are widely used in the food industry. They provide functionalities similar to natural hydrocolloids and offer benefits in terms of stability and consistency.

How do hydrocolloids act as thickening agents and stabilizers?

Hydrocolloids can absorb and retain water, which results in thickening of liquids and providing stability to sauces, dressings, and other culinary preparations.

What are gelling agents?

Gelling agents are hydrocolloids that can form gels when hydrated. They are commonly used in making jellies, desserts, and pie fillings, offering varying gel textures and firmness.

How do hydrocolloids act as emulsifiers?

Hydrocolloids can act as emulsifiers, enabling the blending of oil and water in recipes such as dressings and sauces. They help create and maintain stable emulsions.

FREQUENTLY ASKED QUESTIONS (FAQS) CONTINUED

What are water-binding agents in hydrocolloids?

Hydrocolloids with water-binding properties are excellent at retaining moisture in food preparations. They help improve baked goods' texture, shelf life, and quality.

Where can hydrocolloids be used in cooking and baking?

Hydrocolloids have a wide range of applications in cooking and baking. They can be used in soups, sauces, bread, pastries, and other culinary creations to elevate texture and sensory experience.

What are some tips for using hydrocolloids effectively?

Dosages, hydration methods, and understanding the different properties of hydrocolloids are important for successful results. Experimentation and careful measurement are key to incorporating hydrocolloids effectively into recipes.

PART 5.4

How Hydrocolloids Improve Mouthfeel

Hydrocolloids are essential components of many food and beverage products, vital in enhancing mouthfeel and texture. Mouthfeel, the tactile sensation created by food or beverage in the mouth, is a critical aspect of the sensory experience that affects how consumers perceive a product's quality and flavor.

Hydrocolloids, or ***gums***, are polysaccharides or proteins that can modify food and beverages' texture and physical properties. They can be derived from natural or synthetic sources and provide functional benefits such as thickening, gelling, stabilizing, and emulsifying.

Adding hydrocolloids to food and beverage formulations allows manufacturers to create products with desirable sensory attributes, such as a smooth, creamy, or thick texture. Hydrocolloids can also improve shelf life and prevent unwanted changes in texture, such as separation or syneresis.

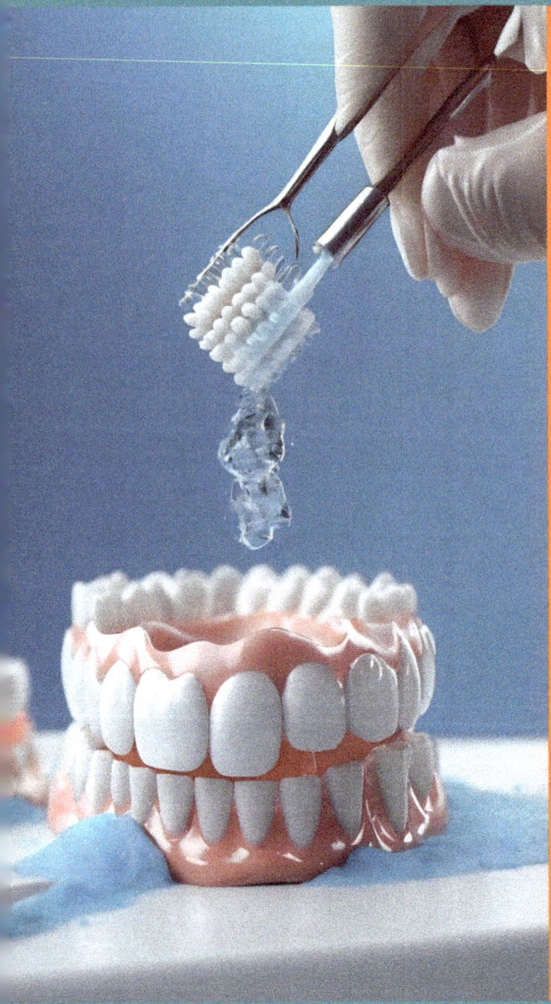

Core Insights:

- ☑ Hydrocolloids are essential components in food and beverage products that enhance mouthfeel and texture.
- ☑ Mouthfeel is a crucial aspect of the sensory experience that affects consumer perception of a product's quality and flavor.
- ☑ Hydrocolloids provide functional benefits such as thickening, gelling, stabilizing, and emulsifying, allowing manufacturers to create products with desirable sensory attributes.

5.4.1 The Uses of Hydrocolloids

Hydrocolloids are commonly used for various applications in the food and beverage industry. These natural or synthetic polymers have unique properties that contribute to a product's texture, viscosity, and stability.

One of the primary uses of hydrocolloids is as stabilizers for emulsions and suspensions. Hydrocolloids such as xanthan gum, carrageenan, and guar gum are often used to prevent separation and improve the shelf life of products like salad dressings, sauces, and beverages.

Hydrocolloids also function as thickeners, enhancing the texture and viscosity of soups, puddings, and dairy products. Commonly used thickeners include carrageenan, pectin, and agar-agar.

Gelling agents are another common application for hydrocolloids. Hydrocolloids like agar-agar, carrageenan, and gelatin create stable gels in products like jams, jellies, and desserts.

Hydrocolloids can also be used as film-forming agents, creating a protective coating around food products. This technique is frequently used to extend the shelf life of fruits and vegetables.

Finally, meat and meat analogs can benefit from using hydrocolloids as binders and stabilizers. Hydrocolloids like carrageenan and Gellan gum can improve the texture and moisture retention of products like sausages and burgers.

5.4.2 The Benefits of Hydrocolloids

Hydrocolloids offer a range of benefits beyond their ability to enhance texture and mouthfeel in food and beverages. These benefits include:

Table 45: Benefits of Hydrocolloids

BENEFIT	DESCRIPTION
Shelf stability	Hydrocolloids can help prevent ingredient separation and maintain stability over time, ensuring consistent quality and appearance.
Improved nutrition	Hydrocolloids enable the production of low-fat, low-sugar, and low-calorie foods without sacrificing taste and texture.
Extended product life	Hydrocolloids can help extend the shelf life of food and beverage products, reducing waste and increasing profitability.
Clean-label options	Natural hydrocolloids can replace artificial additives, allowing for cleaner and simpler ingredient labels.

The functional properties of different hydrocolloids also contribute to their benefits. For example, xanthan gum is known for its excellent stabilizing and emulsifying properties, while carrageenan is valued for its ability to form gels and thicken liquids.

Overall, hydrocolloids can improve the quality, appearance, and sensory experience of food and beverages while providing additional benefits to manufacturers and consumers.

5.4.3 Types of Hydrocolloids

Hydrocolloids are classified based on their origin, structure, and functionality. Numerous types of hydrocolloids are available in the market, each with unique properties and applications.

Plant-based Hydrocolloids

The following are examples of plant-based hydrocolloids:

Table 46: Hydrocolloids Sources and Functionality

HYDROCOLLOID	SOURCE	FUNCTIONALITY
Agar-agar	Red Seaweed	Thickener, stabilizer, gelling agent
Pectin	Fruits and vegetables	Gelling agent, thickener, stabilizer
Carrageenan	Red Seaweed	Stabilizer, thickener, gelling agent, emulsifier
Guar gum	Guar bean	Thickener, stabilizer, emulsifier, texturizer

Microbial-based Hydrocolloids

The following are examples of microbial-based hydrocolloids:

Table 47: Examples of microbial-based hydrocolloids

HYDROCOLLOID	SOURCE	FUNCTIONALITY
Xanthan gum	Bacteria (Xanthomonas campestris)	Thickener, stabilizer, emulsifier, suspending agent
Gellan gum	Bacteria (Sphingomonas elodea)	Gelling agent, thickener
Pullulan	Bacteria (Aureobasidium pullulans)	Thickener, film-former, encapsulating agent

Animal-based Hydrocolloids

The following are examples of animal-based hydrocolloids:

Table 48: Examples of animal-based hydrocolloids

HYDROCOLLOID	SOURCE	FUNCTIONALITY
Gelatin	Collagen from animal bones, skin, and connective tissue	Gelling agent, stabilizer, emulsifier
Casein	Milk protein	Emulsifier, stabilizer, thickener

These hydrocolloids have different functionalities and are used in various food and beverage applications to improve texture, viscosity, and stability.

5.4.4 The Hydrocolloid Industry and Market

The hydrocolloid industry has experienced significant growth over the past few years, largely driven by the increasing demand for processed foods and beverages. Hydrocolloids are widely used in the food industry as stabilizers, emulsifiers, thickeners, and gelling agents, among other applications. They play a vital role in improving the texture and mouthfeel of food and beverage products, making them more attractive to consumers.

The hydrocolloid market is expected to grow in the coming years, driven by the expanding food and beverage industry. The demand for natural and clean-label products has also increased, leading to a shift towards natural hydrocolloids and away from synthetic ones. As a result, manufacturers and suppliers of hydrocolloids are focusing on developing and offering a range of natural and functional hydrocolloids to meet consumers' evolving needs and preferences.

Table 49: Key Players in the Hydrocolloid Industry

KEY PLAYERS IN THE HYDROCOLLOID INDUSTRY	MARKET SHARE
CP Kelco	20%
DuPont Nutrition & Health	18%
Cargill, Incorporated	15%
Ingredion Incorporated	12%
Ashland Global Holdings Inc.	10%

Hydrocolloid manufacturers and suppliers have a crucial role in ensuring the quality and safety of hydrocolloid ingredients. They must adhere to rigorous standards and regulations to ensure that their products meet the required specifications and do not pose any risk to human health. Buyers of hydrocolloids must also exercise caution in sourcing reliable and trustworthy suppliers to ensure that they receive high-quality and safe products.

5.4.5 Hydrocolloid Ingredients for Mouthfeel Enhancement

Hydrocolloids are a diverse group of ingredients that can significantly impact food and beverages' texture and sensory experience. Certain hydrocolloids have proven particularly effective when it comes to enhancing mouthfeel. Here are some of the most used hydrocolloid ingredients for mouthfeel enhancement:

Table 50: Hydrocolloid ingredients for mouthfeel enhancement

HYDROCOLLOID	FUNCTIONALITY	APPLICATION
Xanthan gum	Stabilizer, thickener	Dressings, sauces, beverages
Carrageenan	Gelling agent, stabilizer	Dairy products, desserts, beverages
Guar gum	Thickener, stabilizer	Baked goods, dairy products, beverages
Agar-agar	Gelling agent, stabilizer	Desserts, jellies, confectionery
Pectin	Gelling agent, thickener	Jams, jellies, dairy products

Xanthan gum is a standard thickening agent that can create a creamy and smooth texture in dressings, sauces, and beverages. Similarly, carrageenan can create a smooth and creamy texture in dairy products, desserts, and beverages. Guar gum can thicken and stabilize a wide range of products and is commonly found in baked goods, dairy products, and beverages. Agar-agar and pectin are gelling agents that create a unique, jelly-like texture in desserts, jams, and other confectionery items.

Other hydrocolloid ingredients that improve mouthfeel include Arabic gum, locust bean gum, and cellulose gum. Each hydrocolloid has a unique functionality that can improve the sensory experience of different food and beverage products.

The right hydrocolloid ingredient(s) for a given application will depend on the specific product, sensory goals, and processing conditions. Working closely with a hydrocolloid supplier is essential to select the optimal hydrocolloid for a particular application.

Hydrocolloid ingredients have transformed the food and beverage industry by improving texture, stability, and sensory perception. By carefully selecting and using the right hydrocolloid ingredients, manufacturers can elevate the mouthfeel of their products, delivering a superior sensory experience that consumers will love.

5.4.6 Understanding Mouthfeel

Mouthfeel is an important aspect of sensory perception that can greatly influence the overall experience of consuming food and beverages. It refers to the physical sensations experienced in the mouth, such as texture, viscosity, smoothness, and the taste and flavor of a product.

Mouthfeel is closely related to texture and is often used interchangeably, but they are distinct concepts. Texture refers to the structural properties of a product, such as hardness, crispness, and chewiness, while mouthfeel encompasses a wider range of sensations associated with the oral and nasal cavities. When describing mouthfeel, it is important to consider a product's physical and chemical properties. For example, the thickness, stickiness, and creaminess of a liquid can greatly impact the mouthfeel and overall sensory experience. Similarly, certain ingredients, such as fats, proteins, and sugars, can affect how a product feels in the mouth.

Describing mouthfeel can be challenging, as it requires a specialized vocabulary that considers the various sensations and experiences associated with consuming different products. Terms such as smooth, gritty, creamy, and chewy are commonly used to describe mouthfeel, but many more can be applied depending on the specific product and context.

Understanding mouthfeel is crucial for developing and evaluating food and beverage products. By considering the different aspects of mouthfeel and how they contribute to the sensory experience, manufacturers can create more pleasing products for consumers and stand out in a crowded market.

5.4.7 Mouthfeel Examples in Foods and Beverages

Hydrocolloids are widely used in the food and beverage industry to enhance mouthfeel. Here are some examples of how hydrocolloids are employed to improve the texture and sensory experience of various food and beverage products:

Table 51: Mouthfeel in Foods and Beverages

PRODUCT	HYDROCOLLOID	FUNCTIONALITY
Ice Cream	Guar Gum	Stabilizer, Thickener
Soup	Carrageenan	Emulsifier, Thickener
Soda	Pectin	Gelling Agent, Texturizer
Salad Dressing	Xanthan Gum	Emulsifier, Thickener

Ice cream is a popular application for hydrocolloids, where guar gum is used to prevent ice crystals and improve the creamy mouthfeel. In soups, carrageenan is added to thicken and stabilize the texture, while pectin is used in soda to provide a smooth, velvety mouthfeel. Xanthan gum is commonly utilized in salad dressings to maintain emulsion and provide a uniform texture.

Other examples of hydrocolloid applications in food and beverages include yogurt, cheese, bakery products, and sauces. In each case, hydrocolloid is chosen based on its functional properties and how it can enhance the desired mouthfeel and texture of the end product.

5.4.8 Texture vs. Mouthfeel: Understanding the Difference

When describing sensory experiences in food and beverages, "texture" and "mouthfeel" are often used interchangeably. However, they refer to distinct aspects of the sensory experience and should be evaluated separately.

Texture refers to the physical properties of a food or beverage, such as its hardness, chewiness, or grittiness. It is determined by factors such as the size and shape of particles, the presence of air pockets, and the moisture content. Texture is important because it affects how a product feels in the mouth and can influence taste perception.

Conversely, mouthfeel refers to the overall sensation in the mouth when consuming a food or beverage. It encompasses texture, viscosity, creaminess, and mouth coating. Mouthfeel is important because it contributes to the overall enjoyment of the product and can influence consumer preferences.

When formulating food and beverage products, it is essential to consider texture and mouthfeel to create a well-rounded sensory experience. Hydrocolloids play a crucial role in enhancing both aspects and can be used to achieve specific sensory profiles.

Conclusion

Hydrocolloids are critical in enhancing texture and mouthfeel in food and beverage products. By acting as stabilizers, thickeners, gelling agents, and emulsifiers, hydrocolloids contribute to improved sensory experiences that drive consumer satisfaction.

Hydrocolloid ingredients are crucial in formulations that require specific texture and viscosity profiles. They enhance the mouthfeel, improve shelf stability, and reduce the need for other additives.

By understanding the various types of hydrocolloids and their functional properties, manufacturers and suppliers can offer tailored solutions that meet the unique needs of their consumers.

Moreover, the hydrocolloid industry shows significant potential for growth and advancement. With technological innovations and increased demand for clean-label products, manufacturers must prioritize quality control and sourcing reliable suppliers to ensure consistent product quality.

As the industry continues to evolve, it is increasingly important to understand the difference between texture and mouthfeel. These two sensory experiences contribute to the overall sensory perception of a product, and manufacturers must consider both when formulating products that meet the desires of their consumers.

In conclusion, hydrocolloid ingredients are crucial in formulating food and beverage products. Texture and mouthfeel enhancement are key aspects of these products, and using hydrocolloids provides the desired properties. The industry shows significant potential for growth, and as consumers continue to seek products with enhanced sensory experiences, hydrocolloids will undoubtedly play a critical role.

FAQ

Q: How do hydrocolloids improve mouthfeel?

Hydrocolloids enhance mouthfeel by improving the texture of food and beverages. They contribute to a smoother, creamier, or thicker consistency, providing a more pleasurable sensory experience.

Q: What is the hydrocolloid industry and market like?

The hydrocolloid industry encompasses manufacturers and suppliers of hydrocolloids. The market for hydrocolloids is constantly evolving, with trends and growth opportunities. Quality control and sourcing reliable suppliers are crucial factors in this industry.

Q: Which hydrocolloid ingredients are commonly used to enhance mouthfeel?

Specific hydrocolloid ingredients, such as carrageenan, xanthan gum, and pectin, are commonly used to enhance mouthfeel. These ingredients improve texture, viscosity, and sensory experience in food and beverages.

Q: What is mouthfeel, and how is it described?

Mouthfeel refers to the physical sensations perceived in the mouth when consuming food and beverages. It includes factors such as texture, viscosity, creaminess, and thickness. Mouthfeel can be described using terms like smooth, creamy, chewy, crunchy, or watery.

Q: Can you explain how hydrocolloids enhance mouthfeel in foods and beverages?

Hydrocolloids can enhance the mouthfeel in various food and beverage products. For example, hydrocolloids contribute to a smoother and creamier texture in ice cream. They help achieve the desired viscosity and mouth coating in sauces and dressings. In bakery products, they improve the softness and moistness.

Q: What is the difference between texture and mouthfeel?

Texture refers to a product's overall consistency and physical properties, including factors like smoothness, crunchiness, and elasticity. Mouthfeel, on the other hand, specifically relates to the sensations experienced in the mouth when consuming a product, such as creaminess, thickness, and coating.

CHAPTER 06

Health and Nutrition

PART 6.1

Understanding Hydrocolloid Digestion and Metabolism

Hydrocolloids are substances commonly used in the food industry to improve product texture, stability, and overall quality. However, not many people understand the digestion and metabolism of these substances and their impact on health. In this section, we will delve into the digestibility and metabolism of hydrocolloids, exploring how they are broken down and utilized by the body.

Core Insights:

- ☑ Hydrocolloids are commonly used in the food industry to improve the texture and stability of products.
- ☑ Understanding how hydrocolloids are digested and metabolized is crucial in recognizing their potential impact on overall health and well-being.
- ☑ Hydrocolloids significantly promote gut health, improve digestion, enhance satiety, and support weight management.
- ☑ Their diverse applications in the food industry, pharmaceuticals, and cosmetics make hydrocolloids a crucial ingredient in various products.
- ☑ Further research is necessary to fully understand the digestion and metabolism of hydrocolloids and their potential benefits to human health.

6.1.1 Hydrocolloids in Baking

Hydrocolloids are essential in improving baked goods' texture, consistency, and shelf life. These substances are commonly used in baking, as they mimic gluten's functional properties, the primary protein in wheat flour.

Gluten is responsible for providing elasticity and structure to baked goods. It allows the dough to rise and expand, giving bread and pastries their characteristic texture. However, many people are intolerant or sensitive to gluten, necessitating alternative ingredients in baking.

Types of Hydrocolloids Used in Baking

Several types of hydrocolloids are commonly used in baking, including:

- Xanthan gum
- Guar gum
- Carrageenan
- Agar
- Locust bean gum

These hydrocolloids have unique properties that make them suitable for different types of baked goods. For example, xanthan gum is often used in gluten-free bread and pastry recipes, as it can mimic the elasticity and structure of gluten. Carrageenan is commonly used in dairy-based baked goods, as it can act as a stabilizer and prevent syneresis (the separation of liquids) in custards and puddings.

Benefits of Hydrocolloids in Baking

Hydrocolloids offer several benefits in baking, including:

- Improving texture and consistency
- Increasing shelf life
- Reducing staling
- Enhancing moisture retention
- Improving freeze-thaw stability

These benefits are especially important in gluten-free baking, as achieving the desired texture and consistency without gluten can be challenging. Using hydrocolloids, bakers can create gluten-free products that are just as delicious and visually appealing as their gluten-containing counterparts.

> "Hydrocolloids are a game-changer in gluten-free baking. They allow us to create products that are just as tasty and visually appealing as traditional baked goods."

6.1.2 Hydrocolloids in Pharmaceuticals

Hydrocolloids have found diverse applications in the pharmaceutical industry due to their unique properties. They are commonly used in drug delivery systems, oral suspensions, and controlled-release formulations. This section will shed light on how pharmaceuticals use hydrocolloids and their potential benefits.

Drug Delivery Systems

Hydrocolloids are frequently used in drug delivery systems due to their ability to control the release of active substances. When added to a drug formulation, hydrocolloids can form a gel-like matrix that slows down drug release and prolongs the drug's action. This is particularly useful for drugs that need to be released over a prolonged period, such as painkillers or antidepressants.

Oral Suspensions

Hydrocolloids are also used in the preparation of oral suspensions. These are liquid formulations that contain finely dispersed drug particles. Hydrocolloids are added to the suspension to prevent the drug particles from settling at the bottom of the container and ensure uniform distribution of the active substance. This improves the efficacy of the drug and makes it easier to administer.

Controlled Release Formulations

Controlled-release formulations are drug formulations that release the active substance in a controlled manner over time. Hydrocolloids are often used in these formulations to control drug release and improve drug stability. They can also help to protect the drug from enzymatic degradation in the gastrointestinal tract and increase drug absorption.

Table 52: Hydrocolloid Applications

HYDROCOLLOID	APPLICATION
Sodium Alginate	Controlled release formulations
Carrageenan	Oral suspensions
Guar gum	Drug delivery systems

Hydrocolloids play a crucial role in the pharmaceutical industry by improving drug delivery and stability. Their ability to form gel-like matrices and control drug release makes them an attractive option for drug formulations.

In conclusion, hydrocolloids are versatile substances that offer numerous benefits in the pharmaceutical industry. They can be used in drug delivery systems, oral suspensions, and controlled-release formulations to improve drug efficacy and stability. The unique properties of hydrocolloids make them an attractive option for pharmaceutical applications, and ongoing research is uncovering new ways to utilize them.

6.1.3 Hydrocolloids in Cosmetics

Hydrocolloids have found numerous applications in the cosmetics industry, particularly in skincare and hair care formulations. These natural substances can provide various benefits, from moisturizing and nourishing to enhancing product texture and stability.

One of the primary uses of hydrocolloids in cosmetics is as a thickener or gelling agent. They can improve the texture of creams, lotions, and other products, making them more spreadable and easier to apply. Some commonly used hydrocolloids in skincare formulations include carrageenan, xanthan gum, and gelatin.

Hydrocolloids, mixtures of oil and water commonly used in cosmetics, can also help stabilize emulsions. Hydrocolloids can enhance the shelf life and overall quality of products by preventing the separation of these two phases. In hair care formulations, hydrocolloids can provide similar benefits by helping to create stable suspensions of ingredients like oils and surfactants.

Another important application of hydrocolloids in cosmetics is their ability to moisturize and soothe the skin. Hydrocolloids like aloe vera, guar gum, and hyaluronic acid can help to hydrate and nourish the skin, reducing dryness and irritation. Some hydrocolloids also have antioxidant properties, protecting against damage from environmental stressors like UV radiation and pollution.

Overall, hydrocolloids are a versatile and valuable ingredient in the cosmetics industry. From improving product texture and stability to providing moisturizing and soothing benefits, they offer a range of advantages that can enhance the effectiveness and appeal of skincare and hair care products.

6.1.4 Benefits of Hydrocolloids for Health

Hydrocolloids are known for their unique properties that offer many health benefits. Let's explore some of the significant benefits that hydrocolloids can provide:

Improved Digestion

Hydrocolloids can improve digestion by slowing down the rate food moves through the gut. This, in turn, increases the time available for nutrient absorption, promoting overall digestive health. Hydrocolloids such as psyllium husk and guar gum can relieve constipation and increase bowel regularity.

Increased Satiety

Hydrocolloids can absorb water and form a gel-like substance in the gut, increasing feelings of fullness and reducing hunger. This property makes hydrocolloids an excellent ingredient for weight management and controlling appetite.

Support for Weight Management

Hydrocolloids can help reduce calorie intake by delaying the rate of digestion and absorption, thus lowering energy intake and promoting weight loss. Additionally, hydrocolloids can lower blood sugar and insulin levels, which can help prevent weight gain.

Improved Gut Health

Hydrocolloids can support gut health by promoting the growth of beneficial gut bacteria. Research has shown that hydrocolloids can improve the gut microbiome by increasing the production of short-chain fatty acids, which are essential for gut health.

Overall, hydrocolloids offer a range of health benefits due to their unique properties. From improving digestion to supporting weight management and promoting gut health, these substances have a valuable role to play in maintaining overall wellness.

6.1.5 Digestibility of Hydrocolloids

Hydrocolloids, as we discussed earlier, are a group of substances widely used in the food industry, pharmaceuticals, and cosmetics owing to their unique properties. Understanding how hydrocolloids are digested in the human body is important in recognizing their potential impact on overall health and well-being.

Hydrocolloids are known for their ability to hold water and form gels, thereby altering the texture and consistency of food. However, due to their complex molecular structure, they are not fully digestible by human enzymes. Instead, they are fermented by gut bacteria in the large intestine, providing a source of nutrition for the microbiota.

Despite their limited digestibility, hydrocolloids have been shown to have a range of positive effects on the digestive system. For instance, they can help regulate bowel movements and alleviate symptoms of constipation. Additionally, they may improve the bioavailability of certain nutrients by slowing down their absorption rate, which is particularly beneficial for individuals with blood sugar imbalances.

Table 53: Comparison of the digestibility of selected hydrocolloids

HYDROCOLLOID	DIGESTIBILITY
Guar Gum	Partially digestible, fermented by gut bacteria
Xanthan Gum	Partially digestible, fermented by gut bacteria
Psyllium	Partially digestible, fermented by gut bacteria
Methylcellulose	Not digestible, ferments very slowly
Agar	Not digestible, ferments very slowly

As shown in the table above, the digestibility of hydrocolloids varies depending on their chemical structure and properties. Some, such as guar gum and xanthan gum, are partially digestible and readily fermented by gut bacteria. In contrast, others, such as methylcellulose and agar, are not digestible and ferment more slowly.

Overall, the digestibility of hydrocolloids plays a significant role in their impact on human health. While they are not fully digestible, they still provide a source of nutrition for the gut microbiota and have been shown to have beneficial effects on digestive health. As with any food or supplement, it is important to consider individual needs and potential sensitivities when incorporating hydrocolloids into your diet.

6.1.6 Metabolism of Hydrocolloids

Hydrocolloids are complex substances that undergo unique digestion and metabolism in the body. Their breakdown and absorption can impact various metabolic processes, energy expenditure, and overall health.

When ingested, hydrocolloids form a gel-like substance in the stomach that slows gastric emptying and delays the release of nutrients into the small intestine. This can lead to increased satiety and decreased appetite, making them beneficial for weight management.

The complex structure of hydrocolloids also affects their absorption in the small intestine. Some hydrocolloids, such as guar gum and pectin, can lower the absorption of nutrients like glucose and cholesterol. This can be beneficial for individuals with diabetes or high cholesterol levels.

Table 54: Comparison of the metabolism of different hydrocolloids

HYDROCOLLOID	METABOLIC IMPACT	FOOD SOURCES
Guar gum	Reduces glucose and cholesterol absorption	Legumes, seeds, and grains
Psyllium	Increases satiety and bowel regularity	Plant seeds
Xanthan gum	Low caloric value and improved texture	Fermented sugars

Overall, the metabolism of hydrocolloids varies depending on their chemical structure and degree of solubility. Research suggests they can influence gut health, lipid metabolism, and glucose homeostasis. Therefore, incorporating hydrocolloids into the diet may benefit overall health.

6.1.7 Hydrocolloids and Gut Health

Hydrocolloids are known for their ability to promote gut health. Hydrocolloids can function as prebiotics, feeding the beneficial bacteria in the gut and supporting a healthy gut microbiota.

Research has shown that certain hydrocolloids, such as inulin and beta-glucan, can improve bowel regularity and alleviate digestive discomfort. These hydrocolloids act as bulking agents, increasing stool volume and promoting food movement through the digestive system.

Furthermore, hydrocolloids can enhance satiety and support weight management. When ingested, hydrocolloids absorb water and expand in the stomach, creating a feeling of fullness that can lead to reduced calorie intake.

Table 55: Hydrocolloids for Gut Health

HYDROCOLLOID	BENEFITS FOR GUT HEALTH
Inulin	Increases bifidobacteria in the gut Improves bowel regularity
Psyllium	Relieves constipation Reduces inflammation in the gut
Beta-glucan	Supports a healthy gut microbiota Regulates the immune system

In conclusion, hydrocolloids have numerous benefits for gut health, including promoting healthy gut microbiota, improving bowel regularity, and enhancing satiety. Incorporating hydrocolloids into your diet can improve digestive health and overall well-being.

Conclusion

Hydrocolloids are essential in various industries due to their unique properties and applications. We have explored how hydrocolloids are utilized in the food industry as stabilizers, emulsifiers, and thickening agents. Hydrocolloids are also used in pharmaceuticals to enhance drug delivery systems and controlled-release formulations. The cosmetics industry plays a vital role in skincare products and hair care formulations, among other preparations.

Aside from their various applications, hydrocolloids offer several benefits for overall health and well-being. They promote gut health by supporting healthy gut microbiota, improving bowel regularity, and alleviating digestive discomfort. Additionally, hydrocolloids can enhance satiety, support weight management, and benefit metabolic processes by impacting energy expenditure.

Understanding the digestibility and metabolism of hydrocolloids is crucial in recognizing their potential impact on overall health and well-being. As we have explored in this article, hydrocolloids are essential in various industries due to their unique properties and applications.

Overall, hydrocolloids play a significant role in enhancing the quality and benefits of many products. With their positive effects on the gut and overall health, hydrocolloids are a valuable ingredient for manufacturers and consumers.

FAQ

Are hydrocolloids easily digestible?

Hydrocolloids are generally considered to be easily digestible. However, digestibility can vary depending on the specific type of hydrocolloid and the individual's digestive system. In most cases, hydrocolloids are broken down and absorbed by the body without any issues.

Do hydrocolloids have any impact on metabolism?

Hydrocolloids can have some impact on metabolism. While they are not a significant source of calories, they may somewhat affect energy expenditure and metabolic processes. However, more research is needed to understand the metabolic effects of hydrocolloids fully.

Can hydrocolloids promote gut health?

Yes, hydrocolloids can promote gut health. They can increase the stool's bulk and viscosity, improving bowel regularity and alleviating digestive discomfort. Additionally, certain hydrocolloids can act as prebiotics, nourishing beneficial gut bacteria.

Are hydrocolloids used in pharmaceuticals and cosmetics?

Yes, hydrocolloids are used in both pharmaceutical and cosmetic products. In the pharmaceutical industry, they are used in drug delivery systems, oral suspensions, and controlled-release formulations. In cosmetics, they can be found in skincare products, hair care formulations, and other cosmetic preparations.

PART 6.2

Hydrocolloid Health Benefits: Dietary Fiber and Gut Health Enhancement

Hydrocolloids are a group of food additives that have gained increasing attention in recent years due to their versatile functionality and potential health benefits. Among their many uses, hydrocolloids are known for their ability to provide dietary fiber and enhance gut health, two key factors in promoting overall wellness. In this article, we will explore the world of hydrocolloids, from their various functions in the food industry to their impact on our digestive system.

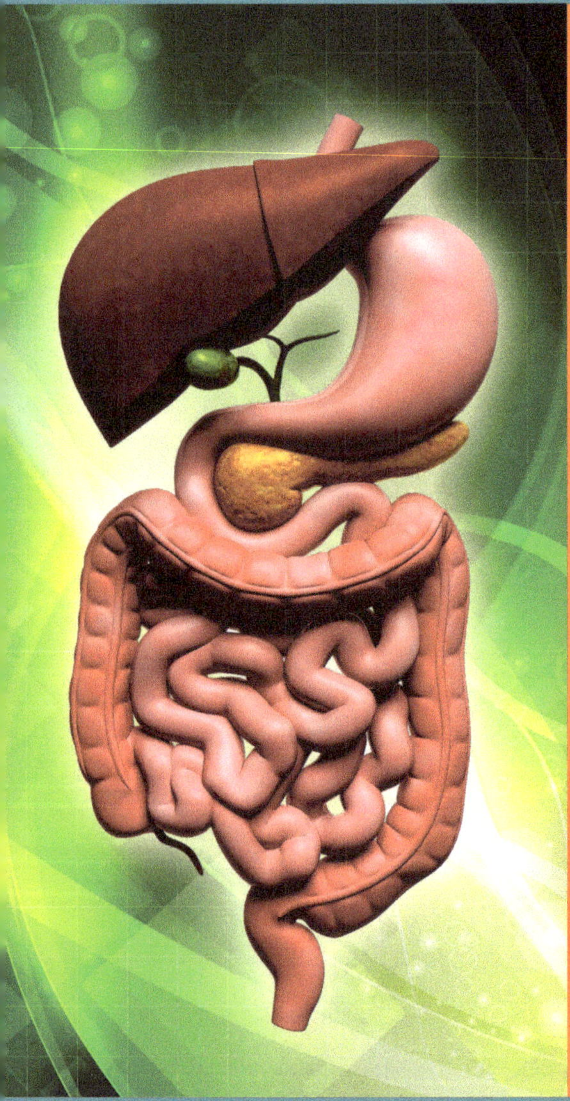

Core Insights:

- ☑ Hydrocolloids are food additives with multiple functions, including thickening, gelling, stabilizing, emulsifying, and enhancing texture.
- ☑ Hydrocolloids can be used as dietary fiber sources in food products, improving digestion and promoting bowel regularity.
- ☑ Hydrocolloids can enhance gut health by supporting a healthy gut microbiome and reducing inflammation.
- ☑ There are different types of hydrocolloids, each with unique functionalities and applications in the food industry and beyond.
- ☑ Hydrocolloids are regulated by guidelines to ensure consumer safety and product quality.

6.2.1 Understanding Hydrocolloids: Food Additives with Multiple Functions

Hydrocolloids are a diverse group of food additives that serve multiple functions in the food industry. They are commonly used as thickening agents, gelling agents, stabilizers, emulsifiers, rheology modifiers, and texture enhancers.

Thickening agents are hydrocolloids that increase the viscosity and consistency of food products. They create a smoother and creamier texture, making them ideal for use in sauces, soups, and dressings. Gelling agents, on the other hand, form gels or semi-solid structures, providing body and texture to products such as puddings, jams, and jellies.

Stabilizers help maintain food products' physical and chemical properties, preventing unwanted changes such as separation, crystallization, and oxidation. Emulsifiers promote the mixing of two immiscible substances, such as oil and water, allowing for the creation of smooth and homogenous products like mayonnaise and ice cream. Meanwhile, rheology modifiers alter food products' flow and deformation characteristics, improving their handling and processing.

Texture enhancers are hydrocolloids that enhance the sensory properties of food products, such as their mouthfeel, creaminess, and smoothness. They are commonly used in products ranging from dairy to bakery and confectionery.

6.2.2 The Role of Hydrocolloids as Dietary Fiber Sources

Hydrocolloids are increasingly recognized as valuable ingredients for the food industry when it comes to improving gut health and enhancing the nutritional value of food products. One of their primary benefits is their role as a source of dietary fiber.

Dietary fiber is an essential nutrient that aids digestion, promotes bowel regularity, and helps maintain a healthy gut microbiome. Unfortunately, many modern diets are deficient in fiber, which can lead to a range of health issues, including constipation, inflammation, and even an increased risk of certain types of cancer.

This is where hydrocolloids come in. Hydrocolloids are polymers that absorb and retain water, making them effective thickeners and stabilizers in food products. They can also act as dietary fiber sources, extracting water as it passes through the digestive tract and adding bulk to stool.

Common hydrocolloids used as dietary fiber sources include:

Table 56: Common hydrocolloids used as dietary fiber

HYDROCOLLOID	SOURCE	DIETARY FIBER CONTENT (PER 100G)
Psyllium Husk	Plantago ovata seeds	88g
Methylcellulose	Cellulose from wood pulp or cotton	71g
Inulin	Chicory root	64g

These hydrocolloids can be added to various food products, from baked goods to dairy to convenience foods, to increase their fiber content and enhance their nutritional value. Many manufacturers now incorporate hydrocolloids into their product formulations to appeal to health-conscious consumers.

Overall, using hydrocolloids as dietary fiber sources is a promising trend for the food industry. By promoting gut health and providing a range of nutritional benefits, they offer a valuable way to improve the quality of our food products and support overall wellness.

6.2.3 Gut Health Enhancement with Hydrocolloids

Hydrocolloids are known for their health benefits, particularly their positive impact on gut health. Hydrocolloids serve as an excellent source of dietary fiber and can also function as texture enhancers in food products.

Dietary fiber is essential for maintaining a healthy gut, as it promotes regular bowel movements, aids digestion, and helps maintain a healthy microbiome. Hydrocolloids, such as psyllium husk and pectin, have been shown to enhance gut health by increasing stool frequency and improving stool consistency.

In addition to their role as dietary fiber sources, hydrocolloids can also serve as texture enhancers in food products. By adding hydrocolloids to food formulations, manufacturers can improve their products' mouthfeel and overall texture while also providing consumers with gut health benefits.

Hydrocolloids as Texture Enhancers

Hydrocolloids can function as texture enhancers in various food products, including dressings, sauces, and beverages. For example, xanthan gum is commonly used as a thickener and stabilizer in salad dressings, while carrageenan stabilizes dairy products such as ice cream and yogurt.

By adding hydrocolloids to food formulations, manufacturers can improve the sensory experience of their products. Hydrocolloids can improve the mouthfeel, increase viscosity, and provide a creamier product texture.

Hydrocolloids and Gut Microbiome

The gut microbiome plays a crucial role in overall health, and research has shown that hydrocolloids can promote a healthy microbiome. Hydrocolloids can help to increase the production of short-chain fatty acids (SCFAs) in the gut, which provide energy to the cells in the colon and help maintain a healthy gut environment.

In addition, hydrocolloids have been shown to have prebiotic properties, meaning that they can feed the beneficial bacteria in the gut and promote their growth.

6.2.4 Types of Hydrocolloids and Their Functionality

Hydrocolloids are diverse food additives with unique functionalities that make them valuable ingredients in various food products. Here are the most used types of hydrocolloids:

Table 57: Types of hydrocolloids and uses

TYPE	FUNCTIONALITY	APPLICATIONS
Agar	Gelling agent	Jellies, puddings, canned meats and fish, confectionery, bakery
Alginates	Thickening agent, gelling agent, stabilizer	Sauces, dressings, bakery, dairy, meat, beverages, confectionery
Carrageenan	Thickening agent, gelling agent, stabilizer	Dairy products, meat products, confectionery, canned foods, sauces, beverages
Gellan gum	Gelling agent, stabilizer	Plant-based meat alternatives, confectionery, desserts, bakery, sauces, beverages
Guar gum	Thickening agent, emulsifier, stabilizer	Bakery, dairy products, sauces, dressings, beverages, meat products
Xanthan gum	Thickening agent, stabilizer	Sauces, dressings, bakery, dairy products, beverages, confectionery, meat products

Hydrocolloids are essential in the food industry, providing unique functionalities that other ingredients cannot match. They can improve food products' texture, stability, and viscosity, allowing manufacturers to create consistently high-quality items. Their versatility and compatibility with various ingredients and manufacturing processes make them ideal for various applications.

Hydrocolloid blends

Hydrocolloid blends, combinations of two or more hydrocolloids, have become increasingly popular in recent years. These blends offer improved functionality and efficiency compared to single hydrocolloids, providing even greater versatility in food product development. Some popular hydrocolloid blends include:

- Carrageenan and xanthan gum for dairy products, such as ice cream and yogurt
- Locust bean gum and xanthan gum for bakery products, such as bread and cakes
- Guar gum and xanthan gum for sauces and dressings

Hydrocolloids will continue to play a crucial role in the food industry as they offer invaluable functionalities and health benefits. Manufacturers are constantly innovating in this space to develop new blends and applications for hydrocolloids, ensuring they remain an essential ingredient in the food products of tomorrow.

6.2.5 Hydrocolloids in Food Formulation and Processing

Hydrocolloids are widely used in food formulation and processing, performing various functions. Texture enhancement is one of the primary reasons for incorporating hydrocolloids in food products. They act as stabilizers, emulsifiers, and texture enhancers, resulting in improved appearance, texture, and shelf-life of the finished product.

Hydrocolloids play a crucial role in the formulation of low-fat and reduced-sugar products, where they act as bulking agents, providing the necessary texture and mouthfeel. They also function as stabilizers, preventing the separation of ingredients in soups, sauces, and dressings.

The Role of Hydrocolloids in Meat Processing

The use of hydrocolloids in meat processing is particularly important, helping to improve the texture, mouthfeel, and sensory attributes of the final product. They act as binders, helping to hold the meat together and ensure adequate water retention. Hydrocolloids also prevent the formation of ice crystals, allowing meat products to retain their texture and shape even after freezing and thawing.

Examples of hydrocolloids used in meat processing include carrageenan, alginate, and sodium carboxymethylcellulose. Carrageenan is popular in processed meat products like luncheon meats and sausages.

The Use of Hydrocolloids in Bakery Products

Hydrocolloids also formulate bakery products like bread, cakes, and pastries. They help to improve the texture and moisture retention of baked goods, resulting in a softer crumb and longer shelf life. Hydrocolloids also act as gluten replacers, allowing gluten-free baked goods to be developed.

Examples of hydrocolloids in bakery products include xanthan gum, guar gum, and locust bean gum. These hydrocolloids are also used in ice creams and frozen desserts to prevent ice crystal formation and improve texture.

The Use of Hydrocolloids in Beverages

Hydrocolloids are increasingly used in the formulation of beverages, where they can act as emulsifiers, stabilizers, and texture enhancers. They help prevent sedimentation and separation of ingredients, improve the product's mouthfeel, and enhance the overall sensory experience. For example, carrageenan is often used in chocolate milk to prevent the cocoa from settling at the bottom of the bottle.

Examples of hydrocolloids used in beverages include pectin, carrageenan, and alginates. In fruit juices, pectin is used as a thickener and stabilizer, preventing the separation of pulp and juice. Carrageenan and alginates are commonly used in dairy-based beverages to improve texture and stability.

6.2.6 Safety and Regulations of Hydrocolloids

Hydrocolloids are widely used as food additives to enhance texture, stability, and emulsification, as well as to increase the dietary fiber content of food products. However, as with all food additives, it is important to ensure hydrocolloids' safety and regulatory compliance.

The safety of hydrocolloids is evaluated by regulatory agencies such as the U.S. Food and Drug Administration (FDA) and the European Food Safety Authority (EFSA). These agencies review the available scientific data on the safety of individual hydrocolloids and establish acceptable daily intake levels (ADI) for each one. ADI is the amount of a food additive that is considered safe for daily consumption over a lifetime.

In addition to safety evaluations, regulatory agencies also establish standards for hydrocolloid identity, purity, and quality. These standards ensure that consumers receive products that are accurately labeled and free from contaminants.

Regulations for Hydrocolloids in the United States

In the United States, the FDA regulates hydrocolloids as food additives under the Federal Food, Drug, and Cosmetic Act. Hydrocolloids must be approved for use by the FDA before they can be used in food products. The FDA maintains a list of approved food additives, including hydrocolloids, in the Code of Federal Regulations.

Hydrocolloids are classified by the FDA as either Generally Recognized as Safe (GRAS) or food additives subject to pre-market approval. GRAS hydrocolloids have a long history of safe use in food without pre-market approval. Hydrocolloids that are not GRAS must undergo a thorough safety evaluation by the FDA before they can be used in food products.

Regulations for Hydrocolloids in Europe

In Europe, hydrocolloids are regulated as food additives by the EFSA under the European Union's Food Additives Regulation. The EFSA must evaluate hydrocolloids before they can be approved in food products. The EFSA establishes ADI levels for each hydrocolloid and sets maximum usage levels in specific food categories.

In addition to safety evaluations, the EFSA establishes specifications for hydrocolloid identity, purity, and quality. These specifications ensure that hydrocolloids used in food products meet the necessary standards for consumer safety and product quality.

> "The safety of hydrocolloids is of utmost importance, and regulatory agencies work to ensure that these food additives are safe for consumption and comply with established standards."

Overall, the safety and regulations of hydrocolloids are critical considerations when using these ingredients in food products. By following established guidelines and ensuring regulatory compliance, manufacturers can provide safe, high-quality food products that meet consumer demands for texture, stability, and enhancement of gut health.

6.2.7 Applications of Hydrocolloids in Various Industries

Hydrocolloids are useful in the food industry and other industries where texture enhancement, stabilization, and emulsification are important. Let's take a closer look at some of the applications of hydrocolloids in various industries:

Pharmaceuticals

Hydrocolloids, substances used as carriers for active ingredients in medications, are widely used in the pharmaceutical industry as excipients. Hydrocolloids can be used as stabilizers and thickeners in liquid and semi-solid medications and as disintegrants in tablets and capsules. They can also improve the taste and texture of oral medications, making them easier to swallow.

Cosmetics

Hydrocolloids are used in cosmetics and personal care products as gelling agents, emulsifiers, and thickeners. They can improve formulations' texture, viscosity, and stability, making them more appealing to consumers. Hydrocolloids are also used in topical medications and wound dressings to retain moisture and promote healing.

Industrial Applications

Hydrocolloids are utilized in various industrial applications, from paper production to oil drilling. In paper production, hydrocolloids are used as binders and coatings to improve the strength and smoothness of paper. The oil industry uses hydrocolloids as thickeners and gelling agents in drilling fluids to improve their viscosity and prevent fluid loss.

6.2.8 Future Trends and Innovations in Hydrocolloids

The future of hydrocolloids looks promising, with ongoing research and innovations. As the demand for functional food ingredients continues to grow, hydrocolloids are well-positioned to play a significant role in meeting these needs. This section will look closely at some emerging trends and innovations in hydrocolloids.

Plant-Based Hydrocolloids

With the rising popularity of plant-based diets, there has been an increased demand for plant-based ingredients in the food industry. Hydrocolloids derived from plants, such as agar, pectin, and carrageenan, are already widely used. However, there is ongoing research to explore the functionalities and applications of other plant-based hydrocolloids, such as chia seeds, flax seeds, and psyllium.

Hydrocolloids for Meat Alternatives

As the demand for meat alternatives continues to grow, there is an opportunity for hydrocolloids to play a significant role in shaping the texture and flavor of these products. Hydrocolloids can be used to create the same texture and mouthfeel as meat. They can also improve the water-holding capacity of meat substitutes, ensuring they remain juicy and tender.

Nanocellulose Hydrocolloids

Nanocellulose hydrocolloids are a new and exciting area of research in the field of hydrocolloids. These hydrocolloids are derived from cellulose, the most abundant biopolymer on earth. They have unique functionalities that can be tailored to suit specific applications, including food and pharmaceuticals. Nanocellulose hydrocolloids can be used as emulsifiers, stabilizers, and thickening agents.

Table 58: Benefits of Nanocellulose Hydrocolloids and Applications

BENEFITS OF NANOCELLULOSE HYDROCOLLOIDS	APPLICATIONS
Biodegradable	Food packaging
Non-toxic	Pharmaceuticals
Thermally stable	Oil drilling fluids

Hydrocolloids for Clean-Label Products

Consumers are increasingly looking for clean-label products that are free from artificial additives. Hydrocolloids offer a natural solution to this demand, as they can be used as alternatives to synthetic ingredients. For example, hydrocolloids can create stable emulsions without synthetic emulsifiers. They can also thicken and stabilize products without relying on artificial thickeners.

- Homemade salad dressings made with xanthan gum instead of synthetic emulsifiers
- Ice cream made with carrageenan instead of artificial thickeners

The future of hydrocolloids is exciting, with ongoing research and innovations. From plant-based hydrocolloids to nanocellulose hydrocolloids, there are numerous opportunities for hydrocolloids to play a significant role in shaping the future of the food industry. As the demand for functional food ingredients continues to grow, hydrocolloids are well-positioned to meet these needs in innovative and sustainable ways.

Conclusion

In conclusion, hydrocolloids are versatile food additives with multiple functions, including thickening agents, gelling agents, stabilizers, emulsifiers, rheology modifiers, texture enhancers, and, most importantly, sources of dietary fiber that can enhance gut health. As consumers become increasingly health-conscious, the demand for food products containing hydrocolloids is rising.

The use of hydrocolloids in the food industry provides many benefits, including improving the texture, stability, and overall quality of food products. Additionally, hydrocolloids provide a cost-effective solution for increasing the fiber content in food products, a crucial dietary component for maintaining good health and preventing chronic diseases.

It is important to note that hydrocolloids' safety and regulatory aspects are strictly monitored, ensuring their safe use in the food industry. Moreover, hydrocolloids have applications beyond the food industry, including pharmaceuticals, cosmetics, and other sectors where texture enhancement, stabilization, and emulsification are important.

The future of hydrocolloids is promising, with emerging trends and innovations in the field that will continue to expand their functionality and applications. As such, the food industry will continue to benefit from using hydrocolloids in food formulation and processing, ultimately providing consumers with healthy and enjoyable food products.

FAQ

How do hydrocolloids enhance gut health?

Hydrocolloids, when consumed as a source of dietary fiber, can improve digestion, promote bowel regularity, and support a healthy gut microbiome.

What types of hydrocolloids are commonly used in the food industry?

Common types of hydrocolloids include agar, carrageenan, pectin, xanthan gum, and guar gum, each with unique functionalities and applications in food formulation.

Are hydrocolloids safe to consume?

Yes, hydrocolloids are considered safe for consumption and are regulated by guidelines and regulations in the food industry to ensure consumer safety and product quality.

In which industries are hydrocolloids used?

Hydrocolloids find applications in the food industry and pharmaceuticals, cosmetics, and other sectors where texture enhancement, stabilization, and emulsification are important.

What are the future trends and innovations in hydrocolloids?

The field of hydrocolloids is witnessing emerging trends and innovations, including new applications, research developments, and potential advancements in hydrocolloid-based products.

PART 6.3

Potential Side Effects and Concerns from Hydrocolloids

Hydrocolloids are becoming increasingly popular in health and wellness due to their wide benefits. However, it's important to note that these substances can also come with potential side effects and concerns.

Hydrocolloids are commonly used in food production, pharmaceuticals, and cosmetics because they thicken, stabilize, and emulsify. While they can positively impact product quality, potential risks are involved if not used properly.

In this article, we will explore the potential side effects and concerns of hydrocolloids and their various benefits, properties, and applications. By the end of this article, you will better understand the role hydrocolloids play in different industries and how they may impact your health regime.

Core Insights:

- ☑ Hydrocolloids have many benefits, but they can also come with potential side effects and concerns
- ☑ Hydrocolloids are used in various industries and can positively impact product quality
- ☑ This article will provide important information to help you make informed decisions about incorporating hydrocolloids into your health regime
- ☑ It is important to understand the potential risks involved if hydrocolloids are not used properly
- ☑ By the end of this article, you will have a better understanding of the role hydrocolloids play in different industries and how they may impact your health

6.3.1 Understanding Hydrocolloid Properties

Hydrocolloids are complex substances with distinctive properties that make them valuable in many applications. Understanding these properties is crucial to appreciate their usefulness and potential side effects in different formulations.

Gelling Properties

Hydrocolloids have the unique ability to form gels with water or other liquids. When a hydrocolloid solution is heated, it can form a gel-like structure as it cools. This property allows hydrocolloids to be used as thickeners and stabilizers in many food products.

Thickening Properties

Hydrocolloids are often added to food products to increase their viscosity or thickness. This property is especially useful in sauces, salad dressings, and other products with desired thicker textures. Hydrocolloids, such as xanthan gum and carrageenan, are commonly used as thickeners in many food products.

Stabilizing Properties

Hydrocolloids can also stabilize emulsions, preventing oil and water from separating. This property benefits many food products, such as salad dressings and ice cream, where a smooth and creamy texture is desired. Hydrocolloids such as guar gum and carrageenan are commonly used for this purpose.

Emulsifying Properties

Hydrocolloids can also emulsify and disperse fat and oil in water-based products. This property is useful in many food products, such as mayonnaise and salad dressings, where oil and water must be mixed. Hydrocolloids, such as lecithin and gum Arabic, are used as emulsifiers in food products.

Overall, understanding the unique properties of hydrocolloids is important to appreciate their usefulness and potential side effects. Hydrocolloids have multiple benefits in various formulations and products, but it is important to consider their potential side effects and use them in moderation.

6.3.2 Potential Side Effects of Hydrocolloids

While hydrocolloids offer numerous benefits for various applications, it's important to be aware of potential side effects that may arise from their use.

One of the most common side effects of hydrocolloids is gastrointestinal disturbances, such as bloating, gas, or diarrhea. These symptoms may occur due to the high fiber content of some hydrocolloids or the body's difficulty digesting them.

Allergic reactions are another possible side effect of hydrocolloids. Individuals with allergies to specific substances, such as soy or gluten, should be cautious when using hydrocolloids that contain these ingredients.

It's crucial to monitor your body's response to hydrocolloid consumption and to stop use immediately if you experience any adverse reactions.

As with any supplement or ingredient, it's important to consult with a healthcare professional before incorporating hydrocolloids into your health regime, particularly if you have existing health conditions or are taking medications that may interact with hydrocolloids.

6.3.3 Concerns and Precautions with Hydrocolloids

While hydrocolloids offer numerous benefits and are widely used in various applications, it's essential to consider any potential concerns or precautions before incorporating them into your health regime.

One potential concern is the possibility of gastrointestinal disturbances, such as bloating, gas, or diarrhea, especially with higher doses or certain types of hydrocolloids like guar gum or xanthan gum. If you experience these symptoms, you must reduce your intake or discontinue use and consult with a healthcare professional.

Another concern is the potential for allergic reactions. While rare, some individuals may be allergic to certain hydrocolloids, such as carrageenan or locust bean gum. If you develop symptoms such as itching, hives, or swelling, seek medical attention immediately.

If you're taking any medications, it's also essential to consider potential interactions. Hydrocolloids may affect the absorption or effectiveness of certain medications, such as antibiotics or diabetes. Consult with a healthcare professional if you're currently taking medication and considering the use of hydrocolloids.

Safety Measures and Precautions

To minimize potential risks and ensure safe use, it's crucial to follow recommended dosages and usage guidelines provided by the manufacturer. Start with small amounts and gradually increase your intake to avoid adverse reactions. Monitoring your body's response and discontinuing use if you experience any negative effects is also essential.

Individuals with pre-existing medical conditions, such as gastrointestinal disorders or allergies, should consult with a healthcare professional before incorporating hydrocolloids into their health regime. Pregnant or breastfeeding women should also exercise caution and seek medical advice before using hydrocolloids.

By considering potential concerns and taking appropriate precautions, you can safely reap the benefits of hydrocolloids without compromising your health.

6.3.4 Expert Perspectives on Hydrocolloid Use

When considering incorporating hydrocolloids into your health regime, seeking expert perspectives on their usage is essential. Nutritionists, healthcare practitioners, and industry professionals can provide valuable insight into the benefits, potential side effects, and precautions when using hydrocolloids.

According to experts, some hydrocolloids, such as xanthan gum, psyllium husk, or guar gum, may cause gastrointestinal discomfort, particularly in high doses. It's crucial to start with smaller amounts and gradually increase consumption while monitoring your body's reaction.

Experts also caution against combining hydrocolloids with certain medications, such as lithium, carbamazepine, or digoxin, as they may reduce absorption or increase the risk of toxicity. Consulting with healthcare professionals before incorporating hydrocolloids into your health regime is recommended.

Reputable hydrocolloid suppliers can guide safe usage and dosage, ensuring you don't exceed the recommended intake levels. It's also essential to follow storage instructions adequately, as improper storage conditions may affect the properties and safety of hydrocolloids.

Expert perspectives on hydrocolloid use can help you make informed decisions and avoid potential side effects and risks. Consider seeking advice from industry professionals, nutritionists, or healthcare practitioners when using hydrocolloids for optimal results.

6.3.5 Techniques for Incorporating Hydrocolloids

Using different techniques, hydrocolloids can be incorporated into various formulations, such as food products, cosmetics, and pharmaceuticals. Here are some popular methods:

Bloom your Gelatin

Gelatin is a commonly used hydrocolloid in the food industry. One of the best methods to add gelatin to a recipe is to bloom it in a small amount of cold water. This process softens the gelatin and prepares it for use in the recipe. Once the gelatin has bloomed, it can be added to the recipe as directed.

Perhydrate your Hydrocolloid

Prehydrating hydrocolloids such as xanthan gum, carrageenan, and pectin can improve their functionality in a product. Hydration allows the hydrocolloid to dissolve more easily, preventing clumping and promoting proper dispersion in the final product.

Emulsify with Hydrocolloids

Hydrocolloids like lecithin and gum Arabic are excellent emulsifiers and can help stabilize emulsions. Typically, these hydrocolloids are added to the oil phase of a formulation to help disperse the oil into the water phase.

Control Texture with Hydrocolloids

Many hydrocolloids, such as agar-agar, gelatin, and carrageenan, can control a product's texture. These hydrocolloids can help stabilize foams, thicken sauces and gravies, and even create elastic gels.

It is important to note that the appropriate technique and amount of hydrocolloid to use may vary depending on the recipe and the desired outcome. Start with small amounts and adjust as necessary to achieve your final product's desired texture and stability.

6.3.6 Understanding the Current Research Landscape

Hydrocolloids have been extensively studied in recent years, with research focused on their potential benefits, safety profile, and possible side effects. Ongoing clinical trials and studies are helping to shed light on the mechanisms by which hydrocolloids exert their effects and their interactions with other compounds and medications.

One area of active research is using hydrocolloids in food formulations, with studies exploring their effects on satiety, glycemic control, and overall nutrient absorption. Researchers are also investigating the use of hydrocolloids in pharmaceuticals, looking at their potential as drug carriers and their potential interactions with other medications.

Another area of interest in hydrocolloid research is their use in cosmetics and personal care products, with studies exploring their texture-modifying properties and ability to improve product stability and shelf life.

Recent studies have also looked at the possible side effects of hydrocolloids, such as gastrointestinal disturbances, allergic reactions, and other adverse reactions. These studies provide valuable information to help consumers make informed decisions about incorporating hydrocolloids into their health regime.

Overall, the research landscape surrounding hydrocolloids is continually evolving, with ongoing studies and clinical trials providing valuable insights into their benefits, safety, and potential side effects. As new findings emerge, staying updated with the latest research is important to ensure the safe and effective use of hydrocolloids.

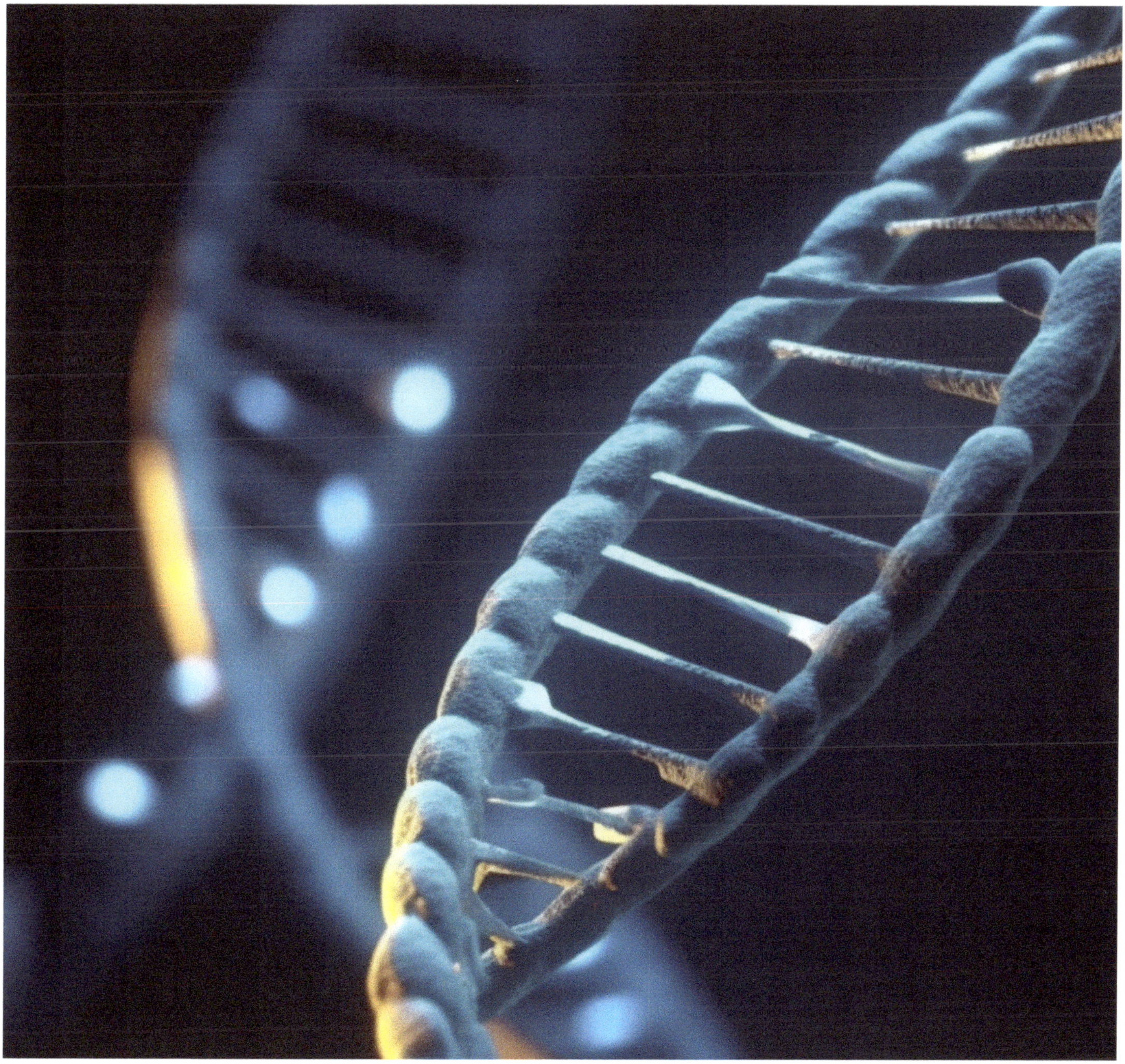

Conclusion

In conclusion, hydrocolloids can benefit various industries, including food, pharmaceuticals, and cosmetics. However, being aware of potential side effects and concerns associated with their use is essential. As you consider incorporating hydrocolloids into your health regime, consult with healthcare professionals to ensure safe and appropriate usage. Additionally, it is crucial to monitor your body's response to hydrocolloid consumption and follow dosage recommendations carefully.

Expert perspectives and ongoing research provide valuable insights into the safe and effective use of hydrocolloids. By staying informed and utilizing proper techniques for incorporating hydrocolloids, you can harness their beneficial properties and achieve optimal results in your formulations.

Overall, hydrocolloids are a promising ingredient with many applications and benefits. With proper precautions and expert guidance, you can safely and effectively incorporate hydrocolloids into your products and take advantage of their unique properties and potential benefits.

FAQ

Q: Are there any potential side effects of using hydrocolloids?

While hydrocolloids are generally considered safe for consumption, some individuals may experience gastrointestinal disturbances or allergic reactions. Monitoring your body's response when incorporating hydrocolloids into your diet or skincare routine is important.

Q: What are some concerns and precautions with hydrocolloids?

It is important to follow recommended dosage guidelines and be aware of any potential interactions with medications. Certain individuals with specific health conditions may need caution when consuming hydrocolloids. Consulting with a healthcare professional is advised.

Q: What do experts say about hydrocolloid use?

Experts, including industry professionals, nutritionists, and healthcare practitioners, consider hydrocolloids safe when used appropriately. It is recommended to source hydrocolloids from reputable suppliers and follow their guidelines for safe usage.

Q: How can I incorporate hydrocolloids into my formulations?

Various techniques for incorporating hydrocolloids depend on the desired product and application. Following recommended usage levels and guidelines provided by hydrocolloid suppliers is important for optimal results.

Q: What is the current research saying about hydrocolloids?

Ongoing research explores hydrocolloids' benefits, safety, and potential side effects. Recent studies and clinical trials contribute to understanding their efficacy and applications across different industries.

CHAPTER 07

Sourcing and Sustainability

PART 7.1

How to Source of Natural Hydrocolloids

Hydrocolloids are an essential component in the food industry, playing a significant role in achieving the desired texture, stability, and taste of various food products. These natural, plant-based additives are used widely in food processing, from creating the perfect viscosity in yogurt to stabilizing dressings and sauces.

However, with the increasing demand for hydrocolloids, the importance of sourcing them sustainably and ethically cannot be overstated. Ethical sourcing considers hydrocolloid making's social, environmental, and economic impact, ensuring a responsible and sustainable supply chain. It is crucial for protecting the natural resources and ecosystems affected by hydrocolloid sourcing.

In this article, we will dive deep into the world of hydrocolloids, exploring their importance in the food industry, types of plant-based alternatives, and the impact of sustainable sourcing practices. We will also address hydrocolloids' potential health and safety concerns and discuss the regulatory oversight of their use in food products. Finally, we will highlight industry initiatives and programs to promote responsible hydrocolloid sourcing and their presence in various consumer products.

Core Insights:

- ☑ Hydrocolloids are natural, plant-based additives extensively used in the food industry to achieve desired textures in various products.
- ☑ Ethical sourcing of hydrocolloids is critical to ensure responsible and sustainable practices and protect natural resources affected by hydrocolloid production.
- ☑ Plant-based alternatives and sustainable sourcing practices can mitigate the environmental impact of hydrocolloid production.
- ☑ Consumers should be aware of any potential health concerns associated with hydrocolloid consumption and the regulatory oversight of their use in food products.
- ☑ Industry initiatives and programs promote the responsible sourcing of hydrocolloids and their presence in consumer products beyond food.

7.1.1 Understanding Hydrocolloids and Their Role in the Food Industry

Hydrocolloids are an essential ingredient in the food industry, playing a crucial role in achieving the desired texture, consistency, and stability in various food products. They are commonly used as gelling agents, thickening agents, stabilizers, and emulsifiers, ensuring the required viscosity, providing structure, and preventing separation.

Hydrocolloids absorb water and create a gel-like substance, contributing to the final product's texture. They are derived from various sources, including plants, seaweed, and animal products, with each type offering distinctive properties and characteristics that suit different applications.

Gelling agents, such as agar and carrageenan, are hydrocolloids that create a gel-like texture in foods, adding to the sensory experience and enhancing consistency. Thickening agents, including xanthan gum and guar gum, increase the viscosity of liquids, resulting in a thicker, more stable product. Stabilizers, such as locust bean gum and cellulose gum, maintain the structure of a food product, preventing it from breaking down or separating. Emulsifiers, such as soy lecithin and mono- and diglycerides, help to mix oil and water-based ingredients, creating a smooth, homogenous texture.

Hydrocolloids are crucial in achieving the desired texture, consistency, and stability in various food products.

"Hydrocolloids are incredibly versatile ingredients that enable us to create a wide range of food products with unique textures and properties", says

-food scientist Dr Mary Lewis.

Function and Applications

Hydrocolloids are used in various food products, including dairy, confectionery, bakery, and meat. In dairy products, hydrocolloids prevent whey separation in yogurt, stabilize cheese, and improve texture and mouthfeel. In confectionery, hydrocolloids create gummies, marshmallows, and other chewy candies. In bakery products, hydrocolloids improve the texture of cakes and muffins, provide moisture to bread, and extend their shelf life.

Hydrocolloids are also used in meat products, where they can improve texture and moisture retention and prevent fat separation. Adding hydrocolloids to sauces and dressings can provide better stability, prevent separation, and improve overall appearance and texture.

"Hydrocolloids are an essential part of our food system, allowing us to create appealing, stable, and nutritious food products that meet consumer demand",says

-food industry expert Dr. John Smith.

Conclusion

Hydrocolloids provide essential functions in the food industry, allowing us to achieve the desired texture, consistency, and stability in various food products. Gelling agents, thickening agents, stabilizers, and emulsifiers are among the main hydrocolloids used in the food industry, each with unique properties and characteristics. Using hydrocolloids responsibly and sustainably, we can continue to meet consumer demand for appealing and nutritious food products while ensuring the industry's long-term viability.

7.1.2 Types of Hydrocolloids: A Look at Plant-Based Alternatives

Hydrocolloids are commonly used in the food industry as thickeners, stabilizers, emulsifiers, and gelling agents. Traditionally, many hydrocolloids were derived from animal sources, such as gelatin. However, with the growing demand for plant-based and vegan options, there has been a shift towards using plant-based hydrocolloids as gelatin substitutes.

One popular plant-based hydrocolloid is agar agar, derived from seaweed and used as a gelling agent in desserts and other foods. Another option is carrageenan, extracted from red seaweed and commonly used as a thickener in dairy and non-dairy products.

Table 59: Hydrocolloid Source and Function

HYDROCOLLOID	SOURCE	FUNCTION
Agar Agar	Seaweed	Gelling agent
Carrageenan	Red Seaweed	Thickener
Xanthan gum	Fermented corn sugar	Stabilizer, thickener

Other plant-based hydrocolloids include guar gum, which is derived from the guar plant and used as a thickener and stabilizer, and xanthan gum, which is made by fermenting corn sugar and commonly used as a thickener and stabilizer in sauces, dressings, and baked goods.

These plant-based hydrocolloids are becoming increasingly popular in the food industry as consumers seek more sustainable and ethical options. In fact, according to a report by Grand View Research, the global plant-based hydrocolloids market is projected to reach $5.32 billion by 2025, driven by increasing demand for plant-based products and growing awareness of their environmental and health benefits.

The trend towards plant-based hydrocolloids not only benefits consumers, but also the environment. Plant-based alternatives are generally more sustainable and ethical than animal-derived options, as they require less water and land to produce and do not involve the use of animals.

As the food industry continues to use, plant-based hydrocolloids are expected to become increasingly common. With their versatility, sustainability, and ethical production methods, plant-based hydrocolloids offer a promising alternative to traditional animal-derived hydrocolloids.

7.1.3 The Sustainable Sourcing of Hydrocolloids

The demand for hydrocolloids in the food industry is increasing, with an increasing emphasis on natural and sustainable sourcing practices. Natural hydrocolloids are obtained from plants, animals, and microbial sources, with plant-based alternatives gaining popularity due to their sustainable and ethical sourcing. Sourcing natural hydrocolloids is crucial for preserving biodiversity, ensuring fair trade practices, and promoting sustainability in the food industry.

Table 60: Benefits of Sustainable Sourcing and Industry Impact

BENEFITS OF SUSTAINABLE SOURCING	IMPACT ON THE FOOD INDUSTRY
Preserve biodiversity: **Sustainable sourcing practices help preserve the natural habitats and ecosystems from which hydrocolloids are obtained. It also promotes the conservation of endangered species and fosters biodiversity.**	*Meets consumer demand:* With the increasing demand for natural and sustainable products, responsible sourcing of hydrocolloids has become crucial for companies to remain competitive.
Fosters fair trade practices: **Sustainable sourcing practices ensure that farmers and workers receive fair wages and work under safe conditions. In turn, promotes ethical labor practices, social responsibility, and community economic growth.**	*Supports industry growth:* Companies that adopt sustainable sourcing practices are likely to experience positive growth due to increased consumer demand, enhanced corporate social responsibility, and promotion of environmental practices.
Promotes transparency: **Sustainable sourcing practices promote transparency in the supply chain, ensuring the product is traceable from its source to the final product. In turn, enhances food safety and quality assurance.**	*Reduces environmental impact:* Sustainable sourcing of hydrocolloids helps mitigate the **environmental impact** of large-scale production of these products. It promotes eco-friendly practices, reduces carbon footprint, and preserves natural resources for future generations.

The food industry increasingly recognizes the importance of responsible and sustainable sourcing practices. Several companies have initiated sustainability programs to promote ethical and sustainable sourcing of hydrocolloids. Collaborations, certifications, and industry initiatives ensure responsible sourcing practices and support sustainable hydrocolloid production. The industry aims to promote environmental protection, social responsibility, and ethical practices in hydrocolloid sourcing through such efforts.

7.1.4 The Environmental Impact of Hydrocolloid Production

As the use of hydrocolloids continues to grow in the food and consumer products industries. It is important to consider their potential environmental impact. Large-scale production of hydrocolloids can lead to ecosystem degradation and pollution, as well as the depletion of natural resources.

However, there are sustainable practices that can mitigate these concerns. Some hydrocolloid manufacturers implement closed-loop systems, where water and resources are recycled, and waste is minimized. Others are sourcing raw materials from sustainable producers and implementing reforestation programs to offset their environmental impact.

One example of sustainable hydrocolloid production is the use of seaweed-based hydrocolloids. Unlike land-based crops, seaweed requires no fresh water, fertilizer, or pesticides, making it a more environmentally friendly option. Additionally, seaweed can help absorb carbon dioxide and improve water quality in marine ecosystems.

7.1.5 The Environmental Impact of Carrageenan Production

Carrageenan is a popular hydrocolloid derived from red seaweed. While it has many functional benefits in food, pharmaceutical, and personal care products, there are concerns about its environmental impact.

Table 61: Environmental Concerns and Sustainable Solutions

ENVIRONMENTAL CONCERNS	SUSTAINABLE SOLUTIONS
Overharvesting of red seaweed	Sourcing from sustainable seaweed farms and implementing responsible harvesting practices
Marine ecosystem damage from farming practices	Implementing eco-friendly seaweed farming techniques, such as using biodegradable ropes and avoiding farming in sensitive areas
Chemical use in refining carrageenan	Limiting the use of chemicals in refining and implementing eco-friendly refining processes

While hydrocolloid production can have a negative environmental impact, some sustainable practices and alternatives can help mitigate these concerns. As consumers, it is important to be aware of the impact of our products and support companies that prioritize sustainable and responsible sourcing.

7.1.6 Health and Safety Considerations of Hydrocolloids

Hydrocolloids are generally considered safe for consumption, but there may be potential side effects or health concerns to be aware of for certain individuals.

For example, some hydrocolloids, such as carrageenan, have been linked to digestive issues in some people, including bloating, gas, and diarrhea. However, the research on these effects is conflicting and inconclusive, and regulatory bodies like the FDA continue to consider carrageenan safe for use in food products.

It is also important to note that some hydrocolloid products may contain trace amounts of allergens, such as soy or gluten, which could pose a risk to individuals with allergies or sensitivities.

However, hydrocolloids are generally considered safe for human consumption when used properly and in moderation. The FDA and other regulatory bodies have established safety guidelines and limits for their use in food products to ensure consumer safety.

7.1.7 Potential Side Effects

The potential side effects of hydrocolloids are generally mild and temporary, but may include:

- Bloating
- Gas
- Diarrhea
- Constipation
- Stomach discomfort

It is worth noting that these side effects are not common and are typically only experienced by individuals with pre-existing digestive issues.

7.1.8 Health Concerns

While hydrocolloids themselves are not considered harmful to human health, there may be certain health concerns, particularly for people who consume large amounts of processed foods containing hydrocolloids.

The high levels of added sugars, salt, and other additives commonly found in these products may contribute to health issues like obesity, diabetes, and hypertension. Additionally, consuming large amounts of processed foods may displace healthier, whole foods from the diet, leading to nutrient deficiencies and other health problems.

7.1.9 Safety

When using hydrocolloids in food products, it's important to follow all safety guidelines and regulations established by regulatory bodies like the FDA. Safety includes maintaining proper sanitation and hygiene practices during production, labeling products accurately, and adhering to dosage and usage guidelines to prevent potential health risks.

> "It's also important for individuals with food allergies or sensitivities to read labels carefully and avoid products that contain hydrocolloids or other ingredients that may trigger an allergic reaction."

Hydrocolloids can be a safe and effective way to enhance food products' texture, stability, and nutritional value when used responsibly and in moderation.

7.1.10 Regulatory Oversight of Hydrocolloids in the Food Industry

The use of hydrocolloids in food products is subject to regulatory oversight and strict industry standards. Regulatory bodies ensure that the production and use of hydrocolloids are safe and adhere to sustainability guidelines. These hydrocolloids are categorized as food additives and must comply with regulatory standards and labeling requirements.

The U.S. Food and Drug Administration (FDA) has established guidelines for using hydrocolloids in food products. The FDA regulates the use of hydrocolloids through the Code of Federal Regulations (CFR) Title 21, which sets permissible levels of use for each hydrocolloid in specific food categories. These regulations ensure consumers' safety and protect against harmful substances or contaminants in food products.

The European Union Regulation (EU) No 1333/2008 also sets out rules for using hydrocolloids in food, including authorized food categories, maximum use levels, and labeling requirements. The regulation also provides guidelines for the safety and quality of hydrocolloids, including their impact on human health and the environment.

Food manufacturers are responsible for ensuring their products comply with industry standards and regulatory guidelines, including using hydrocolloids. It includes testing and labeling their products accurately with information on the use of hydrocolloids and their respective quantities, ensuring transparency for consumers.

7.1.11 Consumer Awareness and Education on Hydrocolloids

As consumers, we have the right to know what goes into our food products. Hydrocolloids are commonly used in the food industry, but not everyone knows what they are and their potential impact.

Education is key to increasing consumer awareness and promoting transparency in the use of hydrocolloids. It includes understanding the various types of hydrocolloids and their functions in food products. For example, some hydrocolloids serve as gelling agents to give products like jelly their texture, while others act as emulsifiers to keep oil and water from separating in salad dressing.

Food labeling plays a significant role in providing information to consumers. By law, manufacturers must list all ingredients used in their products, including hydrocolloids. It allows consumers with allergies or dietary restrictions to make informed choices when purchasing food products. However, unclear or confusing labeling can hinder consumer understanding.

Table 62: Hydrocolloid Labeling on Food Products

PRODUCT	LABELING EXAMPLE
Jam	Ingredients: Strawberries, Sugar, Pectin, Citric Acid
Ice Cream	Ingredients: Milk, Cream, Sugar, Corn Syrup, Guar Gum, Carrageenan
Dressing	Ingredients: Water, Soybean Oil, Vinegar, High Fructose Corn Syrup, Modified Food Starch, Xanthan Gum, Spice

Consumer education initiatives can also help raise awareness of hydrocolloids. It includes public information campaigns, workshops, and educational events. By providing consumers with the knowledge and tools to make informed choices, we can promote responsible use of hydrocolloids in the food industry.

7.1.12 Industry Initiatives Towards Responsible Hydrocolloid Sourcing

The food industry has recognized the need for sustainable and responsible sourcing of hydrocolloids. As such, industry initiatives and sustainability programs have been growing, promoting responsible sourcing practices. By collaborating with stakeholders across the supply chain, these initiatives aim to ensure transparency and ethical practices in hydrocolloid production.

One such initiative is the Roundtable on Sustainable Palm Oil (RSPO), which has established a certification program for sustainable palm oil production. As palm oil is a common ingredient in many hydrocolloids, this initiative helps to ensure that the palm oil used in hydrocolloid production is sustainably sourced.

Industry Collaboration

Industry collaborations have also played a crucial role in promoting responsible hydrocolloid sourcing. The Sustainable Agriculture Initiative Platform (SAI) is one such collaboration, bringing together companies across the food industry to establish sustainable sourcing practices.

Another collaboration is the Global Agri-business Alliance (GAA), which aims to promote sustainable and responsible agricultural practices, including in the production of hydrocolloids. These industry collaborations ensure that responsible hydrocolloid sourcing is a shared priority by working together towards common sustainability goals.

Sustainability Programs

Several sustainability programs have also been established to promote responsible hydrocolloid sourcing. The Sustainable Food Policy Alliance (SFPA) is one example of four major food industry companies committed to responsible sourcing practices across their supply chains.

Another program is the Natural Resources Stewardship Circle (NRSC), which promotes sustainability in the natural resources used in food production, including hydrocolloids. By sustainability standards and guidelines, these programs encourage responsible sourcing practices in hydrocolloid production.

In conclusion, industry initiatives and sustainability programs are vital in promoting responsible hydrocolloid sourcing. By collaborating and establishing sustainability standards, these initiatives ensure that hydrocolloid production is conducted ethically and sustainably, supporting the long-term health of our planet and the food industry.

7.1.13 Hydrocolloids in Everyday Life: From Food Products to Beyond

Hydrocolloids are not just limited to the food industry but are also widely used in various consumer products daily. Hydrocolloids, from personal care products to household items, have diverse applications that showcase their versatility and impact.

Table 63: Hydrocolloids in Applications in Personal Care Products

HYDROCOLLOID	PERSONAL CARE PRODUCT
Xanthan Gum	Shampoo, Toothpaste
Carrageenan	Face Masks, Moisturizers
Guar Gum	Hair Gel, Conditioner

Hydrocolloids such as xanthan gum, carrageenan, and guar gum are commonly used in personal care products to provide stabilizing and thickening properties. Xanthan gum is found in shampoos and toothpaste, while carrageenan is used in face masks and moisturizers. Guar gum is often included in hair gel and conditioner, providing a thicker consistency without weighing hair down.

Applications in Household Items

- Xanthan Gum: Used as a thickening agent in household cleaning products
- Carrageenan: Used as a stabilizer in air fresheners and room sprays
- Agar Agar: Used as a gelling agent in household jelly and jam-making

Hydrocolloids are also used in household items such as cleaning products, air fresheners, and even homemade jams and jellies. Xanthan gum is often used as a thickening agent in cleaning products, while carrageenan stabilizes air fresheners and room sprays. Agar agar, a plant-based hydrocolloid, is commonly used as a gelling agent in homemade jams and jellies.

Hydrocolloids play a significant role not only in the food industry but in our daily lives as well. Understanding their functions and applications is crucial in promoting sustainable and responsible sourcing practices.

Conclusion

As explored in this article, hydrocolloids play a crucial role in the food industry, providing gelling, thickening, stabilizing, and emulsifying functions in many products. It is essential to source natural hydrocolloids sustainably and responsibly, and industry initiatives and collaborations have been established to promote ethical practices and support sustainable production.

While hydrocolloids have many benefits in food production, it is also important to be aware of potential health and safety concerns associated with their consumption. Regulatory bodies oversee the use of hydrocolloids in food products, and food labeling provides valuable information to consumers.

Hydrocolloids have a broader range of applications beyond the food industry, highlighting their versatility and impact on our daily lives. However, it is crucial to consider the potential environmental impact of hydrocolloid production and to use sustainable practices to mitigate any negative effects on ecosystems.

In conclusion, the ethical sourcing of natural hydrocolloids is vital for a sustainable future, and industry efforts towards responsible sourcing and production must continue to support this goal. As consumers, we can educate ourselves on the use of hydrocolloids and make informed choices about the products we purchase, ensuring a safe and sustainable future for all.

FAQ

Why is ethical sourcing of natural hydrocolloids important?

Ethical sourcing of natural hydrocolloids ensures that they are obtained sustainably and responsibly. It helps protect the environment, supports local communities, and promotes fair trade practices.

How does hydrocolloid production impact the environment?

Large-scale hydrocolloid production can have environmental implications, including land and water degradation, deforestation, and greenhouse gas emissions. Sustainable practices are essential to mitigate these impacts.

Are there any health concerns associated with consuming hydrocolloids?

While hydrocolloids are generally considered safe, some individuals may experience digestive issues or allergic reactions. It is important to use them in moderation and consult with a healthcare professional if there are any concerns.

How are hydrocolloids regulated in the food industry?

Regulatory bodies set standards and guidelines to ensure the safe and responsible use of hydrocolloids in food products. These regulations help protect consumer health and promote industry transparency.

How can consumers be more aware of hydrocolloid usage?

Consumer awareness can be increased through clear food labeling that provides information about the presence and purpose of hydrocolloids in food products. It promotes informed choices and transparency in the marketplace.

What initiatives promote responsible sourcing of hydrocolloids?

Various industry initiatives and sustainability programs aim to support responsible sourcing of hydrocolloids. Collaborations and certifications ensure ethical practices and contribute to sustainable hydrocolloid production.

Where else are hydrocolloids used besides the food industry?

Hydrocolloids are found in various consumer products beyond the food industry. Due to their versatile properties, they are used in personal care products, pharmaceuticals, textiles, and other applications.

PART 7.2

Environmental Impact of Hydrocolloid Production

Hydrocolloids are used in the food industry as thickening agents, stabilizers, gelling agents, and emulsifiers. While they are essential to food production, producing hydrocolloids can have a significant environmental impact.

It is crucial to consider the impact of hydrocolloid production on the environment, including the resources required, water and energy consumption, and waste generation and disposal. By adopting sustainable practices, we can ensure the food industry remains sustainable and minimize our environmental impact.

Core Insights:

- ☑ Hydrocolloid production can have a significant environmental impact.
- ☑ Water, energy consumption, waste generation, and disposal are important, considering the resources required.
- ☑ Adopting sustainable practices can minimize the impact of hydrocolloid production on the environment.
- ☑ Certification and labeling schemes can help consumers make informed and sustainable choices.
- ☑ Consumer awareness and responsibility are vital in promoting a greener food industry.

7.2.1 Hydrocolloid Production Process

Hydrocolloid production involves a complex process that requires significant resources, including water and energy. The extraction of hydrocolloids from natural sources involves several steps, including harvesting, cleaning, and processing. On the other hand, synthetic hydrocolloids are produced through chemical reactions and require even more resources.

Hydrocolloid production involves collecting raw materials, including cassava, seaweed, and other plant sources. These raw materials undergo several processing stages for extracting the hydrocolloid, including grinding, soaking, and filtration. The extracted hydrocolloid is then further processed to produce the desired characteristics, such as viscosity and gelling properties.

During the hydrocolloid production process, there is a significant amount of water and energy consumption. The filtration process alone requires large amounts of water, and the energy needed to operate the equipment is substantial. Additionally, some hydrocolloid production processes involve chemicals, posing environmental risks if not managed properly.

7.2.2 Ensuring Sustainable Hydrocolloid Production

As awareness of the environmental impact of hydrocolloid production grows, many manufacturers are implementing sustainable practices to minimize their environmental footprint. These practices can include recycling water used in the production process, using renewable energy sources, and reducing the use of chemicals.

Several technologies are also being developed to help reduce the environmental impact of hydrocolloid production. For example, some companies are exploring using enzymes to extract hydrocolloids, which can significantly reduce water usage. Others are exploring alternative raw materials, such as bacteria and fungi, which can also reduce the environmental impact of hydrocolloid production.

Hydrocolloid manufacturers can help reduce the environmental impact of their production processes by adopting sustainable practices and utilizing new technologies. As a consumer, you can also make a difference by choosing products made by manufacturers committed to sustainability and looking for certification and labeling to ensure that the hydrocolloids used in your food come from sustainable sources.

7.2.3 Water Usage in Hydrocolloid Production

Hydrocolloid production is water-intensive, requiring significant amounts for various production stages. The water usage in hydrocolloid production is a significant concern, as it can lead to water scarcity and various environmental consequences.

The production of hydrocolloids typically involves washing, soaking, and boiling raw materials, such as seeds or roots, to extract the hydrocolloid. The extracted material is then further processed, with water used in each stage, including drying and packaging.

The amount of water used in hydrocolloid production varies depending on the type of hydrocolloid and the production process. For example, the production of carrageenan, a popular hydrocolloid used in the food industry, can require up to 100 times its weight in water. The total water usage in hydrocolloid production significantly affects their environmental impact.

The high-water consumption during hydrocolloid production can negatively affect local water resources, particularly in areas where water scarcity is already an issue. When water is diverted from natural sources, it can lead to the depletion of groundwater reserves, soil degradation, and reduced water availability for communities and ecosystems.

Reducing water usage in hydrocolloid production is critical in promoting sustainability in the food industry. Several initiatives have been implemented to address this issue, including developing more efficient production processes and using recycled water. However, more action is needed to ensure that hydrocolloid production is environmentally responsible.

7.2.4 Energy Consumption in Hydrocolloid Production

Hydrocolloid production is energy-intensive and requires significant electricity, steam, and fuel. The production process involves grinding, mixing, heating, and drying raw materials, which consumes considerable energy.

The energy consumption associated with hydrocolloid production has various environmental impacts, including greenhouse gas emissions, air pollution, and resource depletion. Using non-renewable energy sources such as oil, coal, and natural gas in hydrocolloid production exacerbates the negative impact on the environment.

Sustainable practices such as using renewable energy sources, energy-efficient technologies, and process optimization can be adopted by reducing the environmental impact of energy consumption in hydrocolloid production. For instance, some hydrocolloid producers have invested in renewable energy sources such as solar and wind to power their production lines.

Moreover, technological advancements have led to more energy-efficient production processes that reduce electricity and fuel consumption. Automation of certain processes and high-pressure homogenizers have also been shown to reduce energy consumption in hydrocolloid production.

As a consumer, you can contribute to reducing energy consumption in hydrocolloid production by choosing products from manufacturers who have adopted sustainable practices in their production processes. Look for certifications and labels that indicate the product was produced using renewable energy sources or manufactured using energy-efficient technologies.

7.2.5 Waste Generation and Disposal

One significant environmental impact associated with hydrocolloid production is the generation of waste and the challenges involved in its proper disposal. Hydrocolloid production generates various forms of waste, including solid waste, wastewater, and air emissions. Disposing of this waste can negatively impact the environment, including soil and water pollution and greenhouse gas emissions.

Disposing of hydrocolloid wastes is complex, and disposing of them improperly can have severe consequences. Solid wastes generated during hydrocolloid production often contain high levels of organic matter and other chemicals, making them unsuitable for traditional waste disposal methods. Wastewater from hydrocolloid production may contain high levels of organic matter, suspended solids, and other contaminants, requiring specialized treatment and disposal methods.

The disposal of hydrocolloid production waste remains a significant challenge for manufacturers. The environmental impact of hydrocolloid production waste can be reduced through sustainable practices, such as implementing waste reduction strategies and investing in sustainable technologies.

7.2.6 Sustainable Practices in Hydrocolloid Production

Hydrocolloid production processes can have a significant environmental impact. However, sustainable practices can be adopted to minimize this impact. By implementing these practices, companies can reduce water and energy consumption, waste generation, and greenhouse gas emissions.

7.2.7 Examples of Sustainable Practices

One example of a sustainable practice is using renewable energy sources, such as solar or wind power, to reduce reliance on non-renewable sources. Another is adopting closed-loop systems that recycle water and minimize waste generation. Additionally, some companies invest in research and development to create new hydrocolloids that are more environmentally friendly or produced using sustainable methods.

Other sustainable practices include using locally sourced materials to reduce transportation emissions, optimizing production processes to reduce energy consumption, and implementing waste reduction strategies, such as composting or recycling.

7.2.8 Benefits of Sustainable Practices

Implementing sustainable practices in hydrocolloid production can have numerous benefits. Companies can save money by reducing water and energy consumption and minimizing waste generation. Additionally, these practices can improve a company's reputation as a socially responsible and environmentally conscious organization.

Moreover, sustainable practices benefit the environment and have positive social impacts, such as contributing to water conservation and reducing carbon emissions.

Certification and Labeling

Several certification and labeling schemes exist to promote sustainable hydrocolloid production and consumption. For example, the Rainforest Alliance Sustainable Agriculture Standard provides certification for environmentally and socially responsible practices, while the Marine Stewardship Council certifies sustainable seafood products that use hydrocolloids.

By seeking certified and labeled products, consumers can make informed and sustainable choices, supporting companies committed to sustainable practices.

In conclusion, sustainable practices in hydrocolloid production can have numerous benefits for the environment, society, and companies. Adopting these practices and seeking certified and labeled products can contribute to a greener and more sustainable future.

7.2.9 Certification and Labeling: Making Sustainable Choices with Hydrocolloids

As a consumer, you have the power to make a difference by choosing products made with sustainably sourced hydrocolloids. But how do you determine which products meet these standards? Look for products that carry certification and labeling from trusted organizations.

Certification: Certification schemes, such as Fairtrade or Rainforest Alliance, ensure that products are ethically and sustainably sourced. Look for products with certification logos on their packaging, indicating that the hydrocolloid has been produced in compliance with specific standards.

Table 64: Certification schemes and their Description

CERTIFICATION SCHEME	DESCRIPTION
Fairtrade	A certification scheme that guarantees fair and sustainable trading conditions for small-scale farmers and workers.
Rainforest Alliance	A certification program that supports sustainable farming and conservation of natural resources.

Labeling: Some companies have also begun introducing labeling systems to indicate that their products are produced with sustainably sourced hydrocolloids. Look for labels such as "responsibly sourced" or "sustainably sourced" on product packaging to make informed choices.

Choosing products with certification and labeling can support sustainable hydrocolloid production and promote a greener food industry.

7.2.10 Promoting a Greener Food Industry

The production of hydrocolloids is just one aspect of the food industry that has a significant environmental impact. To truly promote a greener food industry, it is essential to consider the entire supply chain, from farm to table.

One way to reduce the environmental impact of the food industry is by implementing sustainable practices throughout the supply chain. It includes reducing waste, conserving energy, and minimizing water usage. By adopting sustainable practices, companies can reduce their environmental impact and save money in the long run.

Another approach is to promote a circular economy in the food industry. It involves finding uses for all waste products rather than disposing of them. For example, waste from hydrocolloid production could be repurposed as fertilizer or used to generate biogas.

As a consumer, you can also play a role in promoting a greener food industry. By choosing products that use sustainable hydrocolloids, you can support companies taking steps to reduce their environmental impact. Look for certifications and labels that indicate sustainable practices, such as the USDA Organic or Rainforest Alliance certification.

Finally, it is essential to hold companies accountable for their environmental impact and to demand transparency in the supply chain. Being informed and conscious of your choices can contribute to a greener food industry and a healthier planet.

7.2.11 Consumer Awareness and Responsibility

As a consumer, you can drive positive change in the food industry by making informed and sustainable choices when purchasing products that contain hydrocolloids.

By educating yourself about the environmental impact of hydrocolloid production, you can make more conscious decisions about the food you consume and its impact on the planet. Consider reading labels and looking for certifications that indicate sustainable and responsible practices were used in producing the hydrocolloids in your food.

However, it's essential to recognize that your actions are just a small piece of the puzzle. Collective action is necessary to create significant change in the industry. Consider advocating for sustainable practices with local businesses and government officials and supporting companies prioritizing environmentally responsible practices.

The responsibility to reduce the environmental impact of the food industry is not only on producers but consumers as well. By taking an active role in promoting sustainability, together, we can help create a healthier planet for future generations.

Conclusion

Hydrocolloid production has a significant environmental impact, with high water consumption, energy usage, and waste generation. However, some sustainable practices and technologies can be adopted to minimize this impact. As a consumer, you are responsible for making informed and sustainable choices when purchasing hydrocolloid-containing products. Look for certifications and labels that indicate sustainable practices have been utilized in production.

Promoting a greener food industry requires collective action and a commitment to changing practices for a healthier planet. By raising awareness and advocating for sustainable choices, you can contribute to this effort. Remember that small changes can make a big difference.

Sustainability Starts with You

Together, we can make a difference by considering the environmental impact of hydrocolloid production. By choosing sustainable options and promoting a greener food industry, we can protect our planet for future generations. It all starts with you.

FAQ

What is the production process of hydrocolloids?

The production process of hydrocolloids involves several steps, including extraction, purification, and drying. It requires resources such as water, energy, and raw materials.

How does hydrocolloid production contribute to water usage?

Hydrocolloid production requires significant water usage due to the extraction and purification processes. This high-water consumption can have consequences for water resources.

What impact does hydrocolloid production have on energy consumption?

Hydrocolloid production involves energy-intensive processes, which can contribute to increased energy consumption and carbon emissions.

What are the environmental consequences of waste generation during hydrocolloid production?

Waste generated during hydrocolloid production can have environmental impacts if improperly managed and disposed of. It includes potential pollution and resource depletion.

Are there sustainable practices available for hydrocolloid production?

Yes, sustainable practices can be adopted in hydrocolloid production to minimize the environmental impact. These practices include efficient resource management and the use of renewable energy sources.

Are there certification and labeling schemes for hydrocolloids?

Yes, certifications and labels related to hydrocolloids exist to help consumers make informed and sustainable choices. Look for specific certifications and labels indicating sustainable production practices.

FREQUENTLY ASKED QUESTIONS (FAQS) CONTINUED

How can the food industry promote greener practices in hydrocolloid production?

The food industry can promote greener practices in hydrocolloid production by adopting sustainable initiatives and technologies, reducing waste, and prioritizing environmental responsibility.

What role do consumers play in promoting sustainable hydrocolloid choices?

Consumers are essential in driving positive change by increasing their awareness and making responsible choices regarding hydrocolloid-containing products. Individual actions can contribute to a greener food industry.

CHAPTER 08

Innovations and Research

PART 8.1

Current Findings on Hydrocolloids

Hydrocolloids are vital in several food industries to improve texture, stability, and quality. These substances are primarily polysaccharides, which provide the necessary viscosity, gelation, and thickening properties. Researchers are continuously discovering new natural sources and applications for hydrocolloids, making it an exciting field of study.

By exploring the latest research findings on hydrocolloids, we will better understand how these substances play a crucial role in the food industry. Specifically, we will delve into natural sources of hydrocolloids and their benefits, hydrocolloid properties and their role in food preservation, and a comprehensive overview of the hydrocolloid market.

Core Insights:

- ☑ Hydrocolloids are essential ingredients in the food industry, providing essential texture, stability, and quality to food products.
- ☑ Researchers are continually discovering new natural sources of hydrocolloids and their potential applications.
- ☑ By studying hydrocolloid properties, we can better understand how these substances contribute to food preservation and texture.
- ☑ The hydrocolloid market continually expands and evolves, with key players shaping its growth and direction.

8.1.1 Understanding Hydrocolloids: Properties and Benefits

Hydrocolloids are a group of substances that possess unique properties, making them valuable in a variety of applications. These natural polymers are characterized by their ability to form gels and provide viscosity, stability, and texture to various products. Understanding the properties and benefits of hydrocolloids is essential to appreciate their importance in food and other industries.

Hydrocolloids have a range of properties that make them useful in a broad range of applications. Here are some of their most notable properties:

Table 65: Hydrocolloids Properties with Description

PROPERTY	DESCRIPTION
Thickening	Hydrocolloids increase the viscosity of liquids, suspensions, and emulsions, helping stabilize and improve their texture.
Gelling	Hydrocolloids can form gels when heated or exposed to specific conditions, allowing them to shape, mold, and create firm structures.
Emulsifying	Hydrocolloids can help to stabilize emulsions by preventing the separation of oil and water, providing a smooth texture and appearance.
Water-binding	Hydrocolloids can bind water, improving the moisture retention of products and preventing them from drying out or becoming stale.

8.1.2 Hydrocolloid benefits

The unique properties of hydrocolloids provide a wide range of benefits in various applications, including:

- **Improved texture:** Hydrocolloids can improve the texture of food products, making them smoother, creamier, and more appealing.
- **Extended shelf-life:** Hydrocolloids can help to keep food products fresh for longer, reducing waste and increasing profitability.
- **Reduced fat content:** Hydrocolloids can replace fat, providing a healthier alternative to traditional ingredients while maintaining the desired consistency and texture.
- **Improved stability:** Hydrocolloids can help stabilize products by preventing separation, sedimentation, and crystal formation, ensuring consistent quality.

8.1.3 Hydrocolloid Applications

Hydrocolloids find extensive use in various industries, including food, pharmaceutical, personal care, and industrial applications. These diverse applications make hydrocolloids essential in many modern products.

Some common applications of hydrocolloids in the food industry include:

- Improving the texture of ice cream, yogurt, and other dairy products.
- Preventing syneresis in salad dressings and sauces.
- Stabilizing whipped cream and mousses.
- Increasing water retention in baked goods.

The pharmaceutical industry also uses hydrocolloids in various applications, including:

- Drug delivery systems
- Wound dressings
- Capsules and tablets for oral administration

The personal care industry uses hydrocolloids in the production of:

- Toothpaste and mouthwash
- Shampoo and shower gel
- Face masks and makeup

The industrial applications of hydrocolloids include:

- Oil drilling
- Paper manufacturing
- Textile printing

The versatility of hydrocolloids is truly impressive, making them an essential ingredient in many modern products.

8.1.4 Hydrocolloid Market: Current Trends and Future Outlook

The global hydrocolloid market has witnessed steady growth in recent years, primarily driven by the rising demand for natural and organic ingredients in food products. Hydrocolloids, derived from natural sources such as seaweed, plant exudates, and microbial fermentation, have gained popularity as an alternative to synthetic additives due to their clean label status and functional properties.

In 2020, the global hydrocolloid market was valued at $8.1 billion, and it is expected to reach $10.9 billion by 2025, growing at a CAGR of 6% during the forecast period. Europe and North America are the largest hydrocolloid markets, accounting for more than half of the global demand. However, the Asia-Pacific region is expected to show the highest growth rate due to the increasing consumption of processed foods and the growing awareness of the benefits of hydrocolloids.

Key Players in the Hydrocolloid Market

The hydrocolloid market is highly fragmented, with numerous small and large players operating globally. Some of the major players in the market include:

- Cargill, Inc.
- Kerry Group
- Ingredion Incorporated
- DowDupont
- CP Kelco
- Archer Daniels Midland Company

These companies invest heavily in research and development activities to introduce new and innovative hydrocolloid products and expand their portfolio. They also collaborate with local players to strengthen their distribution networks and increase their market share.

8.1.5 Hydrocolloid Suppliers

Several suppliers are operating in the market to meet the growing hydrocolloid demand. Some of the top suppliers of hydrocolloids are:

Table 66: Hydrocolloid Suppliers

COMPANY NAME	COUNTRY
CP Kelco	USA
Ingredion Inc.	USA
Kerry Group	Ireland
Archer Daniels Midland Company	USA
Cargill, Inc.	USA

These suppliers offer a wide range of hydrocolloid products that cater to the diverse needs of customers across various industries. They are also expanding their production capacities and investing in sustainable sourcing practices to ensure a stable supply of raw materials and meet the growing demand for natural and organic hydrocolloids.

Overall, the hydrocolloid market is poised for significant growth in the coming years, driven by the increasing demand for healthy and sustainable food products and the growing applications of hydrocolloids in various industries.

8.1.6 Innovations in Hydrocolloid Research

Hydrocolloid research has gained significant momentum in recent years, with scientists and researchers constantly exploring new applications and discovering novel properties of these natural substances.

One area of significant interest is the development of hydrocolloids with improved functional properties, such as increased stability, solubility, and emulsification capabilities. Researchers are also working to enhance the sensory properties of hydrocolloid-based products, such as taste and aroma.

Examples of Innovative Hydrocolloid Research

- **Nanotechnology:** Scientists are exploring using nanotechnology to create hydrocolloid-based materials with unique properties and functions. For instance, nano-fibrillated cellulose (NFC), a hydrocolloid derived from wood pulp, has enhanced food products' texture and stability while reducing fat content.
- **Encapsulation:** Hydrocolloid-based encapsulation systems are being developed to protect sensitive ingredients, such as probiotics and antioxidants, from degradation during processing and storage. These systems can improve the shelf life and efficacy of functional food products.
- **Biodegradable Hydrocolloids:** With the growing concern over environmental sustainability, researchers are studying the development of biodegradable hydrocolloids that can replace synthetic materials in applications such as packaging and textiles. For instance, chitosan, a hydrocolloid derived from crustacean shells, has been shown to possess antimicrobial properties and can be used as a natural alternative to synthetic polymers.

Overall, the innovations in hydrocolloid research are pointing towards a future where these versatile substances play an even more significant role in various industries. As researchers continue to uncover new applications and properties, hydrocolloid potential is limitless.

Conclusion

Hydrocolloids are a versatile group of natural substances with numerous applications across various industries. Hydrocolloids are a key ingredient from the food industry to pharmaceuticals, textiles, and personal care products, contributing significantly to the end product's texture, stability, and quality.

Through our exploration of hydrocolloids, we have highlighted the latest research findings, delved into their unique properties and benefits, and discussed their diverse applications in the food industry and beyond. We have also examined the different types of natural hydrocolloids available and discussed the current trends and outlook of the hydrocolloid market.

FAQ

What is the current state of the hydrocolloid market?

The hydrocolloid market is experiencing steady growth, driven by increasing demand for processed foods and convenience products. Major players in the industry include suppliers such as Cargill, DuPont, and CP Kelco. The market is projected to expand further in the coming years, fueled by advancements in research and development and emerging trends in clean-label and plant-based products.

Apart from the food industry, where else are hydrocolloids used?

Hydrocolloids have diverse applications beyond the food industry. They are used in pharmaceuticals as binders, film formers, and controlled-release agents. In the textile industry, they are utilized as sizing agents and thickeners for fabric printing. Hydrocolloids also find applications in personal care products, such as lotions, creams, and gels, due to their ability to enhance viscosity and stability.

How do hydrocolloids contribute to food texture and structure?

Hydrocolloids play a crucial role in determining the texture and structure of food products. They can create a smooth, creamy mouthfeel, provide a pleasant gelatinous consistency, or impart elasticity and chewiness. Hydrocolloids help achieve desired attributes like thickness, suspension, and bite resistance by controlling water mobility and interacting with other ingredients.

Are there any recent innovations in hydrocolloid research?

Yes, there have been significant advancements in hydrocolloid research. Scientists are exploring new sources of natural hydrocolloids, such as microorganisms and agricultural by-products, to expand the available options. Additionally, research is focused on optimizing the properties and functionalities of hydrocolloids through modifications and combinations, as well as investigating their potential applications in novel areas, such as drug delivery systems and tissue engineering.

PART 8.2

Emerging Applications and Technologies of Hydrocolloids

Hydrocolloids are a class of molecules that can bind with water and form a gel-like substance, giving them a wide range of functional properties. Hydrocolloids have been used for many years in various industries, but their importance has increased significantly recently, particularly in the food industry.

Hydrocolloids have become vital in the food industry because they improve different formulations' texture, stability, and quality. They are used as thickening, stabilizers, and gelling agents, providing significant benefits to food products. Hydrocolloids play a key role in developing new food products with improved functionality and sensory attributes.

This article will explore the emerging applications and technologies of hydrocolloids in different industries. We will explore the science behind hydrocolloid networks and the latest breakthrough technologies revolutionizing hydrocolloid research and development.

We will also analyze the expanding applications of hydrocolloids beyond the food industry, highlighting their growing importance in sectors such as pharmaceuticals, cosmetics, and personal care. Finally, we will discuss the potential of natural and biodegradable hydrocolloids in promoting sustainability and eco-friendly solutions.

Core Insights:

- ☑ Hydrocolloids are a class of molecules that can bind with water and form a gel-like substance, providing a wide range of functional properties.
- ☑ Hydrocolloids have become increasingly important in the food industry because they improve different formulations' texture, stability, and quality.
- ☑ Hydrocolloids are used as thickening, stabilizers, and gelling agents, providing significant benefits to food products.
- ☑ Hydrocolloids have expanded applications beyond the food industry, including pharmaceuticals, cosmetics, and personal care.
- ☑ Natural and biodegradable hydrocolloids have the potential to promote sustainability and eco-friendly solutions in various industries.

8.2.1 Hydrocolloids in the Food Industry: Revolutionizing Product Formulations

Hydrocolloids have become increasingly popular in the food industry due to their multifunctional properties, contributing to food formulations' texture, viscosity control, and water binding. These ingredients are widely used in various food products, including baked goods, dairy products, sauces, dressings, and beverages.

Hydrocolloids act as stabilizers, thickeners, and gelling agents, providing the desired textural characteristics and overall food quality. They enhance flavor release, prolong shelf-life, and improve the product's appearance. Hydrocolloids are also versatile, as many different hydrocolloids can be used together to create a synergistic effect, further enhancing their functionality.

The use of hydrocolloids in the food industry has revolutionized product formulations. They have helped to address the growing demand for convenient and high-quality food products while maintaining consistency and stability in formulations. Hydrocolloids make it possible to create a range of products, from low-fat spreads to sugar-free baked goods to gluten-free products, without compromising quality, sensory attributes, or shelf-life.

8.2.2 Advantages of Hydrocolloids

The use of hydrocolloids in food products has several advantages. Firstly, they help to reduce the use of artificial additives, which may have negative health consequences. Secondly, they can improve the texture and mouthfeel of low-fat or low-sugar products, making them more palatable. Thirdly, hydrocolloids can help to reduce food waste by improving the shelf-life of products, reducing spoilage and the need for preservatives.

8.2.3 Gel Formation: Unveiling the Science Behind Hydrocolloid Networks

Hydrocolloids have gained popularity for their ability to form gels, widely used in the food industry as thickening agents, stabilizers, and gelling agents. Gel formation is a complex process that involves the interaction of hydrocolloids with water molecules and other substances present in the formulation.

The ability of hydrocolloids to form gels is due to their unique chemical composition, which allows them to create a three-dimensional network structure in the presence of water. This network forms gcules, forming a gel with desired properties such as texture, consistency, and stability.

Several factors, including the type and concentration of hydrocolloids, the pH, temperature, and the presence of other ingredients, influence the gel properties of hydrocolloids. For instance, hydrocolloid gels' gel strength and elasticity increase with higher concentrations of hydrocolloids and lower pH levels.

Table 67: Factors Influencing Gel Formation

FACTORS	INFLUENCE ON GEL FORMATION
Hydrocolloid concentration	Higher concentration increases gel strength.
pH	Lower pH levels increase gel strength and elasticity.
Temperature	Higher temperature accelerates gelation.
Presence of other ingredients	May enhance or inhibit gel formation

Understanding the science behind hydrocolloid gel formation is crucial in designing and formulating food products with desired properties. Knowledge of the factors influencing gel properties can help food manufacturers optimize and formulate their products'' quality, texture, taste, and shelf life.

> "Several factors, including the type and concentration of hydrocolloid, the pH, temperature, and the presence of other ingredients influence the gel properties of hydrocolloids."

Recent advancements in hydrocolloid research have led to the development of innovative approaches for controlling gel properties. For instance, molecular modification of hydrocolloids can significantly alter their gelling behavior, resulting in gels with unique properties. Furthermore, combining different hydrocolloids can lead to hybrid gels with improved properties.

Moreover, in addition to their applications in the food industry, hydrocolloid gels are increasingly used in the pharmaceutical and cosmetic industries, as they offer a versatile and effective means of achieving desired product properties.

In summary, understanding the science behind hydrocolloid gel formation is crucial in optimizing formulations and enhancing the properties of food products. The factors influencing gel properties are diverse and complex, but their knowledge enables scientists to design hydrocolloid-based products with desired properties, leading to improved consumer satisfaction.

8.2.4 Innovations in Hydrocolloid Applications: Beyond the Food Industry

Hydrocolloids have expanded beyond their traditional applications in the food industry and are now being utilized innovatively across different sectors.

Hydrocolloids are used as excipients in the pharmaceutical industry to improve drug delivery systems. They control drug release, increase bioavailability, and improve patient compliance. Hydrocolloids are also being explored as potential wound dressings because they create a moist environment that promotes healing.

In the cosmetics industry, hydrocolloids create high-performance formulations that provide superior texture and stability. They are found in lotions, creams, and hair gels, enhancing the sensory profile and improving product efficacy.

The personal care industry is also benefitting from hydrocolloid innovations. Hydrocolloids are used to produce toothpaste and mouthwash because they provide viscosity control, stabilize the product, and enhance mouthfeel. Additionally, hydrocolloids are used in diapers to improve absorption and increase comfort.

The Versatility of Hydrocolloids

Hydrocolloids are incredibly versatile and can be tailored to meet specific product requirements, making them ideal for diverse applications.

A prime example of this versatility is the use of carrageenan in fire-fighting foam. The hydrocolloid's unique gelling and thickening properties make it ideal for use in foam production, creating a stable foam that is highly effective in extinguishing fires.

In the oil and gas industry, hydrocolloids are used in drilling fluids to maintain viscosity, reduce friction, and improve lubrication. They are also used in secondary oil recovery, where they help to increase oil production by controlling the flow of water and improving reservoir sweep efficiency.

8.2.5 The Future of Hydrocolloid Applications

As more industries discover the benefits of hydrocolloids, their use is expected to increase. Research and development efforts are focused on exploring the potential of new and underutilized hydrocolloids, improving extraction and processing techniques, and developing innovative applications.

With the growing demand for sustainable and eco-friendly solutions, natural and biodegradable hydrocolloids are also gaining popularity. These hydrocolloids are derived from renewable sources and offer improved environmental benefits over synthetic ones.

The Bottom Line

The versatility and potential of hydrocolloids in different industries are vast and continually expanding. The innovative applications of these ingredients showcase their ability to enhance product performance, improve sustainability, and offer new solutions to old problems.

8.2.6 Breakthrough Technologies: Advancing Hydrocolloid Research and Development

The potential of hydrocolloids as multifunctional ingredients is vast, and with breakthrough technologies, research and development are advancing quickly.

> "The development in extraction, modification, and characterization techniques has enabled us to create new hydrocolloid functionalities and tailor their performance to specific applications."

One of the most promising innovations is the use of nanotechnology to enhance the properties of hydrocolloids. Nanoparticles can improve solubility, stability, and dispersion, allowing for more precise control over the texture and functionality of hydrocolloid networks.

The use of enzymes to modify hydrocolloids is also gaining popularity. Enzymatic modification can increase the solubility and viscosity of hydrocolloids, making them more functional in a wider range of applications.

Table 68: Breakthrough Technology and Impact on Hydrocolloids

BREAKTHROUGH TECHNOLOGY	IMPACT ON HYDROCOLLOID DEVELOPMENT
Nanotechnology	Enhances hydrocolloid solubility, stability, and dispersion, allowing for more precise control over texture and functionality.
Enzymatic modification	Increases the solubility and viscosity of hydrocolloids, making them more functional in a wider range of applications.

Another important development is the exploration of hydrocolloids derived from sustainable and natural sources. It includes seaweed-based hydrocolloids, which have emerged as a promising alternative to traditional thickeners and gelling agents.

The future of hydrocolloid research and development looks bright, with new technologies and innovative applications constantly emerging. As the demand for clean labels and sustainable ingredients continues to grow, hydrocolloids will likely play a crucial role in shaping the future of food and other industries.

8.2.7 Hydrocolloids and Sustainability: A Step Towards Eco-friendly Solutions

Hydrocolloids are crucial in promoting sustainability and eco-friendly solutions in various industries. Natural and biodegradable hydrocolloids offer a viable alternative to synthetic materials that harm the environment. As such, their importance continues to grow, especially in industries such as personal care, pharmaceuticals, and food.

Using biodegradable hydrocolloids in packaging reduces environmental waste, as they can break down easily and safely in the environment without leaving harmful residues. It is a significant step forward towards a more sustainable and eco-friendly future.

Commitment to Sustainable Sourcing

Hydrocolloid manufacturers increasingly commit to sustainable sourcing practices to ensure eco-friendly and ethical solutions. Sourcing raw materials from certified sustainable sources ensures the environment is protected and the communities producing these raw materials receive fair treatment.

Moreover, sustainable sourcing practices ensure that the supply of raw materials is sufficient, reducing the risk of shortages that can result in price fluctuations and market volatility.

Advancements in Sustainable Hydrocolloids

The pursuit of eco-friendly solutions has led to the development of sustainable hydrocolloids. These ingredients offer unique functionalities while still promoting sustainability. For instance, researchers have developed hydrocolloid-based packaging that prolongs the shelf life of food products while still being biodegradable and environmentally friendly.

Table 69: Examples of sustainable hydrocolloids and their eco-friendly features

HYDROCOLLOID	SUSTAINABILITY FEATURE	INDUSTRY APPLICATION
Carrageenan	Reduces the need for preservatives	Food industry
Agar	Vegetarian and vegan alternatives to gelatin	Food industry
Guar gum	Superior water-binding properties	Food industry
Gellan gum	Excellent gelling agent	Pharmaceuticals and personal care

The Way Forward

The growing demand for eco-friendly and sustainable solutions continues to fuel the development of natural and biodegradable hydrocolloids. The future looks promising, with researchers and manufacturers working towards improving the functionality and performance of hydrocolloids while still embracing sustainability and environmental protection.

As a consumer, choosing products that use natural and biodegradable hydrocolloids is a step towards promoting sustainable development and protecting the environment from harmful pollutants. The future of hydrocolloids and sustainability looks bright, with research and development paving the way for innovative, eco-friendly solutions in various industries.

8.2.8 Future Trends and Outlook: Hydrocolloids in the Evolving Landscape

As hydrocolloids continue to gain popularity in various industries, the future trends and outlook for their applications remain promising. The following trends are expected to shape the use of hydrocolloids in the coming years:

- **Increased emphasis on clean labels and natural ingredients:** With consumers becoming more health-conscious and environmentally aware, there is a growing demand for natural and clean-label products. This trend is expected to drive the use of natural hydrocolloids, such as carrageenan, agar, and pectin, in food and personal care products.
- **Growing popularity of vegan and plant-based products:** The rise of veganism and plant-based diets is expected to drive the demand for hydrocolloids as substitutes for animal-derived ingredients. Hydrocolloids such as xanthan gum, guar gum, and carrageenan offer functional properties that make them suitable for vegan products.
- **Advancements in extraction and modification techniques:** The development of new extraction and modification technologies is expected to enhance the functionality and performance of hydrocolloids, allowing for more versatile applications. For example, microfluidization and aggregation techniques are being used to improve the solubility and dispersibility of hydrocolloids in food and personal care products.
- **Increasing pharmaceutical use:** Hydrocolloids are gaining popularity as excipients in the pharmaceutical industry due to their functional properties, such as binding, thickening, and gelling. The use of hydrocolloids in pharmaceutical formulations is expected to increase in the coming years, driven by the growing demand for novel drug delivery systems.
- **Exploring new applications:** Hydrocolloids have already found widespread use in the food, pharmaceutical, and personal care industries, but there is potential for their applications to expand further. New research is exploring the use of hydrocolloids in areas such as tissue engineering, coating, and packaging.

Overall, the future of hydrocolloids looks bright, with growing demand, increased research, and expanding applications. The versatility and functionality of hydrocolloids make them an essential ingredient for improving the quality, sustainability, and healthiness of products across various industries.

Conclusion

In conclusion, hydrocolloids are versatile and transformative ingredients with various applications in various industries, especially the food industry. They act as gelling agents, thickening agents, and stabilizers, contributing to texture, viscosity control, and water binding, ultimately enhancing the overall quality of food formulations.

Hydrocolloids revolutionize product formulations by improving texture, shelf life, and sensory attributes, ultimately enhancing consumer satisfaction. They have expanded beyond the food industry and are gaining importance in pharmaceuticals, cosmetics, and personal care.

Research and development in hydrocolloid extraction, modification, and characterization techniques have paved the way for improved hydrocolloid functionality, performance, and sustainability. The potential of natural and biodegradable hydrocolloids is significant in promoting sustainability and eco-friendly solutions.

The future trends of hydrocolloid applications look promising, with emerging technologies, market trends, and consumer preferences shaping the use of hydrocolloids. The potential for further advancements and innovations in hydrocolloid applications is enormous.

FAQ

How do hydrocolloids contribute to gel formation?

Hydrocolloids contribute to gel formation by forming networks that trap water molecules, resulting in a gel-like structure. The properties and functionality of these gels can be influenced by various factors, such as the type of hydrocolloid used and the concentration of the gel.

What are some breakthrough technologies in hydrocolloid research?

Breakthrough technologies in hydrocolloid research include advancements in extraction, modification, and characterization techniques. These innovations contribute to hydrocolloids' improved functionality and performance in various applications.

How do hydrocolloids promote sustainability?

Hydrocolloids can promote sustainability by offering natural and biodegradable alternatives. They reduce waste and contribute to a more sustainable future in the food, pharmaceutical, and cosmetics industries.

What are the future trends and outlook for hydrocolloid applications?

The future trends for hydrocolloid applications include emerging technologies, market trends, and consumer preferences. These factors will shape the use of hydrocolloids and lead to further advancements and innovations in their applications.

CHAPTER 09

Hydrocolloids in Vegan and Vegetarian Foods

PART 9.1

Hydrocolloid Alternatives to Animal-Based Gelling Agents

Recently, there has been a growing concern over using animal-derived gelling agents in the food industry and pharmaceuticals. Apart from ethical considerations, these gelling agents pose health risks due to their potential contamination with bacteria and viruses and the possibility of allergic reactions in sensitive individuals. As a result, there has been a shift towards sustainable and healthier alternatives such as hydrocolloids.

Hydrocolloids are a diverse group of naturally occurring and synthetic compounds that can form gels and thicken solutions. They are derived from various sources, including plants, seaweed, microbial fermentation, and synthetic modification of natural polymers. Hydrocolloids offer numerous benefits over animal-based gelling agents, such as improved functionality, versatility, and safety.

Core Insights:

- ☑ Hydrocolloids offer a sustainable and healthier alternative to animal-derived gelling agents.
- ☑ Hydrocolloids are a diverse group of compounds that can form gels and thicken solutions.
- ☑ Hydrocolloids are derived from various sources, including plants, seaweed, and microbial fermentation.
- ☑ Hydrocolloids offer improved functionality, versatility, and safety over animal-based gelling agents.
- ☑ Hydrocolloids are widely used in the food industry and pharmaceuticals to enhance product texture, stability, and shelf life.

9.1.1 Hydrocolloids role in the development of vegan and vegetarian foods

Hydrocolloids are substances that form a gel in the presence of water and are widely used in the food industry for various purposes, such as thickening, gelling, stabilizing, and improving texture. They play a crucial role in developing vegan and vegetarian foods by helping replicate the texture and mouthfeel of traditional animal-based products.

9.1.2 Types of Hydrocolloids

There are numerous hydrocolloids, each with unique properties and applications in the food industry. Here are some common hydrocolloids used in vegan and vegetarian foods:

Agar-Agar

Agar-agar is an excellent gelling agent, forming a stable gel at relatively low concentrations. It is often used in vegan desserts, fruit preserves, and confectionery. Its gelling properties are useful in creating firm textures in products like vegan gummy candies, custards, and jellies.

Carrageenan

Carrageenan is versatile and comes in different forms with varying textures. Kappa carrageenan forms strong, brittle gels, making it suitable for applications like vegan cheeses and deli-style slices. Iota carrageenan creates soft and elastic gels, ideal for dairy-free yogurt and desserts. Lambda carrageenan is often used as a thickening agent in products like plant-based milk.

Table 70: Carrageenan Applications and Dosage

APPLICATION	DOSAGE
Dairy products	0.05-0.2%
Meat products	0.1-0.5%
Other processed foods	0.05-0.2%

Guar Gum

Guar gum is a potent thickening agent, and it is used to enhance the viscosity of liquid-based products. In vegan ice cream, guar gum helps prevent the formation of ice crystals, contributing to a smoother texture. It's also commonly found in sauces, dressings, and gluten-free baked goods. Guar gum has good viscosity and stabilizing properties and can be used as a fat replacer in certain applications.

Table 71: Guar Gum Applications and Dosage

APPLICATION	DOSAGE
Dairy products	0.05-0.5%
Sauces	0.1-0.5%
Bakery products	0.5-2%

Xanthan Gum

Xanthan gum is a powerful thickener and stabilizer. In gluten-free baking, it mimics the viscosity and elasticity of gluten, providing structure to bread and other baked goods. Vegan salad dressings and sauces prevent ingredient separation and enhance the overall mouthfeel. Xanthan gum has excellent shear stability to maintain its viscosity under high shear conditions, such as during mixing or pumping.

Table 72: Xanthan gum Applications and Dosage

APPLICATION	DOSAGE
Sauces and dressings	0.1-0.5%
Bakery products	0.1-0.3%

Locust Bean Gum (Carob Gum)

Locust bean gum is often used in vegan ice cream to improve texture and prevent ice crystal formation. It works well in combination with other hydrocolloids, enhancing stability and creaminess. It's also used in dairy-free yogurts and plant-based cheeses.

Konjac Gum (Konjac Flour)

Konjac gum is known for its ability to form a gel with very few calories. In vegan and vegetarian foods, it thickens and gels, contributing to the desired texture in products like plant-based jellies, gummy candies, and noodles.

Pectin

Pectin is widely used in producing jams, jellies, and fruit spreads. It can also be incorporated into vegan baking desserts like fruit pies and gels. Pectin is valued for its ability to create a gel with a good balance of firmness and spreadability.

Alginate

Alginate is often used in molecular gastronomy and food innovation. In vegan cuisine, it can be employed to create unique textures, such as vegan caviar or spheres. Alginate forms a gel when it comes into contact with calcium ions, creating visually interesting and texturally appealing dishes.

These hydrocolloids are crucial in developing vegan and vegetarian foods, as they contribute not only to texture but also to the overall quality and stability of the products. Formulators often combine different hydrocolloids to achieve the desired characteristics in plant-based alternatives to traditional animal-based foods.

9.1.3 Hydrocolloids in Pharmaceuticals

Hydrocolloids have significant applications in the pharmaceutical industry, owing to their versatile functionalities and compatibility with various drug delivery systems. They are commonly used as viscosity modifiers, emulsifiers, and stabilizers in pharmaceutical formulations, such as tablets, capsules, and gels. The unique gelling properties of certain hydrocolloids also make them suitable for controlled-release drug delivery systems, which provide sustained drug release, avoid dose dumping, and enhance patient compliance.

One of the most used pharmaceutical hydrocolloids is sodium alginate, derived from seaweed. Sodium alginate is often used with other hydrocolloids, such as xanthan gum and guar gum, to improve the rheological properties of gels and suspensions, enhancing their bioavailability and efficacy. Sodium alginate is also used in wound dressings, forming a protective gel that promotes healing and prevents infections.

Table 73: Hydrocolloid Applications in Pharmaceuticals

HYDROCOLLOID	APPLICATION
Agar	Thickening agent; capsule formulation
Alginate	Gelling agent; controlled-release drug delivery; wound dressings
Carrageenan	Emulsifier; stabilizer; suspending agent; capsule formulation
Gellan gum	Gelling agent; suspending agent; controlled-release drug delivery
Guar gum	Thickening agent; emulsifier; stabilizer; capsule formulation
Xanthan gum	Thickening agent; stabilizer; emulsifier; suspending agent

Other hydrocolloids commonly used in pharmaceuticals include carrageenan, gellan gum, and xanthan gum. Carrageenan is often used as an emulsifier and stabilizer in suspensions, while gellan gum is a suitable gelling and suspending agent for liquid and semisolid formulations. Xanthan gum is a versatile hydrocolloid commonly used as a thickener, stabilizer, emulsifier, and suspending agent in pharmaceutical formulations.

Hydrocolloids are also used in topical applications, such as eye drops and ointments, where they improve the viscosity and stability of the formulations while enhancing drug absorption. Additionally, due to their moisturizing and healing properties, hydrocolloids, such as carboxymethyl cellulose and hyaluronic acid, are commonly used in cosmetics, personal care products, and wound healing applications.

Benefits of Hydrocolloids in Pharmaceuticals

- Enhance the bioavailability and efficacy of drugs
- Facilitate the development of controlled-release drug delivery systems
- Improve patient compliance and convenience
- Provide viscosity and stability to pharmaceutical formulations
- Offer a range of gelling and thickening properties
- Can be derived from sustainable and renewable sources
- Reduce the reliance on animal-derived gelling agents

Hydrocolloids offer vast potential for innovation and advancement in the pharmaceutical industry, enabling the development of more effective, targeted, and sustainable drug delivery systems. Their functional properties and compatibility with other ingredients make them valuable in diverse applications, ranging from wound healing to controlled drug delivery. With growing awareness and demand for sustainable solutions, hydrocolloids offer a promising solution for a greener, healthier, and more ethical pharmaceutical industry.

9.1.4 Hydrocolloids in Modern Gastronomy

Hydrocolloids have revolutionized modern gastronomy, allowing chefs to create innovative dishes with unique textures and flavors. From foams to gels, hydrocolloids offer a wide range of possibilities for culinary experimentation.

> **"Hydrocolloids have been the most important ingredients to emerge in modern cooking since the introduction of the food processor or immersion blender", says**
>
> **Chef Heston Blumenthal.**

Hydrocolloids are particularly popular in molecular gastronomy, a culinary movement that explores the scientific principles behind food preparation. Chefs use hydrocolloids to create unusual textures and shapes, such as spherical beads and noodles.

One of the most well-known applications of hydrocolloids in modern gastronomy is using sodium alginate and calcium chloride to create spheres known as "caviar" or "pearls." These small, bursting spheres can be filled with various flavors and textures, adding a new dimension to dishes.

Table 74: Hydrocolloid Culinary Functions

HYDROCOLLOID	CULINARY FUNCTION	EXAMPLES OF USE
Agar agar	Thickener, stabilizer, gelling agent	Used in desserts, such as panna cotta and jelly candies
Xanthan Gum	Thickener, stabilizer, emulsifier	Used in sauces, such as gravies and dressings, to improve texture and prevent separation
Carrageenan	Thickener, stabilizer, gelling agent	Used in dairy products, such as ice cream and yogurts, to prevent separation and improve texture

Hydrocolloids in high-end restaurants have found their way into home kitchens. With the growing interest in molecular gastronomy, it is now possible for home cooks to experiment with hydrocolloids and create unique dishes.

Overall, hydrocolloids have opened a whole new world of culinary possibilities, allowing chefs to push the boundaries of traditional cooking and explore the science behind food preparation.

9.1.5 Sustainable Hydrocolloids for a Greener Future

Hydrocolloids offer a healthier and more functional alternative to animal-based gelling agents and are a sustainable option for a greener future. We promote ethical and environmental stewardship by reducing the dependence on animal-derived ingredients.

One of the primary benefits of hydrocolloids is their ability to replace fat in food products, offering a clean-label solution. It allows to production of improved food options without compromising taste and texture. In addition to their functional benefits, hydrocolloids have a smaller carbon footprint than traditional gelling agents. For example:

Table 75: Hydrocolloid Carbon footprint and Animal-based alternative

HYDROCOLLOID	CARBON FOOTPRINT (KG CO2 EQ./KG)	ANIMAL-BASED ALTERNATIVE
Xanthan gum	1.56	Gelatin
Carrageenan	2.36	Agar
Agar	2.41	Gelatin

Furthermore, hydrocolloids derived from plant-based sources can also reduce deforestation and promote biodiversity. By replacing animal-derived ingredients with hydrocolloids, we can decrease the demand for land used for grazing and reduce the environmental damage caused by animal agriculture.

Hydrocolloids offer numerous benefits for a more sustainable and ethical food industry. Their functionality, versatility, and environmental advantages make them an excellent choice for creating healthier and more sustainable food products.

Conclusion

In conclusion, hydrocolloids are pivotal in developing vegan and vegetarian foods, contributing to plant-based alternatives' texture, stability, and overall quality. The diverse range of hydrocolloids, including agar-agar, carrageenan, guar gum, xanthan gum, locust bean gum, konjac gum, pectin, and alginate, offers formulators a toolbox for achieving specific characteristics in vegan and vegetarian products.

From providing gelling properties in desserts (agar-agar) to creating stable and elastic textures in dairy alternatives (carrageenan), thickening in ice cream (guar gum), and mimicking gluten in baked goods (xanthan gum), these hydrocolloids are versatile in meeting the varied demands of plant-based formulations. Whether enhancing creaminess in ice cream (locust bean gum), contributing low-calorie gels (konjac gum), creating jams and jellies (pectin), or enabling innovative textures in molecular gastronomy (alginate), each hydrocolloid brings its unique functionalities to the table.

The interaction of these hydrocolloids allows chefs, food scientists, and manufacturers to craft vegan and vegetarian products that meet consumers' sensory expectations and cater to the growing demand for plant-based alternatives. As the world embraces more sustainable and ethical food choices, the role of hydrocolloids continues to be instrumental in expanding the possibilities and improving the overall culinary experience of plant-based cuisine.

FAQ

What are hydrocolloids, and why are they used in vegan and vegetarian foods?

Hydrocolloids are substances that form a gel in the presence of water and are used in the food industry for various purposes. In vegan and vegetarian foods, hydrocolloids play a crucial role in replicating traditional animal-based products' texture, mouthfeel, and stability.

Which hydrocolloids are commonly used in vegan and vegetarian foods, and what functions do they serve?

Common hydrocolloids include agar-agar, carrageenan, guar gum, xanthan gum, locust bean gum, konjac gum, pectin, and alginate. They serve gelling, thickening, stabilizing, and improving texture in plant-based products.

How is agar-agar used in vegan cooking?

Agar-agar is often used as a vegan alternative to gelatin. It is employed in desserts, fruit preserves, and confectionery to create firm textures in products like vegan gummy candies, custards, and jellies.

What role does carrageenan play in plant-based foods, and how does it differ in its various forms?

Carrageenan is used for thickening and stabilizing in vegan foods. Kappa carrageenan is used for strong, brittle gels in products like vegan cheeses, while iota carrageenan is suitable for soft, elastic gels in dairy-free yogurts and desserts. Lambda carrageenan is often used as a thickening agent in plant-based milk.

In what ways does guar gum contribute to the texture of vegan products?

Guar gum is a thickening agent commonly used in vegan ice cream to prevent ice crystal formation and improve texture. It also enhances viscosity in sauces, dressings, and gluten-free baked goods.

FREQUENTLY ASKED QUESTIONS (FAQS) CONTINUED

- **How does xanthan gum benefit gluten-free baking and vegan sauces?**

  Xanthan gum is a powerful thickener and stabilizer used in gluten-free baking to mimic the structure of gluten. Vegan sauces and dressings prevent ingredient separation and enhance mouthfeel.

- **What is the role of locust bean gum in vegan ice cream and plant-based cheeses?**

  Locust bean gum enhances the texture of vegan ice cream, preventing ice crystal formation and contributing to creaminess. It is also used in plant-based cheeses to improve stability.

- **How is konjac gum utilized in vegan and vegetarian foods?**

  Konjac gum is a thickening and gelling agent in vegan and vegetarian foods, contributing to the desired texture in products like plant-based jellies, gummy candies, and noodles.

- **What function does pectin serve in vegan baking and fruit-based products?**

  Pectin is used in vegan baking to create jams, jellies, and fruit spreads. It contributes to the gel formation with a good balance of firmness and spreadability.

- **In what culinary applications is alginate commonly used, especially in vegan cuisine?**

  Alginate is often used in molecular gastronomy and food innovation. In vegan cuisine, it can create unique textures, such as vegan caviar or spheres, by forming gels with calcium ions.

PART 9.2

Hydrocolloid's Role in Plant-Based Meat and Dairy Alternatives

If you're curious about the world of plant-based alternatives to meat and dairy, you might have heard the term «hydrocolloid.» Hydrocolloids are an essential component in creating vegan alternatives for protein, particularly in the case of plant-based meat. These ingredients provide the necessary texture, binding, and moisture retention to replicate the taste and feel of meat. Without hydrocolloids, many of these products wouldn't exist.

The rise of plant-based meat alternatives has also led to the emergence of numerous brands utilizing hydrocolloids in their product formulations. From Impossible Foods to Beyond Meat, these companies rely on hydrocolloids to create products that mimic the taste and texture of meat without the associated animal welfare and environmental concerns.

Core Insights:

- ☑ Hydrocolloids are crucial in creating plant-based alternatives to meat and dairy.
- ☑ These ingredients contribute to texture, binding, and moisture retention in vegan protein sources.
- ☑ Many popular plant-based meat brands use hydrocolloids in their products.
- ☑ Hydrocolloids play a significant role in addressing animal welfare and environmental concerns associated with meat production.
- ☑ Plant-based alternatives to meat and dairy are on the rise as consumers increasingly seek sustainable and ethical food options.

9.2.1 Hydrocolloids in Plant-Based Meat Production

Hydrocolloids are essential ingredients in the production of plant-based meat alternatives. They play a crucial role in providing the texture and consistency of meat, making these substitutes feel and taste like the real thing. Several plant-based meat substitute companies use hydrocolloids in their products to enhance their quality and taste.

Some hydrocolloids used in plant-based meat production include carrageenan, xanthan gum, and methylcellulose. These hydrocolloids contribute to the binding and moisture retention of plant-based meat substitutes, allowing them to mimic the juiciness and tenderness of real meat.

Table 76: Hydrocolloid Function Popular Brand Uses

HYDROCOLLOID	FUNCTION	POPULAR BRAND THAT USES IT
Carrageenan	Thickener and stabilizer	Beyond Meat
Xanthan gum	Thickener and emulsifier	Impossible Foods
Methylcellulose	Emulsifier and gelling agent	Gardein

Hydrocolloids also help plant-based meat alternatives retain their shape during cooking, preventing them from falling apart or becoming too mushy. Plant-based meat substitutes can be cooked like traditional meat without losing texture or flavor.

Moreover, hydrocolloids have contributed to the growing popularity of plant-based meat alternatives. They make these products more accessible to people who want to reduce meat consumption for ethical or health reasons. The result is a range of delicious and nutritious plant-based meat substitutes suitable for vegans and vegetarians.

9.2.2 The Health Benefits of Plant-Based Meat Alternatives

Plant-based meat alternatives have become increasingly popular for their environmental benefits and potential health advantages. By consuming plant-based substitutes, you can reduce your intake of saturated fats while increasing essential nutrients like fiber, vitamins, and minerals. However, not all plant-based meat substitutes are created equal. Consider the nutritional content and ingredients to ensure you get the healthiest option.

If you're concerned about sodium intake, choose low-sodium plant-based meat alternatives. While some plant-based meat products may be high in sodium, many have lower amounts. Be sure to check the nutrition label and compare products before purchasing.

Some plant-based meat substitutes are enriched with essential vitamins and minerals, making them healthier than meat. Look for substitutes that contain vitamin B12, iron, and zinc commonly found in meat products. Incorporating hydrocolloids in these substitutes can also improve their nutritional profiles, enhancing their texture and flavor without sacrificing health benefits.

9.2.3 Addressing Concerns about Plant-Based Meat

If you're considering incorporating plant-based meat into your diet, you may have heard some concerns and drawbacks associated with it. Evaluating the potential risks and considerations before making any dietary changes is important.

One concern is the presence of certain additives, including hydrocolloids, which may be used to enhance texture and moisture retention in plant-based meat products. While hydrocolloids are generally considered safe, some people may have an allergic reaction to them. It's always important to read ingredient labels and be aware of potential allergens.

Another concern is that plant-based meats often contain higher sodium levels than traditional meat. It can disadvantage individuals with high blood pressure or other health concerns. However, low-sodium options are available, and it's crucial to pay attention to nutritional labels and choose products that align with your dietary needs.

Despite these concerns, plant-based meat can offer a range of health benefits, including lower levels of saturated fat and cholesterol. Additionally, using hydrocolloids and other plant-based ingredients can enhance the nutritional profile of these products, offering a healthier alternative to traditional meat.

9.2.4 Exploring Vegan Alternatives for Protein

With the rise of plant-based diets and veganism, finding alternative protein sources has become increasingly important. When you rely on plant-based sources for protein, it's crucial to incorporate diverse options to ensure you receive all the necessary amino acids.

Vegan alternatives for protein can include a variety of foods such as legumes, nuts, seeds, tofu, tempeh, and even certain grains like quinoa. These options offer a range of nutritional benefits, from fiber and healthy fats to vitamins and minerals. It›s important to note that consuming enough protein is key to maintaining overall health and well-being, regardless of whether you follow a vegan diet. Some studies suggest incorporating plant-based protein sources can have additional health benefits, such as improving heart health and lowering the risk of chronic diseases. Regarding vegan protein sources, there are various options, all of which can be incorporated into a balanced, healthy diet. The key is to find the ones that work best for you and that you enjoy eating.

9.2.5 Notable Plant-Based Meat Substitute Brands

You have plenty of options if you are looking for high-quality plant-based meat substitutes. Many top food companies now include plant-based options in their product lineups, catering to a growing demand for ethically and sustainably produced food. Here are some of the most notable plant-based meat substitute brands that incorporate hydrocolloids into their products to create meat-like texture and flavor.

Table 77: Brand, Product Lineup and Hydrocolloid Uses

BRAND	PRODUCT LINEUP	HYDROCOLLOID USE
Beyond Meat	Burgers, sausage, ground beef, meatballs	Xanthan gum, methylcellulose
Impossible Foods	Burgers, sausage, ground beef, meatballs	Methylcellulose
Gardein	Chicken tenders, beef tips, meatballs, fish filets	Methylcellulose, konjac flour, carrageenan
Tofurky	Delicatessen slices, sausages, burgers	Xanthan gum, carrageenan
Lightlife Foods	Hot dogs, sausages, ground beef, chicken tenders	Carrageenan, konjac flour

These brands have gained popularity for their ability to mimic the taste and texture of traditional meat products, making it easier for people to embrace a plant-based diet and reduce their meat consumption.

Beyond Meat's products are soy-free and gluten-free, combining xanthan gum and methylcellulose to create a meat-like texture. Similarly, Impossible Foods uses methylcellulose to bind its plant-based ingredients and create a burger that «bleeds» like real meat. Gardein›s products rely on methylcellulose, konjac flour, and carrageenan to achieve a similar effect, while Tofurky and Lightlife Foods use a combination of carrageenan, konjac flour, and xanthan gum.

If you are new to plant-based eating or looking to expand your options, these brands are a great place to start. Their innovative use of hydrocolloids offers a range of delicious and satisfying alternatives to traditional meat products.

9.2.6 The Role of Hydrocolloids in Dairy Alternatives

Regarding plant-based dairy alternatives, hydrocolloids play a crucial role in creating products that mimic the taste and texture of traditional dairy. Hydrocolloids are food additives that work as thickening, gelling, or stabilizing agents, creating a range of textures and mouthfeel in foods. They are used in plant-based dairy alternatives to create products like vegan cheese, milk, and yogurt.

Hydrocolloids are particularly useful in creating plant-based dairy products similar in texture and flavor to their dairy counterparts. For example, in vegan cheese, hydrocolloids help to create a cheese-like texture and meltability. In plant-based yogurt, hydrocolloids contribute to the thickness and creaminess of the yogurt.

Some of the most commonly used hydrocolloids in plant-based dairy alternatives include agar, carrageenan, and xanthan gum. Agar is a natural gelling agent derived from seaweed and is often used to create vegan cheese and custards. Carrageenan is extracted from red seaweed and is commonly used to thicken and stabilize plant-based milk. Xanthan gum is a soluble fiber that helps to create a creamy texture in plant-based dairy alternatives. Hydrocolloids are also useful in providing nutritional benefits to plant-based dairy alternatives. They can enrich products with vitamins and minerals, improve moisture retention, and increase protein content.

In conclusion, hydrocolloids are essential in creating plant-based dairy alternatives, contributing to texture, flavor, and overall product quality. With more people turning towards plant-based alternatives for health and sustainability reasons, hydrocolloids will continue to play an important role in creating innovative and appealing plant-based substitutes for traditional dairy products.

9.2.7 Best Plant-Based Meat Substitutes

Whether you're a committed vegan, a dedicated vegetarian, or simply looking to incorporate more plant-based options into your diet, there are plenty of delicious and satisfying meat substitutes. While not all plant-based meat alternatives are created equally, some standout products deliver on taste, texture, and nutritional content. Here are some of the best plant-based meat substitutes to try:

Table 78: Best Plant-Based Meat Substitutes

PRODUCT	DESCRIPTION
Beyond Burger	This popular plant-based burger is known for its meat-like texture and juicy flavor. Made with pea protein and coconut oil, the Beyond Burger is also a good source of protein, iron, and other essential nutrients.
Impossible Burger	The Impossible Burger is another high-profile plant-based burger with a loyal following. Made with soy protein and potato protein, this burger is designed to look, taste, and cook like traditional beef.
Gardein Seven Grain Crispy Tenders	If you're looking for something crispy and satisfying, these Gardein tenders are a great option. Made with a blend of soy, wheat, and pea protein, they're also a good source of fiber and iron.
Quorn Meatless Grounds	For a versatile and protein-packed meat substitute, try Quorn Meatless Grounds. Made with mycoprotein, a fungus-based protein high in fiber and low in fat, these grounds can be used in various dishes, from spaghetti Bolognese to tacos.

These are just a few examples of the available plant-based meat substitutes. Whether you're looking for burgers, tenders, or grounds, there are plenty of options to choose from that are tasty, nutritious, and satisfying.

9.2.8 Comparing Plant-Based Meat Brands

If you're looking to try plant-based meat alternatives, you'll find a growing number of brands on the market. But with so many options, deciding which one can be challenging. To help you make an informed decision, we've compiled a list of some of the most popular and highly regarded plant-based meat substitute brands.

Table 79: Comparing Plant-Based Meat Brands

BRAND	DESCRIPTION	HYDROCOLLOID USE
Beyond Meat	Beyond Meat is a well-known brand that offers a wide range of plant-based meat alternatives, including burgers, sausages, and ground meat. Their products are made primarily from pea protein isolate and use potato starch for texture.	Beyond Meat uses methylcellulose as a hydrocolloid in its products, which helps to bind the ingredients together and create a meat-like texture.
Impossible Foods	Impossible Foods developed their signature product, the Impossible Burger, to be as close to real beef as possible. Their burgers are made from a blend of soy and potato protein, and they also use coconut oil and sunflower oil to mimic the fattiness of beef	Impossible Foods uses a combination of methylcellulose and food-grade carrageenan to create a texture that resembles ground beef.
Gardein	Gardein is a brand that offers a variety of plant-based meat alternatives, from chicken tenders to meatless meatballs. Their products are made from a blend of soy, wheat, and pea proteins, and they also use tapioca starch, potato starch, and pea fiber for texture.	Gardein combines methylcellulose and konjac flour to create a meat-like texture.
Tofurky	Tofurky is a brand specializing in vegan products, including plant-based deli slices and sausages. Their products are made from various ingredients, such as tofu, wheat protein, and soy protein.	Tofurky combines carrageenan and konjac flour to create a texture that resembles meat.

These brands are just a few examples of the many plant-based meat alternatives available. When choosing a brand, consider factors such as taste, texture, nutritional content, and the specific hydrocolloids used in the product. You can find the plant-based meat substitute that best suits your preferences by exploring different options.

9.2.9 The Future of Hydrocolloids in Plant-Based Alternatives

As the demand for plant-based meat alternatives grows, food companies are investing in research to find new and innovative ways to create tasty and nutritious substitutes for meat and dairy products. Hydrocolloids have already been proven to be a valuable ingredient in these products and are likely to continue playing a significant role in the future of plant-based food technology.

Researchers are exploring the potential of hydrocolloids in creating more complex textures and flavors in plant-based meat substitutes. One area of research involves using hydrocolloids to create fibrous tissue structures that mimic the texture of animal muscle. Additionally, hydrocolloids may be used to create plant-based fat alternatives that melt and behave similarly to animal fats.

Food companies are also investing in sustainable alternatives to traditional hydrocolloids, such as seaweed-based gelling agents, which could further reduce the environmental impact of plant-based meat production.

As plant-based meat and dairy alternatives become increasingly popular, food companies are also exploring the potential of hydrocolloids in creating new and innovative products. For example, hydrocolloids could be used to create plant-based seafood alternatives, such as vegan shrimp or crab meat, or even to develop edible packaging from plant-based materials.

Overall, the future of hydrocolloids in plant-based food technology is bright, with ongoing research and development promising exciting new developments. As more consumers seek sustainable and ethical food options, hydrocolloids will likely continue to be valuable in creating delicious and nutritious plant-based alternatives to traditional meat and dairy products.

Conclusion

As the demand for more sustainable and ethical food options grows, hydrocolloids have become a pivotal ingredient in creating plant-based meat and dairy alternatives. Their ability to contribute to texture, binding, and overall product quality has made them a favorite among vegan and plant-based food producers.

Research on hydrocolloids as substitutes for animal proteins has also shown promising results, suggesting their potential to revolutionize the food industry. As more consumers turn to plant-based alternatives for their protein needs, the role of hydrocolloids in these products will only become more significant.

Applying hydrocolloids in plant-based meat and dairy alternatives constantly evolves as food producers seek to improve their products' taste, texture, and nutritional profiles. Ongoing research and advancements in hydrocolloid technology pave the way for innovative and appealing plant-based alternatives to traditional meat and dairy products.

As a consumer, you can shape the food industry's future by supporting companies prioritizing sustainability and ethical practices. Plant-based meat alternatives can reduce environmental impact and contribute to a more compassionate food system.

Hydrocolloids are just one ingredient in the complex web of plant-based meat and dairy alternatives, but their impact is unquestionable. In the coming years, we can expect to see more exciting developments in using hydrocolloids in creating sustainable and delicious plant-based options for meat and dairy products.

So, whether you're looking for vegan alternatives for protein, plant-based substitutes for meat, or simply exploring new food options, hydrocolloids are an ingredient to keep an eye on. With their versatility and potential, they are sure to play an important role in the future of the food industry.

FAQ

What is the role of hydrocolloids in plant-based meat and dairy alternatives?

Hydrocolloids play a pivotal role in developing plant-based meat and dairy alternatives. They contribute to texture, binding, and overall product quality.

What are hydrocolloids, and how do they function?

Hydrocolloids are substances that form gels or thicken liquids when mixed with water. They function as stabilizers, emulsifiers, and thickeners in plant-based meat alternatives.

How do hydrocolloids contribute to plant-based meat production?

Hydrocolloids contribute to the texture, binding, and moisture retention in plant-based meat production, creating meat-like alternatives.

What are the health benefits of plant-based meat alternatives?

Plant-based meat alternatives offer various health advantages, including low sodium and nutritional benefits. Hydrocolloids are often used to enhance the nutritional profiles of these products.

What concerns and drawbacks are associated with plant-based meat?

Some concerns associated with plant-based meat include the presence of certain additives, like hydrocolloids, and potential negative health effects. It's important to address these considerations when consuming plant-based meat.

What are vegan alternatives for protein, and why are they important?

Vegan alternatives for protein refer to relying on plant-based sources for protein intake. Incorporating diverse protein sources into a vegan diet is important for meeting nutritional needs.

Which are some notable plant-based meat alternative brands?

Several popular plant-based meat substitute brands utilize hydrocolloids. These brands offer a range of products catering to different consumer preferences.

FREQUENTLY ASKED QUESTIONS (FAQS) CONTINUED

How do hydrocolloids contribute to the creation of plant-based dairy alternatives?

Hydrocolloids play a role in creating plant-based dairy alternatives like vegan cheese, milk, and yogurt by impacting texture, flavor, and overall product quality.

What are the best plant-based meat substitutes available?

There are various plant-based meat substitutes available in the market. The best ones are determined based on factors such as taste, texture, nutritional content, and the role of hydrocolloids in enhancing meat-like qualities.

How do different plant-based meat brands compare?

Different plant-based meat brands can vary in taste, texture, and ingredient composition. Considering these factors, hydrocolloids can contribute to the overall quality of these products.

What does the future hold for hydrocolloids in plant-based alternatives?

Ongoing research and advancements in the use of hydrocolloids for plant-based meat and dairy alternatives are expected. Hydrocolloids will continue to play a vital role in meeting the growing demand for sustainable and ethical food options.

CHAPTER 10

Global Market and Trends

PART 10.1

Global Hydrocolloid Market Trends: Leading Producers

The hydrocolloids market is experiencing robust growth globally, driven by increased demand for functional foods, expanding pharmaceutical and personal care industries, and rising health consciousness among consumers. Hydrocolloids are used extensively in various applications, including food processing, drug delivery, wound healing, and cosmetics.

This article will provide insights into the latest global market trends in the hydrocolloids industry, focusing on the leading producers and their impact. It will also discuss the key drivers and challenges faced by the market and explore the outlook for the industry.

Core Insights:

- ☑ The hydrocolloids market is growing globally
- ☑ Increased demand for functional foods drives growth
- ☑ Hydrocolloids are widely used in various applications
- ☑ Leading producers play a significant role in shaping the industry
- ☑ The future outlook for the industry is promising, with potential growth opportunities

10.1.1 Overview of the Hydrocolloids Market

The global hydrocolloids market has witnessed significant growth in recent years, driven by increasing demand from various industries such as food, pharmaceuticals, and personal care. Hydrocolloids are complex molecules that form a gel-like structure when added to water, which makes them an important ingredient in many products. According to a report by MarketsandMarkets, the hydrocolloids market is expected to reach a value of $9.7 billion by 2025, growing at a CAGR of 6.9% during the forecast period.

Hydrocolloids are used for various processes, including thickening, gelling, emulsification, and stabilization. They are commonly used in the food industry to improve texture, enhance flavor, extend product shelf life, and modify liquids' viscosity. The pharmaceutical industry utilizes hydrocolloids in drug delivery systems, wound dressings, and tissue engineering. The personal care industry uses hydrocolloids in products such as shampoos, lotions, and cosmetics to improve consistency and texture.

Types of Hydrocolloids

There are several types of hydrocolloids available in the market. Some of the most used hydrocolloids include:

- Starch
- Gums
- Pectin
- Agar
- Alginate
- Carrageenan

These hydrocolloids have unique properties that make them suitable for specific applications. For example, carrageenan is commonly used in dairy products due to its ability to form a gel-like structure at room temperature. In contrast, pectin is frequently used in jams and jellies because of its strong gelling properties.

Key Players in the Hydrocolloids Market

The hydrocolloids market is highly competitive, with several major players operating in the industry. Some of the leading companies in the hydrocolloids market include:

- DowDuPont Inc.
- Cargill, Incorporated
- Ingredion Incorporated
- Kerry Group
- BASF SE
- Ashland Global Holdings Inc.

These companies are engaged in strategic initiatives such as mergers and acquisitions, partnerships, and product launches to gain market share and expand their product portfolio.

10.1.2 Key Trends in Hydrocolloid Dressings Market

Hydrocolloid dressings have gained immense popularity owing to their effectiveness in wound healing. According to a recent report, the global hydrocolloid dressings market size is projected to reach USD 824.9 million by 2028, exhibiting a CAGR of 6.7% during the forecast period.

The market growth is primarily attributed to the high prevalence of chronic wounds, such as diabetic foot ulcers and pressure ulcers, coupled with the increasing adoption of advanced wound care products. Hydrocolloid dressings provide a moist environment for the wound, which helps in faster healing and reduces the risk of infection. Furthermore, these dressings can be easily applied and removed without causing pain to the patient, making them ideal for use in home settings.

The market is witnessing notable technological advancements, with manufacturers focusing on developing innovative, advanced hydrocolloid dressings to meet the growing demand. For instance, introducing antimicrobial hydrocolloid dressings that help prevent infections is expected to drive further market growth. Additionally, several companies invest heavily in research and development activities to introduce eco-friendly and cost-effective hydrocolloid dressings.

North America is currently the largest market for hydrocolloid dressings, owing to the high incidence of chronic wounds and the rising geriatric population. The United States dominated the market share in the region, followed by Canada. The Asia Pacific region is expected to exhibit significant growth potential during the forecast period, attributed to the growing healthcare infrastructure and the rising prevalence of chronic diseases.

Overall, the hydrocolloid dressings market is poised for significant growth in the coming years, with manufacturers focusing on introducing innovative products that comply with sustainable and eco-friendly regulations. Increasing investment in research and development, coupled with the rising adoption of advanced wound care products, is expected to drive the market growth further.

10.1.3 Collagen Hydrocolloid Market Analysis

The global collagen hydrocolloid market is expected to grow at a CAGR of 7.5% between 2021 and 2026. Collagen hydrocolloids are widely used in the cosmetics and personal care industry as they offer numerous benefits, including improved skin elasticity and hydration.

The market is highly competitive, with a few key players dominating the industry. The leading producers in the collagen hydrocolloid market include Gelita, Rousselot, Nitta Gelatin, PB Gelatins, and Weishardt Group.

The increasing demand for natural and clean-label products is driving the growth of the collagen hydrocolloid market. Consumers are increasingly looking for products made with natural ingredients and free from artificial additives. This trend is expected to drive the growth of the collagen hydrocolloid market in the coming years.

Applications of Collagen Hydrocolloids

Collagen hydrocolloids have numerous applications in the cosmetics and personal care industry. They are widely used to produce anti-aging creams, moisturizers, and other skincare products. Collagen hydrocolloids are also used in hair care products, as they can improve hair strength and thickness.

In the food industry, collagen hydrocolloids are used for their gelling and emulsifying properties. They commonly produce sausages, meatballs, and other processed meats. Collagen hydrocolloids also produce dairy products, such as yogurt and cheese.

Challenges and Opportunities

One of the major challenges facing the collagen hydrocolloid market is the high cost of production. Collagen hydrocolloids are extracted from animal by-products, such as bones and skin, making the production process expensive. Additionally, the use of animal by-products raises ethical concerns and sustainability issues.

However, there are also opportunities for growth in the collagen hydrocolloid market. Manufacturers are exploring new sources of collagen, such as fish and plant-based sources. It could lead to more sustainable and cost-effective production methods.

The collagen hydrocolloid market is poised for significant growth in the coming years, driven by increasing demand for natural and clean-label products. While there are challenges to overcome, such as the high cost of production and sustainability concerns, there are also significant opportunities for innovation and growth in this dynamic industry.

10.1.4 Emerging Innovations in the Food Hydrocolloids Market

The food hydrocolloids market has witnessed several innovations in recent years, driven by the growing demand for functional ingredients. Hydrocolloids, widely used in the food industry for their thickening, gelling, and emulsifying properties, are increasingly finding new applications in food processing, texture modification, and stability enhancement.

One of the emerging innovations in the food hydrocolloids market is using plant-based hydrocolloids as an alternative to animal-based products. With the rise of veganism and clean-label trends, consumers seek plant-based options free from synthetic additives and animal-derived ingredients. It has led to the development of hydrocolloids such as agar, carrageenan, and gellan gum derived from seaweed and other plant sources.

Another innovation in the food hydrocolloids market is using hydrocolloids for encapsulation and controlled release of active ingredients. Hydrocolloid-based capsules are being used to deliver nutraceuticals, flavors, and even functional microorganisms to the human gut. This technology has opened opportunities for functional food and beverage manufacturers to develop products with targeted health benefits.

Furthermore, there has been a growing interest in using hydrocolloids in developing sustainable packaging materials. Hydrocolloid-based films and coatings effectively preserve food quality and extend shelf life.

The food hydrocolloids market is expected to grow significantly in the coming years, driven by the increasing demand for functional and clean-label ingredients. Plant-based hydrocolloids, encapsulation technology, and sustainable packaging materials are likely to be key drivers of innovation and growth in the market.

10.1.5 Leading Producers in the Hydrocolloids Industry

The global hydrocolloids market is highly competitive, with numerous players vying for market share. A few key producers with a strong presence in multiple regions dominate the market. These leading producers are driving innovation and shaping the industry›s future.

Table 80: Leading Producers in the Hydrocolloids Industry

COMPANY NAME	MARKET SHARE	PRODUCT PORTFOLIO	STRATEGIC INITIATIVES
DuPont Nutrition & Biosciences	25%	Hydrocolloid stabilizers, gelling agents, and thickeners for the food, beverage, and pharmaceutical industries	Investment in R&D to develop sustainable hydrocolloid solutions
Cargill, Inc.	20%	Hydrocolloid stabilizers, emulsifiers, and thickeners for the food, dairy, and beverage industries	Acquisition of companies to expand product portfolio and market presence
Ingredion Incorporated	15%	Hydrocolloid solutions for the food, beverage, and industrial markets	Investment in R&D to create clean-label and organic hydrocolloid solutions

These companies have strong global networks and have made significant investments in research and development to stay ahead of the competition. They offer various hydrocolloid solutions for various industries and continuously seek to improve their products to meet changing market demands.

Emerging hydrocolloid industry players include Ashland Global Holdings Inc., Kerry Group plc, and Tate & Lyle plc. These companies also invest in R&D to develop sustainable and innovative hydrocolloid solutions for different applications.

The leading producers in the hydrocolloids industry are expected to continue driving growth and innovation in the market. They will play a crucial role in meeting the increasing demand for hydrocolloid products across various industries.

10.1.6 Market Challenges and Opportunities in the United States

The hydrocolloids market in the United States faces unique challenges and opportunities amid global market trends. The regulatory framework governing the use of hydrocolloids in food and personal care products is complex, variable, and evolving, requiring manufacturers to stay updated and compliant with new regulations.

Consumer preferences in the United States reflect a growing demand for natural, organic, and sustainable products, leading to the emergence of clean-label and plant-based hydrocolloids. The trend towards healthier and functional foods drives innovation and experimentation with new hydrocolloid formulations offering various health benefits. These trends present opportunities for hydrocolloid producers to capitalize on these evolving market demands.

Market dynamics in the United States vary by region, and manufacturers must understand local market conditions, distribution channels, and consumer preferences. The market in the United States is highly competitive, with established players and numerous new entrants vying for market share.

Despite these challenges, the United States remains a critical market for hydrocolloid producers, accounting for a significant share of global demand. Manufacturers must navigate the complexity of the regulatory environment and stay attuned to consumer preferences while also leveraging opportunities for innovation and growth in this key market.

10.1.7 Future Outlook for the Hydrocolloids Industry

The hydrocolloids industry is expected to witness steady growth in the coming years, driven by increasing demand for food and personal care products and technological advancements in wound care and medical materials. According to industry experts, the global hydrocolloids market is projected to reach USD 11.37 billion by 2025, growing at a CAGR of 6.1% from 2020 to 2025 ***(Global Market Insights, 2020)***. This growth is attributed to the rising popularity of convenience and ready-to-eat foods and a growing emphasis on natural and organic ingredients.

One trend that is expected to shape the future of the hydrocolloid industry is sustainability. Consumers are increasingly demanding products that are eco-friendly and transparently sourced. Hydrocolloid producers are responding with innovations in biodegradable and plant-based materials. For instance, some companies are developing biodegradable food packaging that contains hydrocolloids, which is expected to reduce waste and plastic pollution.

Another trend to watch is the clean-label movement. Consumers increasingly demand products free from artificial colors, flavors, and additives. Hydrocolloid producers respond with clean-label solutions using natural and minimally processed ingredients. It is expected to drive demand for hydrocolloids derived from renewable sources, such as seaweed and plant-based sources.

Emerging Market Segments

One of the emerging market segments in the hydrocolloid industry is medical and wound care applications. Hydrocolloid dressings are gaining popularity due to their advanced moisture management properties, which promote faster healing and reduce the risk of infection. Additionally, the use of collagen hydrocolloids in the cosmetics and personal care industry is expected to grow due to their superior skin moisturizing and anti-aging properties.

Another market segment that is expected to grow is the use of hydrocolloids in processed foods. Hydrocolloids are widely used in food processing to modify texture and enhance stability. They are particularly useful in producing bakery, dairy, and meat products. As consumers demand healthier and more natural foods, hydrocolloid producers are expected to develop clean-label and plant-based solutions that meet these demands.

Technological Advancements

Technological advancements are expected to drive innovation in the hydrocolloids industry. For instance, nanotechnology is expected to revolutionize the production of hydrocolloids by improving their stability, bioavailability, and functional properties. Additionally, advances in fermentation technology are expected to reduce production costs and increase the availability of hydrocolloids derived from natural sources.

The future of the hydrocolloids industry looks promising, with significant growth potential in both the food and medical industries. With emerging trends in sustainability, clean-label, and plant-based solutions, hydrocolloid producers are well-positioned to meet the demands of consumers and drive innovation in the industry. Technological advancements are expected to enhance the functional properties of hydrocolloids further, making them more versatile and cost-effective.

Conclusion

In conclusion, understanding the global market and hydrocolloid industry trends is crucial for businesses to stay competitive. The industry has grown significantly in recent years, driven by leading producers and technological advancements. From a broad perspective, the hydrocolloids market has witnessed increased demand from various industries, ranging from food and beverage to pharmaceuticals. The demand for specialized hydrocolloid products, such as collagen, in the cosmetics and personal care industry has also increased, creating new opportunities for market players. However, the industry faces several challenges, such as regulatory frameworks, consumer preferences, and market dynamics. These challenges vary across different regions, such as the United States, where there are specific market challenges and opportunities.

The outlook for the hydrocolloids industry is promising, with growth projections and emerging market segments. Additionally, industry players focus on sustainability and clean-label trends, which are expected to shape the industry's future. In conclusion, staying up to date with the latest market trends and innovations is crucial for businesses to survive and thrive in the global hydrocolloids market. By understanding the market dynamics and collaborating with leading producers, businesses can take advantage of the opportunities the ever-evolving hydrocolloids industry presents.

FAQ

Q: What are the emerging innovations in the food hydrocolloids market?

The food hydrocolloids market is witnessing innovations such as plant-based hydrocolloids, clean-label solutions, and customized texture modification. There is also a growing interest in using hydrocolloids as encapsulation agents for the controlled release of bioactive compounds.

Q: Who are the leading producers in the hydrocolloids industry?

The hydrocolloids industry is led by major players such as DuPont Nutrition & Biosciences, Cargill Incorporated, CP Kelco, Ashland Global Holdings Inc., and Ingredion Incorporated. These companies have a significant market share and offer a wide range of hydrocolloid products.

Q: What are the key challenges faced by the hydrocolloids market in the United States?

The hydrocolloids market in the United States faces challenges such as regulatory compliance, volatility in raw material prices, and increasing consumer demand for clean-label and natural ingredients. Market players need to navigate these challenges to stay competitive.

Q: What is the outlook for the hydrocolloids industry?

The hydrocolloid industry is expected to witness steady growth in the coming years, driven by increasing consumer awareness regarding health and wellness, the demand for clean-label products, and technological advancements in hydrocolloid production. The industry is also likely to witness the emergence of new market segments and sustainable practices.

PART 10.2

Hydrocolloids Market Dynamics and Future Projections

Hydrocolloids are a vital component of the food industry, serving as essential ingredients in various food products. Market dynamics in the hydrocolloids industry have been evolving, driven by changing consumer preferences and technological advancements in the food sector. Industry players must understand current market trends and future projections to make informed investment decisions.

Core Insights:

- ☑ The hydrocolloids industry is experiencing dynamic growth, driven by changing consumer preferences and technological advancements.
- ☑ Hydrocolloids are hydrophilic polymers used as food additives, providing thickening, gelling, stabilizing, and emulsifying functionalities.
- ☑ Viscosity modifiers play a significant role in enhancing the texture and mouthfeel of food products using hydrocolloids.
- ☑ Hydrocolloids are used as texture enhancers in the food industry, improving sensory perception and stability of food products.
- ☑ Hydrocolloids contribute to various food properties, such as shelf life, emulsion stability, and flavor release, making them crucial ingredients in the food industry.

10.2.1 Understanding Hydrocolloids as Hydrophilic Polymers

Hydrocolloids are food additives widely used in the food industry as thickening agents, gelling agents, stabilizers, and emulsifiers. These additives are hydrophilic polymers, meaning they have an affinity for water and can absorb and retain large amounts.

Hydrocolloids generally consist of long-chain polysaccharides, proteins, or combinations of both. They are characterized by their ability to form hydrogels or viscous solutions when dissolved in water, making them ideal for use in various food products.

Hydrocolloids are essential ingredients in many foods, from bakery products and dairy to meat and poultry. They contribute to these products' texture, mouthfeel, and stability by providing the necessary thickness, viscosity, and elasticity.

One of the unique properties of hydrocolloids is their water-absorbing capacity, which makes them highly effective in controlling the water content and preventing syneresis in food products. This property also makes them ideal for enhancing the juiciness and tenderness of meat products.

Hydrocolloids are also used in various applications outside the food industry, such as pharmaceuticals, cosmetics, and personal care products. In these industries, they are used as thickening agents and stabilizers to improve the texture and stability of products.

10.2.2 The Importance of Viscosity Modifiers in Hydrocolloids

Hydrocolloids are widely used in the food industry as they provide various functionalities, such as thickening, gelling, stabilizing, and emulsifying. The key characteristic of hydrocolloids is their ability to absorb water, which is why they are often referred to as hydrophilic polymers. This property makes them ideal for improving food products' texture, mouthfeel, and stability. However, the effectiveness of hydrocolloids depends on their ability to modify the viscosity of food systems.

Viscosity modifiers play a critical role in hydrocolloids as they determine the thickness and flow of food products. Hydrocolloids that have a high molecular weight and form strong gels tend to have a high viscosity. In contrast, hydrocolloids that have a low molecular weight and form weak gels have a low viscosity. Therefore, adjusting the viscosity modifiers in hydrocolloids is essential to achieve the desired texture and consistency in food products.

The importance of viscosity modifiers can be seen in various food products. For example, salad dressings require a lower viscosity to ensure they can be easily poured, while sauces like ketchup need a higher viscosity to prevent them from dripping off food. Frozen desserts such as ice cream and gelato require a specific viscosity to ensure a smooth and creamy texture. Viscosity modifiers also play an essential role in ensuring the stability of emulsions, preventing separation, and maintaining a consistent texture.

Hydrocolloids that act as viscosity modifiers include guar gum, xanthan gum, and carrageenan. Guar gum is widely used in the food industry because it enhances viscosity and forms gels. Xanthan gum is another popular hydrocolloid that can increase the viscosity of food products while maintaining stability. Carrageenan is primarily used as a thickener and stabilizer in dairy products, such as chocolate milk and ice cream, due to its ability to retain moisture and prevent ice crystal formation.

Viscosity modifiers are crucial in hydrocolloids, as they contribute to food products' texture, stability, and sensory experience. The effectiveness of hydrocolloids in food applications depends on their ability to modify viscosity, making it essential for food manufacturers to understand the role of viscosity modifiers in hydrocolloids.

10.2.3 Applications of Hydrocolloids as Texture Enhancers

Hydrocolloids are widely used in the food industry as texture enhancers. They are highly versatile and can be tailor-made to create various textures, from gummy to creamy. Hydrocolloids such as carrageenan, xanthan gum, and pectin are commonly used as stabilizers and thickeners in many food products, including sauces, dressings, and desserts.

One key application of hydrocolloids as texture enhancers is in producing dairy products. Hydrocolloids are added to improve the body and texture of yogurt, ice cream, cream cheese, and other dairy products. These ingredients help to prevent ice crystal formation, improve mouthfeel, and enhance creaminess. In low-fat versions of these products, hydrocolloids can function as fat replacers, mimicking the texture and mouthfeel of full-fat products.

Another popular application of hydrocolloids is in producing bakery products such as cakes, muffins, and bread. Hydrocolloids can help improve baked goods' texture, volume, and shelf-life. They can also be used to adjust the texture of gluten-free products, providing a more pleasant mouthfeel.

In addition to these applications, hydrocolloids can be used as texture enhancers in meat products. They can help improve water retention, increasing yield and improving juiciness. Hydrocolloids such as carrageenan and alginate are also used in producing plant-based meat substitutes, helping to create a meat-like texture.

Overall, hydrocolloids are vital in enhancing the texture and mouthfeel of a wide range of food products. They offer high functionality and versatility, making them popular for food manufacturers looking to create high-quality products with desirable textures.

10.2.4 The Role of Hydrocolloids in Ingredient Functionality

Hydrocolloids are versatile ingredients that play a crucial role in various food applications. They act as hydrophilic polymers, absorbing water and contributing to product texture, viscosity, and stability. In addition, hydrocolloids serve as vital ingredient functionalities, supporting food properties such as shelf life, emulsion stability, and flavor release.

Hydrocolloids can enhance the sensory experience of food products by improving texture and mouthfeel. They can also help stabilize products, preventing separation and prolonging shelf life. Hydrocolloids function as gelling agents, offering the ability to form and maintain a specific shape in products such as jellies and custards. Hydrocolloids help maintain emulsions' homogeneity as stabilizers, preventing the separation of oil and water-based components.

The functionality of hydrocolloids can vary depending on the type of food product being produced and the desired properties. For example, in dairy products, hydrocolloids can be used to improve the creaminess and texture of ice cream. In meat products, hydrocolloids can help to bind ingredients together, improving the texture and moisture retention of the final product.

Due to their unique properties, hydrocolloids have gained widespread use across the food industry and are found in various products, including baked goods, sauces, dressings, and beverages. Businesses can use hydrocolloids to improve product quality and stand out in a competitive market.

10.2.5 Market Dynamics and Growth Drivers of Hydrocolloids

The hydrocolloids market is expected to experience significant growth in the coming years, primarily driven by the demand for convenience foods and the increasing preference for natural and organic products. The market is expected to reach $11.4 billion by 2025, growing at a CAGR of 5.3%.

The increasing use of hydrocolloids in the food and beverage industry as stabilizers, thickeners, gelling agents, and emulsifiers is expected to drive market growth. The demand for hydrocolloids in the pharmaceutical and cosmetics industries is also expected to contribute to market growth. Hydrocolloids are also preferred over synthetic ingredients due to their natural origin, which is expected to fuel market growth further.

The Asia Pacific region is expected to dominate the hydrocolloids market due to the increasing use of hydrocolloids in the food industry, particularly in countries like India and China. Furthermore, the region's growing demand for convenience and processed foods will drive market growth.

Market Dynamics

The hydrocolloids market is driven by various market dynamics, including:

- Increasing demand for processed and convenience foods
- Growing demand for natural and organic ingredients
- Rising consumption of functional foods
- Advancements in hydrocolloid technology
- Increasing demand for hydrocolloids in pharmaceuticals and cosmetics

However, certain factors may impede market growth, including:

- High cost of hydrocolloids compared to synthetic ingredients
- Concerns regarding the safety and regulatory compliance of hydrocolloids
- Volatility in raw material prices

Overall, the hydrocolloids market is poised for significant growth in the coming years, driven by various factors, including increasing demand for natural and organic products, advancements in hydrocolloid technology, and rising consumption of functional foods.

10.2.6 Regional Analysis of the Hydrocolloids Market

The global hydrocolloids market is expected to grow significantly in the coming years, with strong demand from industries such as food and beverages, pharmaceuticals, cosmetics, and personal care. The market is expected to be dominated by North America, followed closely by Europe and the Asia Pacific region. The increasing demand for processed food and rising health awareness among consumers is expected to drive market growth in these regions.

North America

The North American hydrocolloids market is expected to grow steadily in the coming years, driven by the rising demand for functional food and increasing pharmaceutical and personal care applications. The United States is the largest market for hydrocolloids in the region, owing to the substantial consumption of processed food and beverages and the presence of major market players such as Cargill, DuPont de Nemours, Inc., and Ashland Global Holdings Inc.

Europe

The European hydrocolloids market is expected to grow significantly in the coming years, driven by the increasing demand for natural food additives and functional products. The United Kingdom, Germany, France, and Italy are among the major markets in the region, with strong demand for hydrocolloids from the bakery and confectionery, dairy and frozen desserts, and meat and poultry industries.

Asia Pacific

The Asia Pacific hydrocolloids market is expected to witness the fastest growth in the coming years, driven by the increasing demand for convenience food, especially in China and India. Additionally, the rising health awareness among consumers and the growing trend of clean labeling are expected to drive the demand for natural hydrocolloids in the region. Key market players in the Asia Pacific region include Ingredion, Tate & Lyle PLC, and CP Kelco.

10.2.7 Future Projections and Emerging Trends in the Hydrocolloids Market

The hydrocolloids market is poised for significant growth in the coming years, driven by innovative applications, technological advancements, and shifting consumer preferences. As the food industry evolves, hydrocolloids are becoming increasingly important ingredients that enhance product quality, texture, and stability.

Innovative Applications

Hydrocolloids are increasingly used in innovative applications, such as in developing plant-based and functional food products. As consumers become more health-conscious and eco-friendly, the demand for sustainable and nutritious food products rises. Hydrocolloids play a key role in achieving this, as they offer unique functionalities that improve the sensory experience of plant-based products while increasing their nutritional value.

Technological Advancements

Technological advancements are also shaping the hydrocolloids market. As new technologies emerge, manufacturers are developing more efficient and cost-effective ways of producing hydrocolloid products. For example, fermentation technology advances enable hydrocolloid production from renewable resources, reducing the industry's environmental impact.

Changing Consumer Preferences

Consumer preferences are also driving growth in the hydrocolloids market. As consumers seek out food products that are convenient, healthy, and tasty, hydrocolloids are becoming increasingly important as ingredients that deliver on these attributes. Hydrocolloids improve various food products' texture, flavor, and stability, from ice cream to baked goods. As the food industry continues to evolve, hydrocolloids will play an increasingly vital role in meeting the changing needs of consumers.

The hydrocolloid market's future projections are bright, with significant growth potential and opportunities for innovation. As the food industry continues to evolve, hydrocolloids will remain a key ingredient in developing new and exciting food products that meet the changing needs of consumers.

10.2.8 Key Players and Competitive Landscape in the Hydrocolloids Market

The hydrocolloids market is highly competitive, with many players operating. Some of the key players in the market include Cargill Inc., DowDuPont Inc., Ingredion Incorporated, Ashland Global Holdings Inc., and Kerry Group, among others. These players collectively account for a significant share of the overall market.

Various factors, including technological advancements, changing consumer preferences, and competitive pressures, shape the market dynamics of the hydrocolloids market. In recent years, market players have focused on expanding their product portfolios and geographical presence to strengthen their market position.

For instance, in October 2020, CP Kelco, a leading hydrocolloids producer, announced plans to expand its gellan gum production capacity. The company also launched a new product line of gellan gum, providing innovative solutions to meet the increasing demand for clean-label and plant-based food and beverage products.

Another key player, Cargill Inc., has been focusing on collaborations and investments to expand its market reach. In August 2020, the company announced a partnership with five startups to develop innovative solutions for the food and agriculture industry, including hydrocolloid-based products. The company has also invested in research and development activities to introduce new hydrocolloid-based products.

Overall, the hydrocolloids market is expected to witness intense competition in the coming years, with market players adopting various strategies to gain a competitive edge. Players must focus on product innovation, partnerships, and strategic collaborations to stay ahead.

10.2.9 Regulatory Landscape and Challenges in the Hydrocolloids Market

The hydrocolloids market is subject to various regulatory frameworks and compliance requirements, which can challenge industry players. One of the main challenges is the lack of a unified regulatory framework across different regions and countries. It can result in confusion and inconsistencies in compliance requirements, especially for companies operating on a global scale.

Furthermore, regulatory bodies are increasingly focusing on food safety and consumer protection, which has led to stricter regulations and requirements for hydrocolloid manufacturers and users. Compliance with these regulations can be time-consuming and expensive, leading to additional costs for market players.

Another challenge facing the hydrocolloids market is the growing demand for natural and clean-label ingredients. Many consumers are wary of artificial additives and preservatives in their food and are seeking products with simple and recognizable ingredient lists. It has led to a shift in the market towards natural and plant-based hydrocolloids, which can be more expensive and challenging to source.

Additionally, the global COVID-19 pandemic has impacted the hydrocolloids market, with disruptions in supply chains and changes in consumer demand. It has further highlighted the importance of agility and resilience in the industry, as companies must adapt to changing market dynamics and consumer preferences.

Despite these challenges, the hydrocolloids market is expected to continue its growth trajectory in the coming years, driven by increasing demand for convenience foods, rising disposable incomes, and technological advancements. Companies that can navigate the regulatory landscape and capitalize on emerging trends will be well-positioned for success in the market.

Conclusion

The hydrocolloids market is expected to experience substantial growth in the coming years, driven by factors such as expanding food processing industries, changing consumer preferences, and technological advancements. Hydrocolloids are vital in enhancing food texture, stability, and functionality. With increasing demand for natural and organic food products, the market will likely shift towards plant-based hydrocolloids.

Investing in the hydrocolloids market can be a lucrative opportunity for businesses seeking to expand their product portfolio. However, staying informed about market dynamics and emerging trends is crucial to making informed investment decisions. Key players in the market should focus on product innovation, strategic collaborations, and mergers and acquisitions to strengthen their market position.

Looking Ahead

As the food industry continues to evolve, the demand for hydrocolloids is expected to increase. The market is projected to grow significantly in Asia Pacific and Latin America, driven by increasing urbanization and rising disposable income. The emergence of new applications and product formulations will likely create new market opportunities and challenge existing players.

Overall, the hydrocolloids market presents a promising investment opportunity for businesses seeking to expand their foothold in the food industry. With the right strategy and market insights, businesses can capitalize on the growth prospects and contribute to the market›s overall development.

FAQ

- **Q: How do hydrocolloids contribute to the food industry?**

  Hydrocolloids, as hydrophilic polymers, possess water-absorbing properties that make them suitable for various applications in the food industry.

- **Q: What role do viscosity modifiers play in hydrocolloids?**

  Viscosity modifiers are important in hydrocolloids as they impact the texture and mouthfeel of food products, enhancing their quality.

- **Q: What are the market dynamics and growth drivers of hydrocolloids?**

  Market dynamics of hydrocolloids include factors driving market growth, current trends, market opportunities, and challenges for industry players.

- **Q: Can you provide a regional analysis of the hydrocolloids market?**

  The hydrocolloids market will be analyzed regionally, focusing on key regions such as North America, Europe, Asia Pacific, and Latin America, highlighting market trends and growth potential.

- **Q: What are the future projections and emerging hydrocolloid market trends?**

  Future projections and emerging hydrocolloid market trends will be discussed, exploring innovative applications, technological advancements, and changing consumer preferences.

- **Q: Who are the key players in the hydrocolloids market?**

  An overview of the key players and competitive landscape in the hydrocolloids market will be provided, discussing major industry players, market share, strategies, and recent developments.

- **Q: What is the regulatory landscape and challenges the hydrocolloids market faces?**

  The hydrocolloid market's regulatory landscape and challenges, including regulatory frameworks, certifications, and compliance requirements, will be examined for market players.

- **Q: Can you provide a conclusion summarizing the key insights from the article?**

  The conclusion will summarize key insights from the article, highlighting the hydrocolloid market's growth prospects and future projections emphasizing the importance of informed investment decisions.

CHAPTER 11

DIY and Home Uses

PART 11.1

Making Homemade Gels & Thickeners with Hydrocolloids

Have you ever wondered how to make homemade gels and thickeners that are smooth, stable, and packed with flavor? The answer lies in hydrocolloids, a fascinating ingredient class revolutionizing cooking. Hydrocolloid technology has been used for decades in the food industry to create innovative textures and mouthfeel in a wide range of products. But the potential of hydrocolloids goes well beyond the food industry and extends to pharmaceuticals, cosmetics, and more.

In this section, we will explore the basics of hydrocolloid technology and the benefits and uses of hydrocolloids in creating homemade gels and thickeners that will impress your guests.

Core Insights:

- ☑ Hydrocolloid technology has transformed the food industry by enabling the creation of unique textures and flavors.
- ☑ Hydrocolloids are versatile ingredients that can create homemade gels and thickeners.
- ☑ The benefits of using hydrocolloids include improved texture, stability, and flavor enhancement.
- ☑ Hydrocolloid technology is utilized in various industries, with vast potential applications.
- ☑ Incorporating hydrocolloids into your culinary adventures can unlock your creativity and elevate your dishes to a new level.

11.1.1 Homemade Hydrocolloid Gels

You can create homemade hydrocolloid gels perfect for adding texture and flavor to your culinary creations. The key to making a smooth and stable gel is using the right combination of ingredients and following the proper techniques.

One popular hydrocolloid for gelling is agar-agar, a natural seaweed extract. To make a basic agar gel, you'll need:

Table 81: Homemade Hydrocolloid Gels, Ingredients

INGREDIENT	AMOUNT
Agar-agar powder	2 teaspoons
Water or other liquid	1 cup

Combine the agar-agar powder with the liquid in a small saucepan and simmer. Cook for a few minutes until the agar is fully dissolved. Pour the mixture into a container and let it cool until it becomes a gel.

You can experiment with different liquids and flavorings to create unique gels. Try using fruit juices, purees, or extracts to add flavor to your gel.

Another popular gelling hydrocolloid is gelatin, derived from animal collagen. Gelatin is commonly used in desserts such as panna cotta and marshmallows. To make a basic gelatin gel, you'll need:

Table 82: Basic Gelatin Gel Ingredients

INGREDIENT	AMOUNT
Gelatin powder	1 tablespoon
Water or other liquid	1 cup

Bloom the gelatin by soaking it in cold water for a few minutes. Heat the liquid in a small saucepan until it›s hot but not boiling. Remove from heat and add the bloomed gelatin, stirring until it dissolves. Pour the mixture into a container and let it cool until it sets.

You can also combine agar and gelatin to create a gelled mixture with a unique texture and flavor.

- Experiment with different liquids and flavorings to create unique gels.
- Bloom gelatin by soaking it in cold water before adding hot liquid.
- Combining agar and gelatin can create a unique texture and flavor.

By mastering the art of making homemade hydrocolloid gels, you can add new dimensions to your culinary creations and impress your guests with your creativity.

11.1.2 Enhancing Culinary Creativity

Hydrocolloids offer endless possibilities when it comes to elevating your culinary creativity. With their unique properties and versatile uses, they can create textures and flavors that were once impossible to achieve. Here are some innovative ways to use hydrocolloids in your cooking:

- **Thicken sauces and dressings:** Hydrocolloids such as xanthan gum and guar gum can thicken sauces and dressings without altering their taste. They can also stabilize emulsions, preventing them from separating.
- **Create foam:** By incorporating a small amount of hydrocolloid, such as lecithin, into liquids and blending them, you can create a foam or froth that adds a light and airy texture to dishes.
- **Make noodles:** Using hydrocolloids such as konjac gum and gellan gum, you can make noodles low in calories and carbohydrates, making them a perfect choice for those on a low-carb or gluten-free diet.
- **Produce spherical foods:** Hydrocolloids such as sodium alginate and calcium chloride can create flavorful liquid spheres that burst in your mouth.
- **Add crunch:** Hydrocolloids such as methylcellulose can create a crispy coating on foods such as meats and vegetables, adding a satisfying crunch.

By incorporating these techniques into your cooking, you can take your culinary creations to the next level with the help of hydrocolloids. Don't be afraid to experiment and try new things!

11.1.3 Hydrocolloroids Application in do-it-yourself (DIY) and home settings

Hydrocolloids, often associated with their use in the food industry, also find applications in various do-it-yourself (DIY) and home settings. Here are some DIY and home uses of hydrocolloids:

Gel Air Fresheners

Hydrocolloid: Agar-agar or carrageenan.

DIY Use: Dissolve agar-agar or carrageenan in water, add a few drops of your favorite essential oil for fragrance, and let it set to create a gel air freshener. The hydrocolloid helps maintain the gel consistency, providing a stable medium for the fragrance.

Homemade Jams and Jellies

Hydrocolloid: Pectin.

DIY Use: Pectin is a natural hydrocolloid found in fruits. To make homemade jams and jellies, you can use commercially available pectin or extract it from fruits rich in this substance. Pectin forms a gel with sugar and acid, giving your spreads the desired texture.

DIY Face Masks

Hydrocolloid: Konjac gum.

DIY Use: Konjac gum can create hydrating and soothing DIY face masks. Mix konjac gum with water to form a gel-like consistency and add other beneficial ingredients such as aloe vera or essential oils. Apply the mask to your face for a refreshing experience.

Plant-Based Gel Desserts

Hydrocolloid: Agar-agar or carrageenan.

DIY Use: Experiment with agar-agar or carrageenan to make vegan-friendly gel desserts. Dissolve the hydrocolloid in fruit juices or coconut milk, heat the mixture until it thickens, then let it cool to form delicious plant-based jellies and puddings.

Seed Tape for Gardening

Hydrocolloid: Guar gum.

DIY Use: Create a seed tape by making a paste with guar gum and water. Dip paper strips in the paste, place seeds at intervals, and allow it to dry. The resulting seed tape can be planted directly in the soil, ensuring evenly spaced seedlings.

Gluten-Free Baking

Hydrocolloid: Xanthan gum.

DIY Use: Xanthan gum is a common ingredient in gluten-free baking. It helps improve the texture and structure of baked goods made with gluten-free flour. Add small amounts to your recipes, adjusting until you achieve the desired consistency.

Thickening Homemade Sauces and Soups:

Hydrocolloid: Guar gum or xanthan gum.

DIY Use: Use guar or xanthan gum as thickeners when making sauces or soups. Start with a small amount, as these hydrocolloids are potent, and add it slowly until you reach the desired thickness.

DIY Plant-Based Yogurt

Hydrocolloid: Locust bean gum.

DIY Use: Enhance the texture and stability of homemade plant-based yogurt with locust bean gum. Experiment with plant-based milk and locust bean gum for a creamy and satisfying yogurt alternative.

Crafting Edible Bubbles

Hydrocolloid: Agar-agar.

DIY Use: Create a solution using agar-agar and water to make edible bubbles. Blow the solution into bubbles; enjoy these unique and edible treats once it solidifies. Agar-agar provides the necessary structure for the bubbles.

DIY Gel Ice Packs

Hydrocolloid: Agar-agar or carrageenan.

DIY Use: Mix agar-agar or carrageenan with water to make gel ice packs. These hydrocolloids help create a flexible gel even when frozen, making it ideal for DIY ice packs.

Following recommended proportions and instructions is essential to ensure success when working with hydrocolloids in DIY projects. The unique properties of each hydrocolloid contribute to various applications, adding versatility to your creative and practical projects at home.

11.1.4 Tips and Tricks for Success

Working with hydrocolloids can initially seem daunting, but you can achieve exceptional results with the right techniques and methods. Here are some valuable tips and tricks to keep in mind when using hydrocolloids in your recipes:

Measure Accurately

Hydrocolloids can be sensitive to even small changes in measurements. Measure accurately using a digital scale to ensure your recipes turn out as intended.

Hydrate Properly

Hydrocolloids need to be hydrated properly to work effectively. Follow the manufacturer's instructions on the correct amount of water to hydrate your chosen hydrocolloid.

Use Gradually

When adding hydrocolloids to your recipe, add them gradually while constantly stirring to avoid clumping. It will help you incorporate the hydrocolloid into your mixture more evenly.

Avoid Heat

High temperatures can alter the properties of hydrocolloids, so it's important to avoid overheating. Ensure to follow temperature guidelines specified in the recipe or by the hydrocolloid manufacturer.

Experiment and Have Fun

Hydrocolloids offer endless possibilities for culinary creativity. Don't be afraid to experiment and try different hydrocolloids and techniques to achieve unique and exciting results.

Hydrocolloid Adhesives

Although primarily used in wound care and bandages, hydrocolloid adhesives can also be used in food preparation to create stunning visual effects. You can attach edible decorations to your dishes for a show-stopping presentation using a small amount of hydrocolloid adhesive.

11.1.5 Tips and Tricks for Success

Working with hydrocolloids requires a certain level of skill and knowledge to ensure consistent results. Here are some valuable tips and tricks to help you successfully incorporate hydrocolloids into your culinary adventures:

Measure accurately: Hydrocolloids should be measured by weight, not volume, for precise results.

Use the right tools: An immersion blender must disperse hydrocolloids in liquids evenly.

Hydrate slowly: When adding hydrocolloids to liquids, sprinkle them gently on the surface and allow them to hydrate for a few minutes before blending.

Mix thoroughly: Ensure that hydrocolloids are fully dispersed by blending for several minutes, and then allow them to rest for a few hours to hydrate fully.

Be patient: Hydrocolloids can thicken fully for several hours or overnight, so prepare your recipes beforehand.

Don't overuse: Using too much hydrocolloid can result in a gummy or rubbery texture, so use in moderation.

Experiment: Try different types of hydrocolloids and techniques to find what works best for your recipes.

Hydrocolloid adhesives can also be used for stunning food preparation and presentation effects. For example, you can use a hydrocolloid adhesive to attach edible flowers or decorations to your dishes.

11.1.6 Exploring Beyond the Kitchen

Hydrocolloid technology has revolutionized various industries beyond the culinary world. The unique properties of hydrocolloids make them ideal for use in pharmaceuticals, cosmetics, and other applications.

Hydrocolloid dressings have become a game-changer in wound care, providing a moist and protective environment for faster healing and reduced scarring. Additionally, hydrocolloid-based skincare products have gained popularity for their effectiveness in treating acne and improving skin texture.

The pharmaceutical industry uses hydrocolloids as binders, disintegrants, and suspending agents in tablet formulations. They also serve as a delivery mechanism for controlled-release drugs and vaccines.

Hydrocolloids have significantly created various products, including paints, adhesives, and personal care items. They have also helped reduce waste in industries such as oil and gas by aiding in recovering valuable resources.

The versatility of hydrocolloids makes them an essential component in many industries, from food to cosmetics and beyond.

Conclusion

In summary, incorporating hydrocolloids into your culinary creations can elevate your dishes to a new level. With their unique properties and versatility, hydrocolloids offer countless possibilities for creating homemade gels and thickeners in the kitchen. Remember to explore the different types of hydrocolloids available, their specific applications, and the dos and don›ts of working with these ingredients for optimal results.

Hydrocolloids, commonly known for their role in the food industry, offer versatile applications in DIY and home settings. From crafting homemade jams using pectin, creating plant-based gel desserts with agar-agar or carrageenan, and making DIY face masks using konjac gum, these hydrocolloids add functionality beyond the kitchen. Hydrocolloids showcase their adaptability and value in various creative and practical home applications, whether used in gluten-free baking, thickening sauces, or developing unique projects like edible bubbles or gel ice packs.

Additionally, remember the importance of proper storage and handling to maintain the quality and safety of hydrocolloid products. Beyond the kitchen, hydrocolloid technology is utilized in various industries, such as pharmaceuticals and cosmetics, highlighting their wide range of uses and benefits. So, whether you're a professional chef or a home cook, don't be afraid to experiment with hydrocolloids in your recipes. Your guests will thank you for the exceptional dining experience!

FAQ

- **What are hydrocolloids, and why are they used in DIY and home applications?**

  Hydrocolloids are substances that form gels in the presence of water, commonly used in the food industry. In DIY and home settings, they are employed for their versatile properties, such as creating gels, enhancing textures, and providing stability in various projects.

- **Can hydrocolloids be used in non-food DIY projects?**

  Yes, hydrocolloids have applications beyond the kitchen. They are used in non-food DIY projects such as crafting gel air fresheners, homemade face masks, and even creating unique items like edible bubbles.

- **How can I use hydrocolloids to make homemade jams and jellies?**

  Pectin, a natural hydrocolloid, can make homemade jams and jellies. It can be obtained commercially or extracted from fruits to achieve the desired gel-like consistency.

- **Which hydrocolloid is suitable for DIY face masks, and what benefits does it offer?**

  Konjac gum is often used in DIY face masks due to its ability to form a hydrating and soothing gel-like consistency. When mixed with water, it can enhance the overall quality of the face mask.

- **What hydrocolloids are commonly used in gluten-free baking, and how do they contribute to the texture of baked goods?**

  Xanthan gum is commonly used in gluten-free baking. It helps mimic the elastic properties of gluten, providing structure and improving the texture of baked goods made with gluten-free flour.

- **Can hydrocolloids be used to thicken homemade sauces and soups?**

  Yes, hydrocolloids such as guar gum or xanthan gum can be added gradually to thicken homemade sauces and soups to the desired consistency.

FREQUENTLY ASKED QUESTIONS (FAQS) CONTINUED

How can I create plant-based gel desserts at home using hydrocolloids?

Agar-agar or carrageenan can be dissolved in fruit juices or coconut milk to create the base for plant-based gel desserts such as jellies and puddings.

How can hydrocolloids like guar gum be used to make seed tapes in gardening?

When mixed with water, Guar gum can create a paste for making seed tapes. This paste ensures evenly spaced seedlings when the tape is planted in the soil.

Are there any safety considerations when working with hydrocolloids in DIY projects?

While hydrocolloids are generally safe, following recommended proportions and handling instructions is important. Ensure you use food-grade hydrocolloids for projects involving contact with skin or ingestion.

Can hydrocolloids create edible bubbles, and how is this done?

Yes, agar-agar can be mixed with water to create a solution for crafting edible bubbles. When blown, the solution solidifies, allowing for a unique and enjoyable treat.

PART 11.2

Practical Tips for Using Hydrocolloids in Everyday Cooking

Hydrocolloids are an essential component of the food industry, serving as gelling agents, stabilizers, thickening agents, and water-binding agents. They also act as texture modifiers, enhancing the quality and appearance of various dishes. Understanding the functional properties of hydrocolloids is critical for achieving desired outcomes in cooking, and ingredient technology plays a vital role in their effectiveness. This article will explore practical tips for incorporating hydrocolloids in everyday cooking, from selecting the appropriate hydrocolloid for specific dishes to troubleshooting common issues.

Core Insights:

- ☑ Hydrocolloids are crucial in the food industry as gelling, stabilizers, thickening, and water-binding agents.
- ☑ They can also act as texture modifiers, improving the quality and appearance of various dishes.
- ☑ Understanding the functional properties of hydrocolloids is essential for achieving desired outcomes in cooking.
- ☑ Ingredient technology plays a vital role in the effectiveness of hydrocolloids.
- ☑ This article will provide practical tips for incorporating hydrocolloids in everyday cooking, from selecting the appropriate hydrocolloid for specific dishes to troubleshooting common issues.

11.2.1 Choosing the Right Hydrocolloid for Your Dish

Choosing the right hydrocolloid can be daunting, given the range of options available. It's crucial to understand the characteristics of various hydrocolloids to ensure optimal results in your cooking. Hydrocolloids can be used as gelling, stabilizers, thickening, and water-binding agents, depending on the desired outcome in your recipe.

Let's take a closer look at some of the commonly used hydrocolloids, their functional properties, and suitable applications:

Table 83: Hydrocolloid Functional Properties and Applications

HYDROCOLLOID	FUNCTIONAL PROPERTIES	APPLICATIONS
Gelatin	Gelling, stabilizing, water binding	Gummy candies, mousses, creams, marshmallows
Agar	Gelling, stabilizing, thickening	Jellies, custards, fruit fillings, ice cream, soups
Pectin	Gelling, thickening	Jams, jellies, fruit fillings
Carrageenan	Stabilizing, thickening	Dairy products, chocolate milk, soy milk, ice cream
Xanthan gum	Thickening, stabilizing, water binding	Sauces, salad dressings, gravies, soups

It's important to note that the amount and type of hydrocolloid used can significantly affect the final product's texture. Excess hydrocolloids can produce a slimy or rubbery texture, while too little can lead to a runny or watery consistency.

Food science and ingredient technology also play a significant role in choosing the right hydrocolloid. PH, temperature, and salt concentration can affect a hydrocolloid›s functional properties and performance.

Therefore, when choosing the right hydrocolloid, consider the desired outcome, the recipe's requirements, and the environmental conditions under which it will be prepared.

11.2.2 Enhancing Texture with Hydrocolloids

Hydrocolloids are versatile ingredients that can dramatically improve the texture of various dishes. In this section, we will explore the role of hydrocolloids as texture modifiers in cooking and highlight specific hydrocolloids that excel in improving texture. We will also discuss their functional properties based on food science and ingredient technology.

Why Texture Matters in Cooking

Texture plays a crucial role in the enjoyment and overall quality of food. A dish with the right texture can enhance its flavor and appearance, making it more appealing to the palate. Texture is also important in creating a satisfying eating experience and can impact how we perceive a dish's freshness, quality, and overall appeal.

Hydrocolloids as Texture Modifiers

Hydrocolloids can be used as texture modifiers in cooking to enhance the mouthfeel and texture of various dishes. They can create a wide range of textures, from creamy and smooth to chewy and crunchy. Hydrocolloids alter how water interacts with other ingredients in a recipe, creating the desired consistency and texture.

Some hydrocolloids that excel in improving texture include:

Table 84: Hydrocolloid Texture Properties

HYDROCOLLOID	TEXTURE PROPERTIES
Gellan gum	Creates a firm, brittle gel with a clean break
Agar agar	Forms a firm, elastic gel with a smooth texture
Carrageenan	Produces a soft, elastic gel that can withstand freezing and thawing
Xanthan gum	Provides a creamy, smooth texture and improves mouthfeel

Functional Properties of Hydrocolloids

The functional properties of hydrocolloids are based on their physical and chemical properties, which determine their effectiveness in various applications. Hydrocolloids can act as gelling, stabilizers, thickening, and water-binding agents, among other functions. These properties are based on their ability to influence the movement of water, control the viscosity of a solution, and alter the texture, appearance, and flavor of a dish.

For example, gelling agents like gellan gum and agar agar can form gels that provide structure and stability to a dish. In contrast, thickening agents like xanthan gum can improve the viscosity and mouthfeel of a sauce or dressing. Water-binding agents like carrageenan can retain moisture in a dish, preventing it from drying out, while stabilizers like guar gum can prevent the separation of ingredients in a product.

11.2.3 Tips for Incorporating Hydrocolloids into Sauces and Soups

Hydrocolloids are versatile ingredients that can improve the texture and consistency of sauces and soups. By acting as gelling agents, stabilizers, thickening agents, and water-binding agents, hydrocolloids can help achieve the desired outcomes in your recipes. Here are some tips for incorporating hydrocolloids into your sauces and soups:

Varying Viscosity

Depending on the type of sauce or soup you are preparing, you may want to adjust the viscosity to create the desired texture. For thicker sauces and soups, you may use hydrocolloids like xanthan gum and guar gum, which are powerful thickening agents. For thinner sauces and soups, you may use hydrocolloids like carrageenan and agar, which can help create a smooth and even texture.

Matching Flavor Profiles

When selecting hydrocolloids for your sauces and soups, it is important to consider their flavor profiles. For example, if you are making a savory soup, you may want to use a hydrocolloid like kappa carrageenan, which has a neutral flavor profile. On the other hand, if you are making a sweet dessert sauce, you may want to use a hydrocolloid like pectin, which is derived from fruits and can add a fruity flavor.

Proper Dispersal

Hydrocolloids should be dispersed evenly throughout your sauce or soup to achieve the desired texture. One way to do this is by blending the hydrocolloid with a small amount of liquid before adding it to the recipe. It allows the hydrocolloid to dissolve more easily and mix evenly with the rest of the ingredients. Another technique is to sprinkle the hydrocolloid slowly and evenly into the sauce or soup while whisking vigorously to prevent lumps from forming.

Temperature Considerations

Hydrocolloids can thicken sauces and soups at different temperatures. Some hydrocolloids, like xanthan gum and guar gum, can thicken at room temperature, while others, like agar and carrageenan, require higher temperatures to activate. Please pay attention to the specific hydrocolloid's temperature requirements to ensure it is used correctly.

Sauces and soups can benefit greatly from hydrocolloids to improve texture and consistency. By following these tips, you can achieve the desired outcomes in your recipes and elevate your cooking skills.

11.2.4 Creating Creamy and Rich Desserts with Hydrocolloids

When creating luxurious and indulgent desserts, hydrocolloids can be a game-changer. These versatile ingredients can serve as gelling, stabilizers, thickening, and water-binding agents, resulting in desserts with smooth textures and improved mouthfeel. Here are some tips and techniques for incorporating hydrocolloids into your dessert recipes:

Using Gelatin for Mousse

Gelatin is a popular hydrocolloid used in many dessert recipes. It is an excellent gelling agent and thickener, which makes it perfect for creating smooth and creamy mousses. To use gelatin in your mousse recipe, dissolve it in warm water and mix it with your other ingredients. The resulting mousse will have a light and airy texture that will impress.

Thickening with Xanthan Gum

If you want to thicken your dessert without adding extra flavor or color, xanthan gum might be the perfect hydrocolloid. This white powder is a powerful thickener and stabilizer, which means it can improve the texture of your ice cream, pudding, or custard without altering its taste or appearance. To use xanthan gum, sprinkle it evenly over your dessert mixture and whisk thoroughly until the desired consistency is achieved.

Creating Creamy Vegan Desserts with Agar-Agar

Agar-agar is a plant-based hydrocolloid derived from seaweed. It is a popular alternative to gelatin in vegan cooking and can be used as a gelling agent, thickener, and stabilizer. Agar-agar is particularly useful in creating creamy and rich vegan desserts like cheesecake and panna cotta. To use agar-agar in your recipe, dissolve it in hot liquid and set it in the fridge. The resulting dessert will have a smooth and velvety texture that is sure to please.

Improving Mouthfeel with Guar Gum

Guar gum is a versatile hydrocolloid that can be used as a thickener, stabilizer, and emulsifier. It is particularly useful in improving the mouthfeel of your dessert by creating a smooth and creamy texture. Guar gum is often used in ice cream and other frozen desserts to prevent ice crystallization and improve scoopability. To use guar gum in your recipe, sprinkle it over your dessert mixture and blend until fully incorporated.

Experimenting with Hydrocolloids

Don't be afraid to experiment with hydrocolloids in your dessert recipes. You never know what kind of delicious creation you might remember! Remember to start with small amounts and gradually increase the quantity until you achieve the desired texture consistency. With some trial and error, you'll soon be whipping up decadent and unforgettable desserts that will leave your guests wanting more.

11.2.5 Hydrocolloids: Transforming Beverages and Drinks

Hydrocolloids significantly impact the texture and appearance of beverages and drinks. They are often used as stabilizers, thickening water-binding to create unique and appealing textures. The functional properties of hydrocolloids offer numerous benefits in the food industry, and their application in beverages is no exception. Understanding the role of hydrocolloids in this context can enhance your beverage-making skills and take your creations to the next level.

Hydrocolloids in Beverages

Hydrocolloids are used in various beverages, from carbonated soft drinks to fruit juices. They can be incorporated into beverages to improve mouthfeel, texture, and appearance. For example, gellan gum is often used in energy drinks to create a smooth, thick texture, while pectin is added to fruit juices to enhance their thickness and pulpiness.

Stabilizers like xanthan gum and carrageenan are used in dairy-based beverages to prevent separation and improve the overall texture. Additionally, hydrocolloids such as guar gum and locust bean gum can improve the stability of soy and nut-based beverages.

Creating Unique Textures with Hydrocolloids

Hydrocolloids can create unique textures in beverages, such as foams, gels, and suspensions. For example, agar-agar is often used in molecular gastronomy to create foam and gel-like textures. In contrast, carrageenan can create a smooth, creamy mouthfeel in dairy-based beverages.

To create a suspension, hydrocolloids such as xanthan gum and gum arabic can be added to ingredients like fruit juices or coffee to prevent separation and create a more visually appealing drink.

Practical Tips for Using Hydrocolloids in Beverages

When using hydrocolloids in beverages, it is important to remember their functional properties and how they interact with other ingredients. Here are some practical tips for using hydrocolloids in your drink creations:

- Start with small quantities of hydrocolloids and gradually increase as necessary to achieve the desired texture.
- Consider the pH and temperature of the beverage, as these can affect the effectiveness of certain hydrocolloids.
- Ensure the hydrocolloid is completely dissolved and dispersed throughout the beverage to prevent clumping or uneven texture.
- Experiment with different hydrocolloids to achieve unique textures and flavors.

By following these tips and understanding the properties of hydrocolloids, you can create delicious, visually appealing, and unique beverages that are sure to impress.

11.2.6 Hydrocolloids for Gluten-Free and Vegan Cooking

Hydrocolloids are highly versatile ingredients used for gluten-free and vegan cooking. They act as gelling, stabilizers, thickening, and water-binding agents, ensuring that desired textures and structures are achieved without animal products or gluten.

Gelling agents: Hydrocolloids can create gels in vegan desserts, such as panna cotta or jellies. Agar-agar, carrageenan, and gellan gum are commonly used for this purpose.

Stabilizers: Hydrocolloids can stabilize emulsions in vegan sauces and dressings. Xanthan gum, guar gum, and locust bean gum are popular choices for stabilizing emulsions and preventing separation.

Thickening agents: Hydrocolloids can thicken gluten-free and vegan soups, sauces, and gravies. Cornstarch, arrowroot, and tapioca starch are effective thickening agents that can improve the viscosity and mouthfeel of different dishes.

Water binding agents: Hydrocolloids can bind water and improve the texture of gluten-free and vegan bakery products. Psyllium husk, methylcellulose, and hydroxypropyl methylcellulose can replace the gluten›s water-binding properties, ensuring the dough has the necessary elasticity to rise and create fluffy baked goods.

It's important to consider the functional properties of hydrocolloids when using them in gluten-free and vegan cooking. Some hydrocolloids may affect dishes' taste, texture, or appearance, so it's best to experiment with small amounts before using them in large quantities or critical recipes.

11.2.7 Overcoming Common Challenges When Using Hydrocolloids

While hydrocolloids can enhance your cooking experience, there are some common challenges that you may encounter when using them. Understanding these challenges and learning to overcome them can help you to achieve the desired results with your recipes.

Challenge 1: Inconsistent Results

One common challenge when using hydrocolloids is achieving consistent results. It can be due to several factors, including variations in ingredient quality and environmental conditions (such as temperature). To overcome this challenge:

- Use high-quality ingredients and purchase from reputable suppliers
- Measure ingredients accurately
- Follow recipe instructions carefully
- Keep track of environmental conditions and adjust the recipe as needed

Challenge 2: Clumps or Lumps

Another challenge when using hydrocolloids is avoiding clumps or lumps in your recipes. It is often caused by improper dispersion or hydration of the hydrocolloid. To overcome this challenge:

- Slowly add hydrocolloids to the liquid while mixing vigorously
- Allow the hydrocolloid to hydrate completely before heating or cooling the mixture
- Use a blender, immersion blender, or whisk to disperse the hydrocolloid evenly
- If clumps still occur, strain the mixture through a fine mesh strainer

Challenge 3: Undesired Texture

Hydrocolloids can also create an undesired texture when used improperly. It can include a slimy or gummy texture or a too-thick or too-thin texture. To overcome this challenge:

- Follow recipe instructions carefully
- Use the appropriate hydrocolloid for the desired texture
- Use the right amount of hydrocolloid for the recipe
- Adjust the recipe as needed based on the desired outcome

By overcoming these common challenges when using hydrocolloids, you can achieve the desired results in your recipes and enhance your everyday cooking experience.

11.2.8 Safety Considerations and Storage of Hydrocolloids

While hydrocolloids are safe for consumption and widely used in the food industry, some safety considerations should be considered when handling and storing them. Following the guidelines below will help ensure the effectiveness of hydrocolloids in your cooking and prevent any potential hazards.

Safety Considerations

When handling hydrocolloids, it is essential to take proper precautions to avoid any harm. Here are some safety considerations to keep in mind:

- Hydrocolloids are generally safe, with no known hazards or adverse effects. However, contact with some hydrocolloids can cause mild skin irritation, so gloves are recommended when handling them.
- When adding hydrocolloids to hot liquids, use caution to prevent burns. Slowly sprinkle them into the liquid while stirring to avoid clumping or splashing.
- Some hydrocolloids can create dust particles when poured out of their packaging, which can cause respiratory issues when inhaled. Using a mask or working in a well-ventilated area is recommended to avoid potential harm.

Storage

Proper storage is crucial in maintaining the quality and effectiveness of hydrocolloids. Here are some storage guidelines to keep in mind:

Table 85: Hydrocolloid Storage Conditions and Shelf Life

HYDROCOLLOID	STORAGE CONDITIONS	SHELF LIFE
Agar	Airtight container in a dry place away from direct sunlight	2-3 years
Carrageenan	Airtight container in a dry place away from direct sunlight	2-3 years
Gellan gum	Airtight container in a dry place away from direct sunlight	2-3 years
Xanthan gum	Airtight container in a cool, dry place away from direct sunlight	2 years

Keeping hydrocolloids in an airtight container in a dry, cool place away from direct sunlight is crucial to prevent moisture and heat from affecting their properties. Always check the expiration date on the packaging and discard any expired hydrocolloids.

By following these safety considerations and storage guidelines, you can ensure hydrocolloids› safe and effective use in your everyday cooking.

11.2.9 Exploring the Future Applications of Hydrocolloids

Hydrocolloids continually evolve in line with ingredient technology, and the future looks promising. The food industry can expect to see even more innovative uses of hydrocolloids that will aid in maintaining food quality, reducing costs, and improving consumer experiences. Here are some potential future applications of hydrocolloids:

Improving shelf-life

Hydrocolloids can play a significant role in extending the shelf-life of food products. They can be used to reduce water activity, which limits the growth of bacteria and fungi, leading to increased food safety. They can also work as antioxidizing agents, slowing down the oxidation of foods and preserving their quality for a longer period. Future research will likely focus on developing hydrocolloids that can work in combined effect to improve shelf-life further.

Reducing fat and calorie content

Consumers are increasingly seeking healthy and low-calorie food options. Hydrocolloids can help reduce fat and calorie content in food products while preserving their taste and texture. They can lower the fat content of dairy, meat, and baked goods while maintaining their desired properties. In the future, we could see more extensive use of hydrocolloids to help create healthier food products with lower calories and fat content.

Creating plant-based meat substitutes

Plant-based meat substitutes are becoming increasingly popular, and hydrocolloids could be crucial in their creation. Hydrocolloids can create plant-based meats with similar textures to traditional meats, giving consumers a more familiar experience. They can also help improve plant-based meats' mouthfeel and overall sensory experience, making them more appealing to a wider audience. As the demand for plant-based alternatives grows, we can expect to see even more exciting applications of hydrocolloids in this area.

Enhancing sustainability

Hydrocolloids can be extracted from various sources, including seaweed, algae, and plants, making them a sustainable alternative to animal-derived ingredients. They can also help reduce food waste by preserving food for longer periods, reducing the need to throw away spoiled food. In the future, hydrocolloids may play an essential role in promoting sustainability in food production and reducing its environmental impact.

The future looks bright for hydrocolloids in the food industry. Hydrocolloids are set to revolutionize how we cook and eat, from reducing costs to improving food safety, enhancing sustainability, and creating healthier food products. It will be exciting to see how these applications develop further and contribute to a more sustainable and healthy food industry.

Conclusion

In conclusion, hydrocolloids play a crucial role in everyday cooking, particularly in the food industry. Gelling, stabilizers, thickening, and water-binding agents are essential in achieving the desired textures and improving food quality. Moreover, hydrocolloids function as texture modifiers and provide a range of functional properties based on food science and ingredient technology.

Understanding their science and technology is crucial to making the most of hydrocolloids in cooking. The ability to select the appropriate hydrocolloid for different recipes and overcome common challenges when using them is essential. Furthermore, safety considerations and proper storage are also crucial to maintaining their quality and effectiveness. As we look to the future, hydrocolloids are expected to play an even more significant role in the food industry as emerging trends and advancements in ingredient technology expand their use in innovative ways.

Enhancing Everyday Cooking

Incorporating hydrocolloids into your everyday cooking can enhance your recipes' flavors and textures and bring your dishes to the next level. By delving deeper into hydrocolloids and understanding how they function as gelling, stabilizers, thickening, and water-binding agents, you can select the appropriate hydrocolloid for your recipe and experiment with different textures and structures.

We hope this article has provided practical tips and insights into hydrocolloid use in cooking. Applying this article's knowledge and techniques, you can create delicious and visually appealing dishes that impress your guests and enhance your everyday cooking.

FAQ

How do I choose the right hydrocolloid for my dish?

Selecting the appropriate hydrocolloid for your dish depends on the desired outcome and the characteristics of the hydrocolloid. Consider whether you need a gelling agent, stabilizer, thickening agent, or water-binding agent. Understanding the functional properties of different hydrocolloids and how they interact with other ingredients can help you make informed choices.

Can hydrocolloids be used in gluten-free and vegan cooking?

Yes, hydrocolloids are commonly used in gluten-free and vegan cooking. They serve as gelling, stabilizers, thickening, and water-binding agents, ensuring the desired textures and structures in recipes that adhere to gluten-free and vegan dietary restrictions.

What are some common challenges when using hydrocolloids in cooking?

Some common challenges when using hydrocolloids include achieving the right consistency, preventing clumping, and maintaining stability over time. Following proper techniques and dosage recommendations is important to overcome these challenges. Understanding the science and technology behind hydrocolloids can help troubleshoot any issues.

How should hydrocolloids be stored to maintain their quality?

Hydrocolloids should be stored in a cool, dry place away from direct sunlight. Keeping them in airtight containers is best to prevent moisture and air exposure, which can affect their effectiveness. Following proper storage guidelines will help maintain the quality and longevity of hydrocolloids.

What are the future applications of hydrocolloids in the food industry?

The future applications of hydrocolloids in the food industry are continually evolving. With advancements in ingredient technology, hydrocolloids may be used innovatively to create new textures, improve food quality, and contribute to developing sustainable and plant-based alternatives. Stay tuned for exciting developments in the world of hydrocolloids!

FUNCTIONS OF HYDROCOLLOIDS in Various Food Products

HYDROCOLLOID FUNCTION	APPLICATION
Adhesive	Bakery glazes
Binding agent	Sausage
Bulking agent	Dietetic foods
Clarifying agent	Beer and wine
Cloud agent	Fruit juice
Coating agent	Confectionary
Emulsifying	Salad dressings
Encapsulate	Powdered fixed flavors
Film forming	Sauge casings and protective coatings
Flocculating	Wine
Foam stabilizer	Whipped toppings and beers
Gelling agent	Pudding, desserts, aspics and mousses
Ice Crystal Inhibiting	Ice creams and sugar syrups
Mold release	Confectionary gum drops and jelly candy
Protective colloid	Flavor emulsions
Stabilize	Beer, Mayonnaise
Swelling agent	Processed meats
Suspending agent	Chocolate milk
Syneresis inhibitor	Cheese and frozen foods
Thicken	Jams, pie fillings, sauces and gravies
Whipping	Toppings and icings

CHAPTER 12

Comparison and Substitution

PART 12.1

Why Use One Hydrocolloid Over Another

In food science, hydrocolloids are essential ingredients that enable the creation of various textures and consistencies in processed foods. Hydrocolloids are food additives that function as thickening agents, gelling agents, stabilizers, and emulsifiers, among other things. They modify food products' viscosity, texture, and stability, making them a vital component of the food industry.

Whether you're a professional chef or a home cook, understanding the unique properties of different hydrocolloids is crucial in creating mouth-watering dishes and achieving the desired texture in your recipes. This article will explore the different types of hydrocolloids, their applications in the food industry, and when to use one hydrocolloid over another.

Core Insights:

- ☑ Hydrocolloids are food additives that are essential in the food industry.
- ☑ They function as thickening agents, gelling agents, stabilizers, and emulsifiers.
- ☑ Understanding the unique properties of different hydrocolloids is crucial in creating mouth-watering dishes and achieving the desired texture in your recipes.
- ☑ By selecting the right hydrocolloid for a specific application, you can make informed decisions and achieve optimal results in your culinary creations.
- ☑ Hydrocolloids continue to evolve alongside advancements in food science and technology, creating exciting possibilities for the future of the food industry.

12.1.1 Considerations for choosing one hydrocolloid over another?

The choice of one hydrocolloid over another depends on several factors, including the specific function needed in the product, the desired texture, the processing conditions, and the overall formulation goals. Here are some considerations for choosing one hydrocolloid over another:

Functionality

Gelling Properties: If the primary function is to create a gel, hydrocolloids like agar-agar, carrageenan, and pectin are commonly used. The choice depends on the specific texture and clarity required in the final product.

Thickening: For thickening applications, guar gum, xanthan gum, and locust bean gum are often chosen. The concentration and characteristics of the thickening can vary, influencing the choice.

Stabilizing: Hydrocolloids like carrageenan and xanthan gum are excellent stabilizers, preventing ingredient separation in products like sauces and dressings.

Synergy and Compatibility

Combination of Hydrocolloids: A combination of hydrocolloids is used to achieve a synergistic effect in many formulations. For example, combining xanthan gum with guar gum or locust bean gum can provide stability and enhanced texture.

Processing Conditions

Heat Stability: Some hydrocolloids, like agar-agar and carrageenan, are heat-stable and suitable for products that undergo heat processing, such as jams and hot-fill applications.

Cold Processing: Hydrocolloids like xanthan gum and guar gum are effective in cold applications, making them suitable for salad dressings and cold beverages.

Texture and Mouthfeel

Creaminess: Locust bean gum is often chosen for its ability to contribute to a creamy texture in products like ice cream and dairy alternatives.

Elasticity: Carrageenan, especially iota carrageenan, is known for providing elasticity in products like vegan cheeses and dairy alternatives.

Synergies with Other Ingredients

Interaction with Proteins: Some hydrocolloids interact well with proteins, making them suitable for applications like plant-based meats. Carrageenan, for example, can enhance the texture of vegan deli slices.

Interaction with Starches: Hydrocolloids can also interact with starches, influencing the viscosity and texture of products like sauces and gravies.

pH and Ionic Strength

Acidic vs. Alkaline Environments: The product's pH can influence certain hydrocolloids' performance. For instance, pectin is effective in acidic conditions, while locust bean gum is more stable in alkaline environments.

Ionic Strength: The presence of ions, especially calcium ions, can affect the gelling properties of hydrocolloids. Alginate, for example, forms gels in the presence of calcium ions.

Specific Dietary Requirements

Vegan and Vegetarian Considerations: Agar-agar, carrageenan, and pectin are plant-derived hydrocolloids often chosen for vegan and vegetarian formulations.

Gluten-Free Formulations: Xanthan gum and guar gum are commonly used in gluten-free baking as substitutes for gluten.

Cost Considerations

Economic Factors: The cost of hydrocolloids can vary, and sometimes the choice is influenced by economic factors. Some hydrocolloids are more cost-effective than others, making them suitable for large-scale production.

In summary, selecting one hydrocolloid over another is a multifaceted decision involving the specific functions needed, processing conditions, desired product characteristics, and dietary or cost considerations. A combination of hydrocolloids is often used to achieve the desired results in complex formulations. Formulators need to understand the unique properties of each hydrocolloid and how they interact to optimize the final product's overall performance.

12.1.2 Thickening Agents

Thickening agents are hydrocolloids that enhance the viscosity and texture of food products. They are commonly used in processed foods, such as soups, sauces, and gravies, to achieve the desired thickness and improve mouthfeel.

There are several types of thickening agents used in the food industry, including:

Table 86: Thickening Agent Function and Applications

THICKENING AGENT	FUNCTION	APPLICATIONS
Gums (e.g., xanthan gum, guar gum)	Increases viscosity and enhances texture	Soups, sauces, dressings, ice cream, bakery fillings
Starches (e.g., cornstarch, potato starch)	Thickens and stabilizes	Soups, sauces, gravies, pies, custards, pudding
Proteins (e.g., gelatin, collagen)	Forms rigid gels and stabilizes foams	Gelled desserts, whipped toppings, mousses

When choosing a thickening agent, it is important to consider the type of food product and the desired texture. For example, gums are ideal for creating smooth and uniform textures, while starches are better suited for creating thicker and more opaque liquids.

It is also important to note that overusing thickening agents can result in an unappetizing texture and mouthfeel. Therefore, it is recommended to use the minimum amount required to achieve the desired thickness.

12.1.3 Gelling Agents

Gelling agents belong to the group of hydrocolloids that create a gel-like consistency in food products. They are widely used in the food industry to enhance processed foods' texture, stability, and quality.

Gelling agents form a network of tiny particles that trap liquid and create a semi-solid or solid structure. This structure can be either reversible, such as in the case of a gelatin dessert that melts in the mouth, or irreversible, such as in the case of canned meat that maintains its shape over time.

The choice of gelling agent depends on the desired texture, processing conditions, and compatibility with other ingredients. Some common gelling agents used in the food industry include:

Table 87: Gelling Agent Sources and Applications

GELLING AGENT	SOURCE	APPLICATIONS
Agar	Red Seaweed	Jellies, custards, ice cream, canned meat
Carrageenan	Red Seaweed	Chocolate milk, ice cream, processed meats, toothpaste
Gelatin	Animal bones and skin	Jellies, marshmallows, yogurt, canned meat
Agar	Red Seaweed	Jellies, custards, ice cream, canned meat

Each gelling agent has unique properties that affect its functionality and compatibility with other ingredients. For example, agar requires high temperatures to dissolve and solidify, while gelatin requires specific pH and temperature conditions. Carrageenan has different types that provide different strengths and textures.

Gelling agents can be used in processed foods such as dairy products, confectioneries, and ready-to-eat meals. They offer a cost-effective way to enhance the texture and structure of foods while providing other benefits such as moisture retention and mouthfeel enhancement.

However, gelling agents require careful handling and processing to achieve the desired results. Overuse or misuse can lead to undesirable textures, flavors, and appearances. Therefore, following each gelling agent's recommended usage rates and processing conditions is essential.

12.1.4 Stabilizers

Stabilizers are hydrocolloids used in the food industry to maintain processed foods' texture, consistency, and appearance. They prevent separation and maintain homogeneity in products, contributing to their shelf-life and overall quality.

Common stabilizers include carrageenan, xanthan gum, and guar gum. Carrageenan is extracted from red seaweed and is commonly used in dairy products, such as ice cream, to prevent the formation of ice crystals and provide a smooth texture. Xanthan gum is made from fermented corn sugar and helps stabilize salad dressings and sauces, preventing them from separating. Guar gum, derived from the guar bean, is often used in baked goods to provide structure and improve texture.

Table 88: Common Stabilizer and Applications

STABILIZER	APPLICATIONS
Carrageenan	Dairy products like ice cream
Xanthan gum	Salad dressings and sauces
Guar gum	Baked goods

Stabilizers are especially useful in processed foods, where emulsions are common, and separation can occur due to high shear forces, temperature changes, or pH variations. They can also enhance products' mouthfeel and sensory experience, contributing to their overall appeal.

When selecting a stabilizer, it is important to consider the product's specific needs, including its processing conditions, pH range, and compatibility with other ingredients. The right stabilizer can help ensure the stability and quality of the final product.

12.1.5 Natural Hydrocolloids: A Clean-Label Alternative to Processed Foods

Consumers increasingly demand transparency and simplicity in their food choices, and using natural hydrocolloids as food additives has gained popularity in the food industry. Natural hydrocolloids are derived from plant, animal, or microbial sources and offer a clean-label alternative to synthetic hydrocolloids. They provide functionality and stability to processed foods while maintaining a shorter and cleaner ingredient list.

The Benefits of Natural Hydrocolloids

Natural hydrocolloids offer several benefits to the food industry, including:

- Meeting consumer demand for clean-label, natural ingredients
- Providing texture and stability to processed foods
- Reducing the reliance on synthetic hydrocolloids and other additives
- Offering a sustainable and eco-friendly solution

Table 89: popular natural hydrocolloids used in the food industry

HYDROCOLLOID	SOURCE	FUNCTIONALITY	APPLICATIONS
Agar	Seaweed	Gelling agent	Gelled desserts, soups, sauces, and dairy products
Guar gum	Legume seed	Thickener, stabilizer	Bakery products, sauces, ice cream, and other processed foods
Locust bean gum	Carob tree seed	Thickener, stabilizer	Creamy desserts, sauces, and other processed foods
Xanthan gum	Bacteria fermentation	Thickener, stabilizer, emulsifier	Sauces, dressings, bakery products, and other processed foods

The Role of Natural Hydrocolloids in Processed Foods

Natural hydrocolloids are crucial in processed foods, contributing to texture, stability, and other sensory characteristics. They can be used in a variety of applications, such as:

- Creating smooth and creamy textures in ice cream and other dairy products
- Gelling fruit preserves, jelly candies, and other confections
- Thickening sauces and dressings to improve mouthfeel and cling
- Stabilizing emulsions such as mayonnaise and salad dressings

Natural hydrocolloids offer a clean-label alternative to synthetic hydrocolloids and provide functionality and stability to processed foods while maintaining a shorter, simpler ingredient list. By incorporating natural hydrocolloids into your culinary creations, you can meet the demands of consumers seeking clean, natural ingredients without sacrificing product quality or functionality.

12.1.6 Factors to Consider When Choosing Hydrocolloids

Choosing the right hydrocolloid for a specific application involves considering various factors. The following are key factors to consider:

Desired Texture

The final product's texture is a crucial factor in selecting a hydrocolloid. Some hydrocolloids are better suited for creating a smooth and creamy texture, while others are ideal for a firmer or thicker consistency. Understanding the unique properties of each hydrocolloid is essential in achieving the desired texture.

Processing Conditions

The processing conditions, such as temperature, pH, and shear, can affect the performance of hydrocolloids. For example, some hydrocolloids may degrade or lose their functionality under high temperatures or acidic conditions. Selecting a hydrocolloid that can withstand the processing conditions and maintain its functionality throughout the manufacturing process is crucial.

Compatibility with other ingredients

Hydrocolloids interact differently with other ingredients in a recipe. Some may form unwanted interactions or interfere with the functionality of other ingredients. It is essential to select a hydrocolloid compatible with the other ingredients in the recipe to avoid undesirable outcomes.

Cost

The cost of hydrocolloids can vary depending on the source and quality of the ingredient. It is important to consider the cost of hydrocolloids when selecting the right one for a specific application. Some hydrocolloids offer better functionality at a higher cost, while others may be more affordable but offer lower performance.

Regulatory Environment

The regulatory environment surrounding hydrocolloids can also impact the selection process. Some hydrocolloids may be restricted or banned in certain countries or require additional labeling requirements. Considering the regulatory environment when selecting hydrocolloids for a specific application is essential.

12.1.7 Tips and Tricks for Working with Hydrocolloids

Working with hydrocolloids can be challenging for even the most experienced food industry professionals. Here are some tips and tricks to help you navigate the world of hydrocolloids:

Hydration Techniques

The proper hydration of hydrocolloids is critical to achieving the desired texture and functionality of your recipe. Follow these basic rules for optimal hydration:

- Gradually add hydrocolloids to your liquid while stirring to prevent clumping.
- Allow sufficient hydration time, usually 30 minutes to an hour, before using the mixture in your recipe.
- Adjust the hydration level based on your desired texture. More hydration results in a softer, more fluid texture, while less hydration yields a firmer, more solid texture.

Blending Methods

The method of blending your hydrocolloids with other ingredients can affect the outcome of your recipe. Consider these tips:

- Blend hydrocolloids with other dry ingredients before adding them to liquids to prevent clumping.
- Use a high-speed blender or immersion blender for optimal dispersion of hydrocolloids.
- Gradually add your hydrocolloid mixture to other ingredients while blending to prevent clumping.

Troubleshooting Common Challenges

Even with proper hydration and blending, challenges may arise when working with hydrocolloids. Here are some common issues and their solutions:

Table 90: Common Challenges of Hydrocolloids and Solutions

CHALLENGE	SOLUTION
Clumping	Gradually add hydrocolloids to liquid while stirring, or blend with other dry ingredients before adding to liquid.
Inconsistent texture	Ensure proper hydration and blending, adjust hydration level as needed, and consider the compatibility of hydrocolloids with other ingredients.
Undesirable flavor or odor	Consider the compatibility of hydrocolloids with other ingredients and experiment with different types of hydrocolloids to achieve the desired flavor and aroma.

Following these tips and tricks, you can overcome common challenges and master working with hydrocolloids. You can achieve the perfect texture and stability in your culinary creations with practice and experimentation.

12.1.8 Innovations and Future of Hydrocolloids in the Food Industry

The food industry is constantly evolving, and hydrocolloids are no exception. As demand for clean-label ingredients and natural formulations increases, the role of hydrocolloids is becoming even more critical. This section will explore some of the latest innovations and advancements in hydrocolloids and their potential impact on the food industry.

Natural Hydrocolloids

One of the most significant hydrocolloid trends is the shift toward natural ingredients. Natural hydrocolloids such as agar, carrageenan, and xanthan gum offer clean-label alternatives to synthetic ingredients and are gaining popularity in the food industry. Companies are also exploring new sources of natural hydrocolloids, such as seaweed and fruit extracts, to meet consumer demand for natural and sustainable products.

Novel Functionalities

Hydrocolloids also see advancements in their functionalities beyond the traditional roles of thickening, gelling, stabilizing, and emulsifying. For example, some hydrocolloids have been developed to improve texture and mouthfeel in low-fat foods. Others have been designed to encapsulate flavors, release nutrients, or reduce sugar content in products. These novel functionalities have the potential to revolutionize the food industry and create new opportunities for product development.

Technological Advances

The use of hydrocolloids in the food industry is also benefiting from technological advances in processing and formulation. For example, microencapsulation techniques can provide a controlled release of active ingredients in products. Nanotechnology is also being explored to develop precision delivery systems for hydrocolloid-based ingredients. These advances can improve hydrocolloid-based products' quality, effectiveness, and safety.

Regulations and Standards

Finally, regulatory bodies are becoming more involved in using and labeling hydrocolloids in food products. The FDA has established guidelines for the safe use of hydrocolloids, and the Codex Alimentarius Commission has developed international standards for food additives. These regulations and standards provide a framework for the responsible use of hydrocolloids and ensure consistent product quality and safety for consumers.

Overall, the future of hydrocolloids in the food industry looks promising. As the demand for natural, sustainable, and functional ingredients grows, hydrocolloids are well-positioned to play a critical role in meeting these needs. Technological advances and regulatory standards will continue to shape the industry, and companies that stay ahead of these trends will be best positioned to succeed.

Conclusion

The world of hydrocolloids is vast and complex, but understanding the unique properties of each one can be a valuable asset in the food industry. Hydrocolloids can transform the texture, stability, and overall sensory experience of processed foods, and by utilizing the right type of hydrocolloid for the right application, you can create mouthwatering dishes and successful products.

Factors such as texture, processing conditions, pH, and compatibility should be considered when choosing hydrocolloids. Proper hydration techniques, blending methods, and troubleshooting common challenges are essential for working with hydrocolloids effectively. As the food industry changes and consumer demands shift towards clean, transparent ingredient lists, natural hydrocolloids offer an exciting alternative to synthetic hydrocolloids.

The future of hydrocolloids holds innovative possibilities for creating healthier, tastier, and more sustainable food products. As advancements in food science and technology continue, hydrocolloids will undoubtedly play a critical role in shaping the future of the food industry.

Therefore, understanding the importance of hydrocolloids and how to use them effectively is crucial for success in the food industry. With some experimentation, creativity, and a deep understanding of hydrocolloids, you can create culinary masterpieces that delight the senses and meet the ever-changing demands of consumers.

FAQ

When should I use one hydrocolloid over another?

The choice of hydrocolloid depends on your recipe's desired texture and functionality. Different hydrocolloids have unique properties that make them suitable for specific culinary scenarios. Understanding the characteristics of each hydrocolloid will help you determine which one to use to achieve the desired outcome.

What factors should I consider when choosing hydrocolloids?

When selecting hydrocolloids, you should consider the desired texture, processing conditions, pH, and compatibility with other ingredients. These factors will help you make informed decisions and achieve optimal results in your culinary creations.

What are some tips and tricks for working with hydrocolloids?

Working with hydrocolloids requires some know-how and experimentation. Here are some practical tips and tricks: properly hydrate the hydrocolloid before use, use appropriate blending methods, and be aware of common challenges and how to troubleshoot them.

What is the future of hydrocolloids in the food industry?

Hydrocolloids continue to evolve with advancements in food science and technology. The future holds exciting possibilities, including clean-label solutions and novel functionalities that can contribute to creating healthier, tastier, and more sustainable food products.

PART 12.2

How to Substitute One Hydrocolloid for Another in Recipes

Hydrocolloids are functional ingredients used in food preparation to improve texture, stability, and mouthfeel. Each hydrocolloid has unique properties that suit specific recipes and applications. Understanding these properties is crucial to choosing the right hydrocolloid for your recipe. Experimenting with different hydrocolloids can unlock new possibilities and elevate your culinary creations.

Core Insights:

- ☑ Hydrocolloids are essential for improving texture, stability, and overall recipe quality.
- ☑ Each hydrocolloid has its specific properties and applications.
- ☑ Agar-agar, carrageenan, guar gum, and xanthan gum are common types of hydrocolloids used in recipes.
- ☑ Agar-agar is a versatile hydrocolloid that can create strong, heat-resistant gels.
- ☑ Carrageenan enhances the texture and stability of food products.

12.2.1 The Role of Hydrocolloids in Food Preparation

Hydrocolloids play a crucial role in food preparation by offering a range of functionalities that improve the texture, stability, and overall quality of recipes. Depending on the application, these versatile ingredients can create different textures, such as creamy, smooth, or chewy. They enhance the mouthfeel of dishes, giving them a professional touch that elevates the dining experience.

One of the key benefits of hydrocolloids is their ability to improve stability in food products. They prevent ingredient separation or breakdown during cooking or storage, ensuring the dish remains visually appealing and maintains its quality over time. Whether it's a sauce, dressing, or baked good, adding hydrocolloids helps maintain the desired consistency and appearance.

Furthermore, hydrocolloid gels are widely used in food preparation. These gels can range from soft and delicate to firm and robust, depending on the hydrocolloid used. For example, agar-agar is known for its ability to form strong gels, ideal for making jellies and desserts. Conversely, Carrageenan forms a gel-like substance that enhances dairy products› and sauces› texture and stability.

The Benefits of Hydrocolloids in Food Preparation

- Improved texture and mouthfeel
- Enhanced stability and shelf life
- Ability to create different textures, from creamy to chewy
- Prevention of ingredient separation or breakdown
- Wide range of applications in various dishes

> "Hydrocolloids are the secret ingredient that takes a recipe from good to great. They allow us to create unique textures and enhance the overall eating experience. From sauces to desserts, hydrocolloids play a vital role in food preparation."

Choosing the right hydrocolloid for a recipe requires understanding its unique properties and how they contribute to the desired outcome. Whether agar-agar for a firm gel or xanthan gum for thickening a sauce, each hydrocolloid offers unique benefits and applications. Incorporating hydrocolloids into your culinary creations can unlock new possibilities and take your dishes to the next level.

Table 91: Hydrocolloid, Texture, Stability and Mouthfeel

HYDROCOLLOID	TEXTURE	STABILITY	MOUTHFEEL
Agar-Agar	Firm gel	Good	Smooth, chewy
Carrageenan	Gel-like substance	Excellent	Enhanced creaminess
Guar Gum	Viscous solution	Good	Thick and creamy
Xanthan Gum	Gel-like substance	Stable	Smooth and consistent

12.2.2 Hydrocolloids can be used as an alternate

Various types of hydrocolloids can be used in food preparation, each with unique properties and applications. These natural hydrocolloids are versatile ingredients that enhance various recipes› texture, stability, and quality. Here are some common types of hydrocolloids:

Agar-Agar: A Versatile Hydrocolloid

Agar-agar, derived from seaweed, is a versatile hydrocolloid with unique properties that make it a valuable ingredient in various recipes. Its ability to create strong, heat-resistant gels has made it popular in desserts, jellies, and even savory dishes. Agar-agar can achieve different textures, from light and airy foams to dense and chewy consistencies, allowing you to experiment and elevate your culinary creations.

An interesting feature of agar-agar is its vegetarian-friendly nature, making it an excellent substitute for gelatin. However, it's important to note that agar-agar sets differently than gelatin. While gelatin requires refrigeration to set, agar-agar gels are set at room temperature. This distinction provides unique opportunities for recipe creation and presentation.

To showcase the applications of agar-agar, consider the following example table:

Table 92: Applications of agar-agar in Recipes

RECIPE	APPLICATION
Mango Panna Cotta	Creates a silky-smooth texture with a fruity twist.
Strawberry Jelly	Forms a firm gel that holds the fruit pieces intact.
Vegan Cheesecake	Offers a creamy consistency without the need for dairy products.

As you can see from the table, agar-agar can be used in various recipes to achieve different outcomes. Its versatility and ability to create stable gels make it valuable to any home or professional kitchen.

Carrageenan: Enhancing Stability and Texture

Carrageenan is a hydrocolloid derived from seaweed that plays a vital role in enhancing the texture and stability of food products. With its unique properties, carrageenan is a reliable thickener and stabilizer in various applications, including dairy products, sauces, and processed meats.

One of the key benefits of carrageenan is its ability to form a gel-like substance that provides viscosity and texture to food products. This gel-like nature prevents the formation of ice crystals in ice cream, resulting in a smoother and creamier product. Additionally, carrageenan can improve the mouthfeel of sauces and dressings, giving them a pleasant, velvety texture. When it comes to stability, carrageenan shines. It helps prevent ingredient separation and breakdown during cooking or storage, maintaining your products' quality and appeal. It is especially important in dairy products, where carrageenan acts as a stabilizer, keeping the product homogeneous and preventing creaming.

Carrageenan is a versatile hydrocolloid that enhances food products' stability and texture. By incorporating carrageenan into your recipes, you can achieve the desired viscosity, improve mouthfeel, and ensure the longevity of your culinary creations.

"Carrageenan is a valuable ingredient in our ice cream production. It not only improves the texture and creaminess of the final product but also enhances stability, preventing the formation of unwanted ice crystals. It has become an essential tool in our quest for creating the perfect scoop."

- Chef Lisa Thompson, Ice Cream Innovators

Table 93: Applications of Carrageenan

FOOD PRODUCT	APPLICATION
Ice Cream	Prevents ice crystal formation and enhances creaminess
Sauces and Dressings	Improves mouthfeel and texture
Dairy Products	Acts as a stabilizer to prevent ingredient separation
Processed Meats	Aids in moisture retention and texture improvement

Guar Gum: A Reliable Thickening Agent

Guar gum is a versatile hydrocolloid widely used as a thickening agent in various recipes. It is derived from the guar plant and has excellent thickening properties that enhance food products› texture and consistency. Guar gum forms a viscous solution when mixed with water, creating a smooth and creamy texture in sauces, dressings, and ice creams.

One of the key advantages of guar gum is its ability to hold ingredients together. It acts as a binder, preventing liquids from separating and ensuring a more stable and cohesive mixture. It makes it an ideal ingredient for gluten-free baking, where it can help improve the structure and texture of breads, cakes, and other baked goods.

When using guar gum, it is important to remember that a little goes a long way. It is highly concentrated, so only small amounts are needed to achieve the desired thickening effect. Using too much guar gum can result in a slimy or gummy texture, so following recommended usage guidelines and adjusting quantities based on the specific recipe and desired outcome is important.

Key Uses of Guar Gum

Thickening sauces and dressings: Guar gum can add body and viscosity, improving their texture and mouthfeel.

Enhancing ice cream: Guar gum helps prevent the formation of ice crystals in ice cream, resulting in a smoother and creamier texture.

Improving gluten-free baking: Guar gum acts as a binding agent in gluten-free recipes, helping to improve the structure and texture of baked goods.

Creating smooth beverages: Guar gum can thicken and stabilize beverages, such as smoothies or shakes, giving them a smoother and more enjoyable consistency.

> "Guar gum is a reliable thickening agent that can elevate the texture and stability of various recipes. Its binding properties and ability to create creamy textures make it a valuable ingredient in sweet and savory dishes."

Table 94: Recipe, Amount of Guar Gum, and Resulting Texture

RECIPE	AMOUNT OF GUAR GUM	RESULTING TEXTURE
Sauce	1/4 teaspoon	Thick and velvety
Ice cream	1/2 teaspoon	Creamy and smooth
Bread	1 teaspoon	Improved structure and texture

Xanthan Gum: The Go-To Hydrocolloid

In various recipes, Xanthan gum is a versatile hydrocolloid widely recognized as an excellent thickener and stabilizer. Derived from bacteria, it possesses unique properties that make it a go-to ingredient for enhancing the texture and stability of food products. When dissolved in water, xanthan gum forms a gel-like substance, providing a consistent and smooth consistency to sauces, dressings, and gluten-free baked goods.

One of the key advantages of xanthan gum is its ability to thicken liquids without adding excessive bulk. It is particularly valuable in recipes with a desired texture, such as creamy salad dressings or perfectly balanced sauces. Unlike other hydrocolloids, xanthan gum creates a stable gel that doesn't break down easily under extreme temperatures or during extended storage periods. It makes it an ideal choice for products that require a longer shelf life without compromising quality.

In addition to its thickening and stabilizing properties, xanthan gum provides several other benefits in food preparation. It can improve the mouthfeel of products, giving them a smooth and pleasant texture. Xanthan gum is also known for preventing ingredient separation, ensuring a consistent and visually appealing appearance in products like salad dressings. With its wide range of applications and reliable performance, xanthan gum is a valuable tool for culinary enthusiasts and professional chefs.

Xanthan Gum Uses

- Thickening agent for sauces, dressings, and gravies
- Stabilizer in dairy products, preventing the formation of ice crystals in ice cream and improving creaminess
- Texture enhancer in gluten-free baking, providing structure and preventing crumbly results
- Preventing ingredient separation in emulsified products like salad dressings

Table 95: Xanthan Gum, Benefits and Applications

XANTHAN GUM	BENEFITS	APPLICATIONS
Thickening	Provides viscosity without excessive bulk	Sauces, dressings, gravies
Stabilizing	Prevents ingredient separation and improves shelf life	Ice cream, dairy products
Texture enhancement	Creates a smooth and pleasant mouthfeel	Gluten-free baking
Emulsification	Ensures consistent appearance and texture	Salad dressings, mayonnaise

Conclusion

Substituting one hydrocolloid for another in recipes can be a creative and exploratory process. By understanding the unique properties of each hydrocolloid and experimenting with different options, you can elevate your culinary creations to new levels. Agar-agar, carrageenan, guar gum, and xanthan gum are just a few examples of versatile hydrocolloids that can be used in various recipes.

When substituting hydrocolloids, following recommended usage guidelines, and adjusting quantities based on the desired texture and outcome is important. Each hydrocolloid has its specific characteristics, so it's essential to understand how they interact with other ingredients in your recipe. Through trial and error, you can find the perfect hydrocolloid substitution that suits your preferences and dietary needs.

Remember, mastering the art of hydrocolloid substitutions takes time and practice. Don›t be afraid to experiment and think outside the box. With creativity and an understanding of hydrocolloid properties, you can achieve culinary success and create delicious and visually appealing dishes.

FAQ

How do I substitute one hydrocolloid for another in recipes?

For substituting one hydrocolloid for another in recipes, it's important to understand the unique properties of each hydrocolloid. Experimentation is key. Start using smaller quantities and adjust based on the desired texture and outcome.

What is the role of hydrocolloids in food preparation?

Hydrocolloids have various uses in food preparation. They can act as a thickening, stabilizers, emulsifiers, and gelling agents. They improve texture, stability, and mouthfeel in recipes.

How is carrageenan used in food preparation?

Carrageenan is a hydrocolloid derived from seaweed. It is commonly used as a stabilizer and thickener in dairy products, sauces, and processed meats.

What is guar gum, and what is its purpose?

Guar gum is a popular thickening agent in gluten-free baking, ice creams, and sauces. It creates a viscous solution that holds ingredients together, resulting in thick, creamy textures.

What is xanthan gum, and how is it beneficial?

Xanthan gum is a versatile hydrocolloid thickener and stabilizer in many recipes. It is known for its ability to thicken liquids without adding excessive bulk and is commonly used in dressings, sauces, and gluten-free baking.

What are some tips for hydrocolloid substitutions?

When substituting one hydrocolloid for another, it's important to understand the unique properties of each hydrocolloid. Experimentation and adjusting quantities based on desired texture and outcome are key.

CHAPTER

13

Regulations and Standards

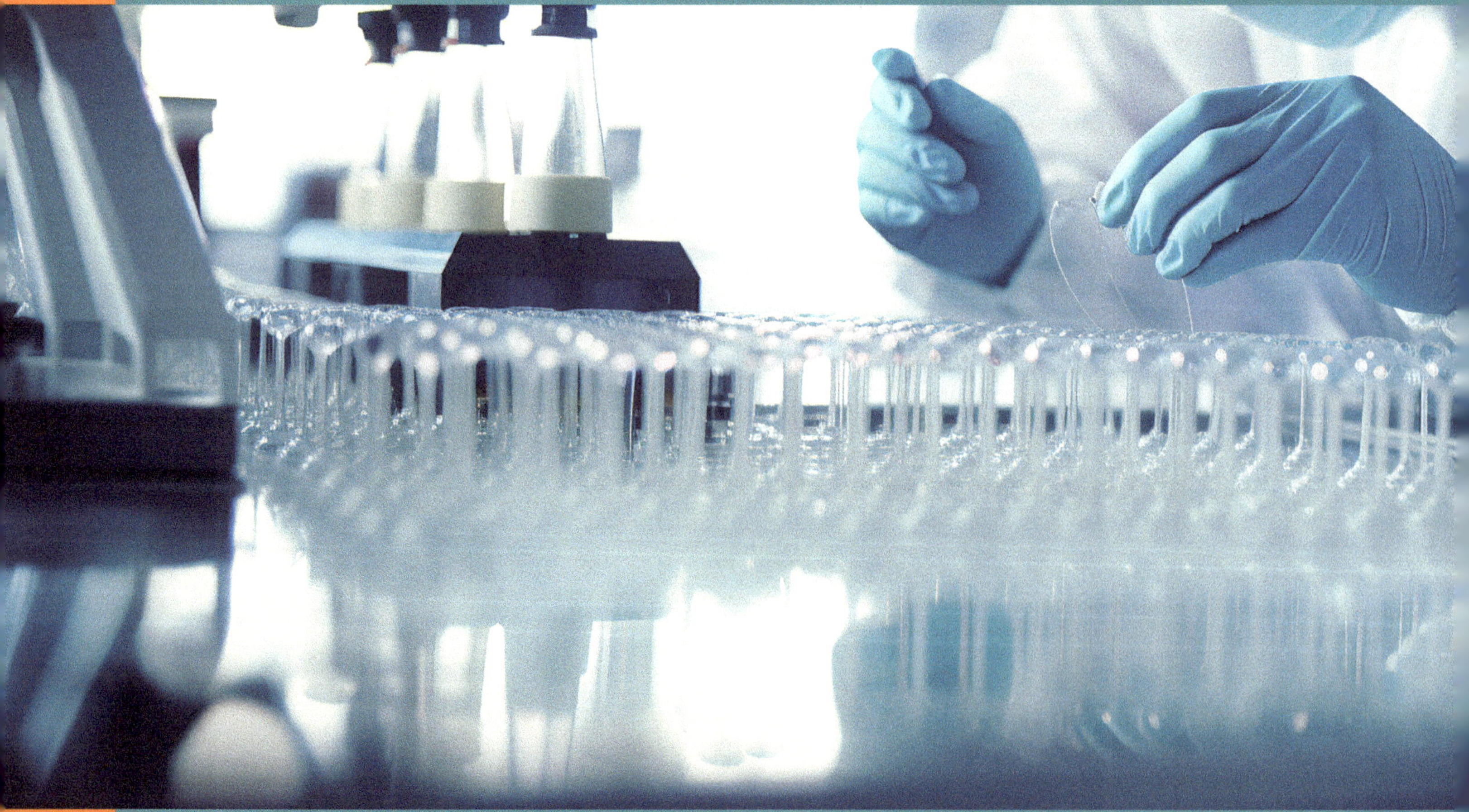

PART 13.1

Hydrocolloid Food Safety and Quality Standards

Hydrocolloids are compounds with unique properties that enable them to function as stabilizers, thickeners, and gelling agents in food applications. These materials are widely used in the food industry because they improve texture, enhance shelf life, and maintain food safety and quality standards. Hydrocolloids can be derived from various sources, including plants, animals, and microorganisms, and are available in a range of forms and grades to suit different food applications.

Hydrocolloid materials are valued for their versatile applications in the food industry, from stabilizing emulsions and suspensions to improving mouthfeel and texture. Their properties can be harnessed to produce various food products, including sauces, dressings, baked goods, dairy products, and meat products. Their ability to hold moisture and control water activity can also help to prevent microbial growth and extend shelf life. Hydrocolloids are often used to create healthier, low-fat, and reduced-sugar food formulations.

Hydrocolloid properties can vary significantly depending on the type of material, its source, and the specific application. They can include viscosity, gel strength, emulsification, and water-binding capacity. Understanding these properties is crucial for creating optimal food formulations. Hydrocolloids must be used in the correct amounts and combined with other ingredients in specific ways to achieve the desired texture, stability, and sensory attributes.

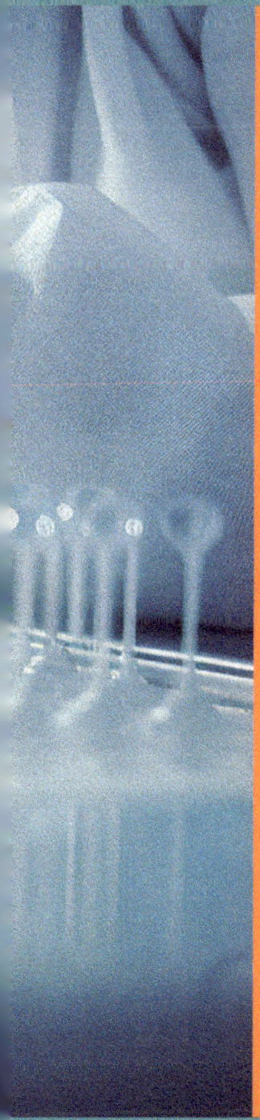

Core Insights:

- ☑ Hydrocolloids play a crucial role in maintaining food safety and quality standards.
- ☑ Hydrocolloid materials are derived from various sources and have a range of applications in the food industry.
- ☑ The properties of hydrocolloids can be harnessed to improve texture, extend shelf life, and create healthier food formulations.
- ☑ Understanding hydrocolloid properties is essential for creating optimal food formulations.
- ☑ Hydrocolloids must be used correctly and combined with other ingredients to achieve desired attributes.

13.1.1 Understanding Hydrocolloids in Food Production

Hydrocolloids are widely used in food production for their unique properties that enhance texture, stability, and sensory attributes. Hydrocolloid manufacturers play a crucial role in ensuring the consistent quality of these materials, making them an essential ingredient in the food industry.

The benefits of hydrocolloids are numerous, and their uses are versatile. They can be used as thickeners, stabilizers, gelling agents, emulsifiers, and fat replacers. Hydrocolloids can improve the shelf life of food products, enhance their appearance, and provide a smoother mouthfeel.

Hydrocolloid manufacturers offer a range of products to cater to the specific needs of food producers. They provide specialized blends of hydrocolloids to meet the functional requirements of each food application. For instance, carrageenan, a seaweed-derived hydrocolloid, is commonly used in dairy products, while xanthan gum is found in salad dressings and sauces.

Understanding the uses and benefits of hydrocolloids is crucial to the food industry's continued success. Hydrocolloid manufacturers provide an essential service by ensuring the consistent quality of these materials, helping producers create high-quality food products that meet consumers' demands for taste, texture, and safety.

13.1.2 Hydrocolloid Standards and Regulations

The hydrocolloid industry is subject to strict standards and regulations to ensure the safety and quality of food products containing hydrocolloid materials. In the United States, the Food and Drug Administration (FDA) regulates the use of hydrocolloids in food products, including their safety and labeling requirements.

Internationally, the Codex Alimentarius Commission (CAC) sets standards and guidelines for food safety and quality, including hydrocolloids. The Global Food Safety Initiative (GFSI) also provides a framework for food safety management systems that incorporate internationally recognized standards.

Compliance with these standards is critical for hydrocolloid manufacturers and food producers to ensure consumer confidence and access to global markets. Adherence to these regulations also helps to prevent foodborne illness and maintain the integrity of food products.

The hydrocolloid market is constantly evolving, and manufacturers must stay current with the latest regulations and standards to remain competitive and succeed in the industry. As new products and applications are developed, industry stakeholders must work together to ensure that safety and quality remain top priorities.

13.1.3 Ensuring Food Safety with Hydrocolloids

Hydrocolloid applications in the food industry extend beyond improving texture and sensory attributes. They also play a crucial role in maintaining food safety. By providing a protective barrier around food particles, hydrocolloids can reduce the risk of microbial contamination. Moreover, they can enhance food stability, prolong shelf life and reduce food waste.

One of the most promising applications of hydrocolloids is in creating healthier food formulations. Hydrocolloids can be used as an alternative to fat, sugar, and salt, which are often associated with negative health effects. With the growing demand for healthier food options, hydrocolloid manufacturers are actively developing new formulations that offer improved nutritional profiles without compromising taste or texture.

The use of hydrocolloids in products like yogurts, dressings, and spreads has been shown to improve the overall quality and consistency of the final product. These hydrocolloid-based products are less prone to separation, have a smoother texture, and are more visually appealing. Moreover, by extending shelf life, they reduce food waste, benefiting both the environment and the bottom line.

Given their versatility and potential benefits, hydrocolloids have a bright future in the food industry. As manufacturers continue to innovate and develop new applications, hydrocolloids will remain an essential ingredient in meeting the evolving needs of consumers.

13.1.4 Quality Control of Hydrocolloid Products

Ensuring consistent quality is a top priority for hydrocolloid manufacturers to maintain food safety and quality standards. Quality control measures are in place to ensure that hydrocolloid products meet specific criteria and perform optimally in various food applications.

Hydrocolloid properties, such as solubility, viscosity, and gelling capacity, are critical parameters manufacturers monitor during quality control testing. Each type of hydrocolloid requires specific tests to ensure that it performs as expected in various food applications.

Manufacturers use high-performance liquid chromatography (HPLC) equipment to analyze hydrocolloid products' molecular weight and purity. Testing for microbial contamination, heavy metals, and other impurities is crucial to quality control testing.

By adhering to strict quality control measures, hydrocolloid manufacturers can ensure that their products consistently meet the needs of the food industry and consumers alike.

13.1.5 Innovations in Hydrocolloid Applications

The hydrocolloid market is constantly evolving, with manufacturers seeking new and innovative ways to enhance the functionality and versatility of these materials. There has been a growing trend towards developing natural hydrocolloid products as consumers become more health-conscious and demand more sustainable food options.

One of the most promising innovations in hydrocolloid applications is using plant-based hydrocolloids, such as agar, carrageenan, and pectin, offering functional benefits without synthetic additives. These natural hydrocolloids can be used as gelling agents, thickeners, stabilizers, and emulsifiers in various food products, from dairy alternatives to meat substitutes and gluten-free baked goods.

Another area of innovation is using hydrocolloids to create cleaner labels and reduce food waste. Hydrocolloid blends, such as modified starches and cellulose derivatives, can replace traditional ingredients like fat and sugar while maintaining the desired texture and sensory attributes. It produces healthier products, reduces the need for preservatives, and extends shelf life, reducing food waste and increasing sustainability in the food industry.

Advances in technology are also driving the development of new hydrocolloid-based products. For example, microencapsulation techniques can protect sensitive ingredients, such as flavors and nutrients, from degradation during processing and storage. It has led to new functional components, such as encapsulated probiotics and omega-3 fatty acids, which can be added to food products to improve health and nutrition.

The hydrocolloid market is expected to grow in the coming years, driven by increasing demand for healthier and more sustainable food options. It presents new opportunities for innovation and collaboration among manufacturers and industry stakeholders as they develop new products and applications that meet consumers' evolving needs and preferences.

13.1.6 Future Trends in the Hydrocolloid Industry

The hydrocolloid industry is poised for significant growth in the coming years. Changing consumer preferences, increasing demand for convenience foods, and the growing popularity of clean-label products are some factors driving the market›s expansion. According to a recent report by Markets and Markets, the global hydrocolloid market size is expected to reach $9.8 billion by 2025, growing at a CAGR of 5.3% during the forecast period.

One of the emerging trends in the hydrocolloid industry is the development of sustainable raw materials. With the increasing focus on environmental sustainability and reducing carbon footprint, manufacturers are exploring alternative sources of hydrocolloids, such as seaweed, plant-based fibers, and microbial fermentation. These new materials offer a more eco-friendly and renewable option for producing hydrocolloids while maintaining the same functional properties.

Another trend that is gaining traction is the use of hydrocolloids in creating plant-based meat and dairy alternatives. Consumers are increasingly adopting a flexitarian lifestyle, seeking meat and dairy alternatives that are healthy, sustainable, and taste like the real thing. Hydrocolloids, with their ability to mimic the texture and mouthfeel of meat and dairy, play a crucial role in creating these innovative products.

Incorporating digital technologies in hydrocolloid manufacturing and supply chain management is also a trend set to transform the industry. With the increasing demand for transparency and traceability in the food sector, manufacturers are leveraging advanced analytics, artificial intelligence, and blockchain technology to improve supply chain visibility, enhance quality control, and ensure compliance with regulatory standards.

Lastly, the hydrocolloid industry is expected to grow significantly in the Asia-Pacific region. Rapid urbanization, changing dietary habits, and increasing disposable income are some factors driving the market's expansion. With the growing demand for processed and convenience foods, there is a significant opportunity for hydrocolloid manufacturers to expand their presence in the Asia-Pacific market and capitalize on the region's growth potential.

13.1.7 The Importance of Collaboration in the Hydrocolloid Sector

The hydrocolloid industry is a complex and highly specialized field that requires collaboration among manufacturers, industry stakeholders, and regulatory bodies. The development, production, and application of hydrocolloids need input from experts and professionals who work together to ensure the highest safety and quality standards.

Hydrocolloid manufacturers play a critical role in this collaboration, as they are responsible for producing high-quality hydrocolloid products that meet industry standards and comply with regulatory requirements. They work closely with food producers, processors, and manufacturers to develop custom solutions for specific applications and help ensure the success of their customers' products.

Collaboration within the hydrocolloid industry extends beyond manufacturers, with industry stakeholders and regulatory bodies playing an essential role in setting standards and guiding industry trends. Industry associations, such as the International Food Hydrocolloids Association (IFHA), serve as a forum for collaboration, education, and research, bringing together hydrocolloid manufacturers, academia, and other stakeholders to promote the use of hydrocolloids in food applications.

The hydrocolloid market is highly competitive, with manufacturers constantly innovating and developing new products to meet consumers' evolving needs and preferences. Collaboration among industry players is key to ensuring that hydrocolloid products remain safe and of the highest quality, even as the industry develops.

Collaboration also plays a crucial role in addressing emerging challenges and opportunities in the hydrocolloid industry. For example, the COVID-19 pandemic has raised concerns about the stability and supply chain of hydrocolloid products, making collaboration more critical than ever to ensure the continued availability of these products.

In conclusion, collaboration is crucial in the hydrocolloid sector to ensure the highest safety and quality standards. Hydrocolloid manufacturers, industry stakeholders, and regulatory bodies must work together to develop and promote best practices, share knowledge and expertise, and address emerging challenges and opportunities in the industry. Only by working together can the hydrocolloid sector continue to deliver safe and high-quality products to consumers.

Conclusion

Hydrocolloids are essential ingredients in the food industry, playing a vital role in maintaining food safety and quality standards. As discussed in this article, hydrocolloid materials are versatile and can improve texture, extend shelf life, and enhance sensory attributes in various food products.

Hydrocolloid manufacturers ensure consistent quality and adhere to global food safety and quality standards. Adherence to these standards is crucial in ensuring consumer confidence and compliance with international regulations in the hydrocolloid market.

Through collaboration among hydrocolloid manufacturers, industry stakeholders, and regulatory bodies, the hydrocolloid industry can continue to innovate and develop new products to meet consumers' evolving needs and preferences. This collaboration can also ensure shared knowledge, best practices, and continuous improvement to maintain the highest food safety and quality standards.

In the future, the hydrocolloid industry will continue to experience growth and face emerging markets, technological advancements, and potential challenges. However, by emphasizing the importance of collaboration and adhering to global standards, the hydrocolloid market can continue to provide safe and high-quality hydrocolloid materials for various applications.

FAQ

What standards and regulations govern the hydrocolloid market?

The hydrocolloid market is governed by global food safety and quality standards. Adhering to these standards ensures consumer confidence and compliance with international regulations.

How do hydrocolloids contribute to food safety?

Hydrocolloids contribute to food safety by improving food stability, reducing the risk of microbial contamination, and extending shelf life. They can also be used to create healthier food formulations.

How is the quality of hydrocolloid products controlled?

The quality of hydrocolloid products is controlled through rigorous testing and quality control measures. Key parameters and tests are used to ensure consistent quality and performance.

What are some innovations in hydrocolloid applications?

There are continuous innovations in hydrocolloid applications. Manufacturers are developing new hydrocolloid-based products to meet consumers' evolving needs and preferences regarding texture, taste, and nutritional profile.

What are the future trends in the hydrocolloid industry?

The hydrocolloid industry is witnessing emerging markets, technological advancements, and potential challenges. Future trends include the exploration of new applications and the development of sustainable and plant-based hydrocolloids.

Why is collaboration important in the hydrocolloid sector?

Collaboration among hydrocolloid manufacturers, industry stakeholders, and regulatory bodies is crucial for sharing knowledge and best practices and ensuring the highest food safety and quality standards.

PART 13.2

Labeling and Claims Related to Hydrocolloids

Hydrocolloids are commonly used as food additives in many processed foods. They serve various functions, including thickening, gelling, stabilizing, emulsifying, water-binding, and modifying viscosity. These ingredients can significantly impact food products› texture, appearance, and mouthfeel, making them an essential component of many formulations.

However, it is crucial to understand the importance of proper labeling and claims related to hydrocolloids. Transparent and accurate labeling is critical to ensuring consumer safety and informed choices. The regulatory framework surrounding hydrocolloid labeling and claims must be considered to comply with the relevant authority›s requirements.

Core Insights:

- ☑ Hydrocolloids are food additives used for thickening, gelling, stabilizing, emulsifying, water-binding, and modifying viscosity.
- ☑ Proper labeling and hydrocolloid claims are essential for consumer safety and informed choices.
- ☑ The regulatory framework surrounding hydrocolloid labeling and claims must be considered to comply with the relevant authorities requirements.
- ☑ Hydrocolloids significantly impact food products' texture, appearance, and mouthfeel, making them an essential component of many formulations.
- ☑ Transparent and accurate labeling is crucial to ensure consumer safety and informed choices.

13.2.1 Understanding Hydrocolloids as Food Additives

Hydrocolloids are commonly used as food additives due to their ability to modify the texture and stability of food products. These ingredients can act as thickening agents, gelling agents, stabilizers, and emulsifiers, among other functions.

Thickening agents increase the viscosity of liquids, while gelling agents form a three-dimensional network that gives food products a semi-solid structure. Stabilizers prevent ingredients from separating; emulsifiers help mix oil and water-based ingredients. Hydrocolloids can also act as water-binding agents and viscosity modifiers, providing additional functionality in food formulations.

Hydrocolloids are typically used in small quantities and are labeled as "gums" on ingredient lists. It is important to understand the specific functions of each hydrocolloid and use them appropriately in food formulations to achieve the desired texture and stability.

The regulatory framework surrounding hydrocolloid labeling and claims is important when using these additives in food products. The U.S. Food and Drug Administration (FDA) sets specific requirements for labeling, and manufacturers must comply with these regulations to ensure consumer safety and prevent misleading claims.

Examples of Hydrocolloids as Food Additives

Some commonly used hydrocolloids as food additives include:

Table 96: Commonly used hydrocolloids as food additives

HYDROCOLLOID	FUNCTION
Xanthan gum	Thickener, stabilizer, emulsifier
Guar gum	Thickener, stabilizer, viscosity modifier
Agar	Gelling agent, stabilizer
Carrageenan	Gelling agent, stabilizer, emulsifier
Pectin	Gelling agent, thickener, stabilizer

Each hydrocolloid has specific functions and is used in different food applications, depending on the desired texture and stability properties.

When using hydrocolloids as food additives, it is important to consider practical aspects such as dosage, processing conditions, and compatibility with other ingredients. Manufacturers must also consider how hydrocolloids impact food texture and the consumer's sensory experience.

Overall, hydrocolloids are an important tool in food product development, providing functional properties that enhance texture and stability. Understanding the specific functions of each hydrocolloid and considering regulatory guidelines and practical considerations when using them is crucial for achieving optimal results in food formulations.

13.2.2 Importance of Proper Labeling for Hydrocolloids

Labeling food products containing hydrocolloids protects consumer safety and ensures informed choices. Hydrocolloids are commonly used as food additives, providing functional properties such as thickening, stabilization, and emulsification. However, without accurate and transparent labeling, consumers may be unaware of the presence of these ingredients and their potential impact on food texture and nutrition.

Labeling hydrocolloids in food products is subject to regulatory guidelines in the United States. The Food and Drug Administration (FDA) requires that all ingredients be listed on the product label, including hydrocolloids and their specific function in the product. It allows consumers to make informed choices based on their dietary needs and preferences.

Claims related to hydrocolloid-containing products must also adhere to regulatory guidelines. Health claims suggest a relationship between a food product and a particular health benefit and are subject to FDA review and approval. Other claims, such as functional or sensory attributes, must be substantiated by scientific evidence and not be misleading or deceptive.

Proper labeling and claims about hydrocolloids in food products are essential for promoting consumer safety and informed choices. Compliance with regulatory guidelines and transparent labeling practices can also help build consumer trust and confidence in the food industry.

13.2.3 Regulatory Guidelines for Hydrocolloid Labeling in the United States

Hydrocolloids are considered food additives in the United States and are subject to strict regulatory guidelines regarding their labeling and claims. The Food and Drug Administration (FDA) regulates the use of hydrocolloids in food products, ensuring their safety for consumer consumption.

The FDA requires that hydrocolloids be listed by their common or usual name and functional class on food product labels. The active type refers to the hydrocolloid's specific function in the food product, such as a thickener, gelling agent, or stabilizer. The title should also indicate the hydrocolloid's percentage by weight per serving of the food product.

The FDA also strictly regulates claims related to hydrocolloids on food product labels. Any health claims regarding hydrocolloids must be backed by scientific evidence and approved by the FDA. Additionally, any claims made regarding the functional properties of hydrocolloids, such as their ability to modify viscosity or improve texture, must be truthful and not misleading to consumers.

Food manufacturers need to comply with these regulatory guidelines to ensure consumer protection and avoid any potential legal and financial consequences resulting from non-compliance.

13.2.4 Practical Considerations for Hydrocolloid Use in Food Products

When formulating food products with hydrocolloids, it's essential to consider several practical factors to achieve the desired texture and stability. It includes the dosage, processing conditions, and compatibility with other ingredients.

The dosage of hydrocolloids should be carefully measured and controlled to avoid over- or under-use. Overuse can lead to a gummy or slimy texture, while underuse can result in instability or poor surface. The optimal dosage varies depending on the specific hydrocolloid and the food application.

The processing conditions also play a crucial role in hydrocolloid performance. For example, heat treatments, such as baking or pasteurization, can affect hydrocolloids› viscosity and gelation properties. Some hydrocolloids, such as hydration or pre-dispersion, may require specific processing conditions to function correctly.

Compatibility with other ingredients is another important consideration. Some hydrocolloids may interact with other components in the food matrix, affecting their functionality or causing undesired effects, such as instability or phase separation. Testing the compatibility of hydrocolloid blends with other ingredients is crucial before use.

13.2.5 Impact of Hydrocolloids on Food Texture

Hydrocolloids have various functions in food products, including their ability to modify texture. The formation of gels and the changes in rheology, mouthfeel, and stability are some ways hydrocolloids can impact food texture.

Texture modifiers, such as hydrocolloids, can alter the properties of food products by changing their rheology. This property affects the flow of liquids and determines whether a product is thick or thin. For example, carrageenan and xanthan gum are commonly used to thicken liquids, such as sauces and dressings, enhancing their viscosity and stability.

Hydrocolloids can also contribute to gel formation, which influences the texture of food products. Gelling agents, such as gelatin and pectin, can create solid-like networks that give food products structure and firmness. This property is desirable in jams, jellies, and dessert gels.

Hydrocolloids can also influence the mouthfeel of a food product. It refers to the sensation of texture experienced in the mouth when consuming a food product. Hydrocolloids can impact the mouthfeel of products by changing their viscosity, thickness, and cohesiveness. For example, guar gum and locust bean gum are commonly used to increase the viscosity and creaminess of dairy products like ice cream.

Lastly, hydrocolloids can affect the stability of food products, preventing ingredients from separating or settling over time. It is particularly important in salad dressings, where it is undesirable to dissolve oils and water-based ingredients.

13.2.6 Consumer Perception and Acceptance of Hydrocolloid-Enhanced Foods

Consumer perception and acceptance of hydrocolloid-enhanced foods play a crucial role in the success of food products that contain these ingredients. Sensory attributes such as texture, taste, appearance, and aroma can be affected by hydrocolloids, which in turn can influence consumer preferences.

Some consumers may be wary of using hydrocolloids in food products, especially health-conscious consumers with dietary restrictions. However, when used appropriately, these ingredients can contribute to developing healthier and more sustainable food products.

Food manufacturers must conduct sensory evaluations to ensure that hydrocolloids do not compromise the product's overall quality and sensory experience. Careful consideration of the type and amount of hydrocolloid used and the processing conditions can help optimize sensory attributes and improve consumer acceptance.

In addition, clear and accurate labeling regarding the presence and number of hydrocolloids can help to build consumer trust and ensure informed choices. Transparency in labeling can also promote awareness and understanding of the role of hydrocolloids in food products and their impact on sensory attributes.

Overall, monitoring consumer perception and acceptance of hydrocolloid-enhanced foods is crucial for the food industry to develop products that meet consumer preferences and needs while maintaining high-quality sensory attributes.

13.2.7 Potential Health and Safety Considerations of Hydrocolloids

As with any food additive, using hydrocolloids in food products raises potential health and safety considerations that must be considered. Regulatory authorities, such as the FDA in the United States, closely monitor the use of hydrocolloids and set specific safety guidelines to ensure consumer protection.

One of the main health concerns associated with hydrocolloids is their potential allergenicity. Some hydrocolloids, such as carrageenan and xanthan gum, have been identified as potential allergens, and their use must be carefully regulated to avoid adverse reactions in susceptible individuals. Regulatory bodies require that food products containing hydrocolloids be clearly labeled to indicate their presence, allowing consumers with allergies or intolerances to make informed choices.

In addition to allergenicity, other health risks may be associated with the consumption of hydrocolloids. However, the available studies have not identified significant safety concerns for most hydrocolloids when used within the recommended dosage ranges. Nevertheless, researchers continue to explore potential health issues related to the long-term consumption of hydrocolloids, and regulatory authorities maintain close oversight to ensure their safety.

Hydrocolloids in food products are generally considered safe when proper labeling, dosage, and regulatory compliance are observed. As with any food additive, evaluating potential health and safety risks and taking appropriate precautions to minimize them is essential.

13.2.8 Future Trends and Innovations in Hydrocolloid Applications

Hydrocolloids have been widely used in the food industry for decades, but ongoing research and innovation reveal exciting new possibilities for these versatile ingredients. As the industry evolves, hydrocolloids will likely play a crucial role in functional food development and formulation. Some of the emerging trends and innovations in hydrocolloid applications include:

- **Plant-based alternatives:** With the increasing demand for plant-based products, hydrocolloids derived from diverse sources such as seaweed, starches, and gums are becoming more popular. These alternatives can provide similar functional properties to traditional hydrocolloids while also meeting the needs of consumers seeking more sustainable and ethical options.
- **Biopolymer blends:** Blending hydrocolloids with other biopolymers, such as proteins and fibers, can result in unique texture and stability properties. This approach allows for developing products that meet specific functional requirements while minimizing the use of synthetic additives.
- **Functional properties beyond Texture:** Hydrocolloids are known for modifying texture, but new research is exploring other useful properties, such as antioxidant and antimicrobial effects. The potential health benefits of hydrocolloid-enriched products could be a key factor in future product development and marketing strategies.
- **New uses in food applications:** Hydrocolloids are currently employed in many food products, but new applications are emerging. For example, hydrocolloids are being investigated as potential fat replacers or as a means of improving the nutritional profile of products.
- **Customized formulations:** As the food industry evolves, the need for tailored product formulations will likely increase. Hydrocolloid suppliers respond to this demand by offering customized blends to meet specific functional and sensory requirements.

These emerging trends and innovations in hydrocolloid applications will likely significantly impact the food industry in the coming years. By leveraging the unique functional properties of hydrocolloids, food manufacturers can develop products that meet evolving consumer needs while also addressing industry challenges.

Conclusion

Labeling and claims related to hydrocolloids are critical for ensuring consumer safety and informed choices. Hydrocolloids serve numerous functions as food additives, such as thickening, gelling, stabilizing, emulsifying, water-binding, and modifying viscosity. Compliance with regulatory guidelines is essential to guarantee consumer protection.

Practical considerations, such as dosage, processing conditions, and compatibility with other ingredients, must be considered to optimize the use of hydrocolloids in food product formulation. Hydrocolloids can modify rheological properties, contribute to gel formation, and influence mouthfeel, which affects consumer perception and acceptance.

Although allergenic risks and potential adverse effects exist, hydrocolloids have been deemed safe for consumption. Ensuring product stability and shelf life, as well as addressing evolving consumer needs, are a priority in the food industry.

The food industry continuously explores new hydrocolloid applications that contribute to product development and formulation advancements. The emergence of functional properties and new hydrocolloid uses are expected to meet evolving consumer demands and tackle industry challenges.

Some promising trends include using hydrocolloids as fat substitutes, coating agents, or edible films that enhance product quality and shelf life. Innovations in encapsulation technology and hybridizing hydrocolloids with other ingredients may lead to new opportunities in food manufacturing.

By staying abreast of these advancements and understanding the role of hydrocolloids, food manufacturers can create enhanced products that meet the evolving needs of consumers while ensuring their safety and satisfaction.

FAQ

- **Why is proper labeling important for hydrocolloids?**

  Proper labeling is important for hydrocolloids to ensure consumer safety and informed choices. Accurate and transparent labeling allows consumers to make informed decisions about the products they purchase and consume.

- **What are the regulatory guidelines for hydrocolloid labeling in the United States?**

  The relevant authorities set the regulatory guidelines for hydrocolloid labeling in the United States. Compliance with these guidelines is important to ensure consumer protection. Specific labeling requirements may exist for health benefits or functional property claims.

- **What factors should be considered when using hydrocolloids in food products?**

  Practical considerations when using hydrocolloids in food product formulation include dosage, processing conditions, and compatibility with other ingredients. Optimizing the use of hydrocolloids can help achieve the desired texture and stability in various food applications.

- **How do consumers perceive hydrocolloid-enhanced foods?**

  Consumer perception and acceptance of hydrocolloid-enhanced foods may be influenced by sensory attributes that hydrocolloids can affect. It can impact consumer preferences and purchasing decisions.

- **Are there any potential health and safety considerations associated with hydrocolloids?**

  Potential health and safety considerations associated with hydrocolloid use include regulatory restrictions or guidelines, allergenic risks, and any known adverse effects or precautions related to their consumption.

- **What are future trends and innovations in hydrocolloid applications within the food industry?**

  Future trends and innovations in hydrocolloid applications include emerging functional properties and new hydrocolloid uses. These advancements contribute to product development and formulation within the food industry, helping to meet evolving consumer demands and address industry challenges.

CHAPTER 14

Research and Innovations in Hydrocolloids

PART 14.1

Hydrocolloids as a research topic

This chapter will provide insights into some general hydrocolloid trends and innovations. However, for the latest developments, It is recommended to check recent scientific literature, industry publications, and news sources for the most up-to-date information.

Core Insights:

- ☑ Hydrocolloids are versatile compounds crucial in food, with diverse sources like plants and seaweeds.
- ☑ Their applications span clean label formulations, plant-based product development, molecular gastronomy, and personalized nutrition.
- ☑ Ongoing innovations include sustainable sourcing, synergistic blends, 3D printing and microencapsulation advancements, reflecting a dynamic intersection of technology and culinary arts.
- ☑ Formulations are tailored to specific dietary needs, and digital tools optimize hydrocolloid use, highlighting their pivotal role in shaping modern food trends.

Here are some areas of research and innovation in hydrocolloids:

Sustainable Sourcing

Research Focus: There is a growing interest in sourcing hydrocolloids sustainably, with a particular emphasis on seaweed-derived hydrocolloids such as agar-agar and carrageenan. Researchers are exploring eco-friendly cultivation practices to ensure a stable supply while minimizing environmental impact.

Clean Label Formulations

Innovation: Formulators increasingly seek clean-label alternatives to traditional hydrocolloids, focusing on natural and minimally processed options. It aligns with consumer preferences for transparent ingredient lists.

Hydrocolloids in Personalized Nutrition

Research Focus: Tailoring hydrocolloid formulations to meet specific dietary needs is an area of ongoing research. It includes developing hydrocolloids suitable for various dietary restrictions, such as gluten-free, vegan, or allergen-free options.

Synergistic Blends of Hydrocolloids

Innovation: Researchers are exploring the synergies between hydrocolloids to enhance their functionalities. Blending hydrocolloids in specific ratios can improve stability, texture, and sensory properties in various food applications.

Microencapsulation Techniques

Research Focus: Microencapsulation of hydrocolloids is being investigated for the controlled release of flavors, colors, and bioactive compounds in food products. This technology enhances the stability and effectiveness of hydrocolloids in specific applications.

Nanostructured Hydrocolloids

Innovation: Nano-structuring hydrocolloids is an emerging area aiming to improve their functional properties. It includes creating nano-sized particles or networks enhancing solubility, gelling properties, and bioavailability.

Responsive Hydrocolloids

Research Focus: Developing hydrocolloids that respond to specific stimuli, such as pH, temperature, or shear, is an area of interest. Responsive hydrocolloids can offer controlled release properties or adapt to changing conditions during processing.

Hydrocolloids in 3D Printing of Food

Innovation: Hydrocolloids are being explored for their potential in the 3D printing of food. Researchers are investigating their role in creating structures and textures in printed foods, opening up possibilities for customized culinary experiences.

Health-Promoting Hydrocolloids

Research Focus: Hydrocolloids with potential health benefits, such as prebiotic properties or the ability to modulate the glycemic response, are areas of active investigation. It aligns with the broader trend of functional and health-promoting ingredients in food.

Digitalization and Hydrocolloid Formulation

Innovation: Computational tools and modeling are increasingly used to optimize hydrocolloid formulations. Digital approaches allow for more efficient screening of formulations, reducing the time and resources required for product development.

It's important to note that the field of hydrocolloids is dynamic, and ongoing research and innovation contribute to the continuous evolution of these ingredients. I recommend checking recent scientific journals, industry conferences, and reports from research institutions and companies involved in food science and technology for the latest developments.

Here are some highlighted research articles published in different journals and magazines.

14.1.1 Recent trends in oil structuring using hydrocolloids

Santiago Bascuas , Pere Morell , Isabel Hernando , Amparo Quiles

In this article, the researchers discuss using hydrocolloids to create healthy and sustainable food structures. The paper highlights the challenges in structuring oil with hydrocolloids, such as the difficulty dispersing hydrophilic polymers in oil to achieve the necessary structure and network formation required for gelation. The article also discusses the indirect methods used to structure oil with hydrocolloids, including the emulsion-template approach, solvent exchange procedure, physical sorption of oil in porous structures, and HIPE gelation.

The article emphasizes the importance of using hydrocolloids to create healthy food structures, as solid fats are associated with adverse chronic health effects. The use of hydrocolloids in creating gelled emulsions and oleogels provides a promising solution to substitute saturated and trans fats. The article also discusses the benefits of using polysaccharides and proteins in creating healthy colloidal systems for food, such as their supramolecular interactions that can produce novel structures with unique sensory, texture, and stability properties. The article references various studies and research papers to support the information presented.

The article's authors also provide their insights and opinions on the topic based on their research and experience in the field. The paper provides a comprehensive overview of recent oil-structuring trends using hydrocolloids and highlights the potential impact on the food industry.

Reference:

Bascuas, Santiago, Pere Morell, Isabel Hernando, and Amparo Quiles. "Recent trends in oil structuring using hydrocolloids." Food Hydrocolloids 118 (2021): 106612.

14.1.2 Hydrocolloids acting as emulsifying agents - How do they do it?

Eric Dickinson

This paper provides a comprehensive overview of the functionality of hydrocolloids as emulsifying agents, focusing on the essential features controlling the formation and stabilization of oil-in-water emulsions. The document delves into the theoretical principles underlying polymer adsorption and steric stabilization by polymer layers, as well as the experimental interfacial functionality of three classes of biopolymer emulsifying agents: gum arabic, pectin, and hydrophobically modified starch (OSA starch).

The document begins by discussing the basic generic principles describing the theory of interfacial stabilization by model polymers with different sorts of molecular architecture. It highlights the relevance and applicability of these basic physicochemical principles to the known molecular structures and colloid stabilizing properties of the three other hydrocolloid ingredients.

The authors aim to illuminate an underlying essence of physical understanding while recognizing the diversity and molecular complexity of the individual systems involved. It also provides insights into the adsorbed polymer layers and colloid stability. It discusses the influence of polymer concentration and molecular structure on the adsorbed layer's capability to confer effective stability to closely approaching droplet surfaces. The document explores the impact of homopolymers, copolymers, and branched block copolymers on the strength of emulsions, shedding light on the structural characteristics and steric stabilizing ability of predominantly hydrophilic copolymers.

Furthermore, the document acknowledges the complexity of food hydrocolloids and the idealized nature of statistical models about the complexity of these biopolymers. It emphasizes the consensus, supported by extensive experimental evidence, that the simplified statistical models provide a sound mechanistic basis for describing the essential features of adsorption and colloidal stabilization by all sorts of macromolecules, including food biopolymers.

The key message from the statistical models is that steric stabilization is most effectively achieved through the adsorption of copolymers, combining anchoring groups that stick directly to the surface and chain regions that extend away from the surface. In conclusion, the document addresses the underlying mechanical question of how food hydrocolloids function as emulsifying agents. It outlines the basic principles of polymer adsorption.

It assesses the accumulated evidence relating to three practically relevant case studies, providing a valuable resource for understanding the molecular mechanisms behind the emulsifying properties of hydrocolloids. Overall, this document offers a detailed and insightful exploration of the functionality of hydrocolloids as emulsifying agents, providing a valuable resource for researchers, students, and professionals in the food science and colloid chemistry fields.

Reference:

Dickinson, Eric. "Hydrocolloids acting as emulsifying agents–How do they do it?." Food Hydrocolloids 78 (2018): 2-14.

14.1.3 Gels, emulsions and application of hydrocolloids at Phillips Hydrocolloids Research Centre

Katsuyoshi Nishinari, Yapeng Fang, Nan Yang, Xaolin Yao, Meng Zhao, Ke Zhang, Zhiming Gao

This paper provides an overview of the research activities at Phillips Hydrocolloids Research Centre in Wrexham, Wales and Wuhan, China. The document covers the historical development, extensively studied topics and the relation between the structure and physicochemical properties of various hydrocolloids and protein-polysaccharide interactions. The paper begins with a retrospective introduction celebrating the 90th birthday of Glyn Phillips, the paternal mentor of the authors.

It then describes the historical development of the research activities in both laboratories, including establishing the Gums and Stabilisers for the Food Industry conferences and the International Hydrocolloids Conference. The extensively studied topics at Phillips Hydrocolloids Research Centre include the rheology and structure of hydrocolloids, the interaction between hydrocolloids and proteins, and the development new hydrocolloid-based products.

The document also discusses the relationship between various hydrocolloids' structure and physicochemical properties, such as xanthan gum, carrageenan, and pectin. The authors also highlight some of their key findings, such as the effect of ionic strength on the rheological properties of hydrocolloid solutions, the role of electrostatic interactions in protein-polysaccharide interactions, and the use of hydrocolloids in the development of low-fat and low-sugar food products. Overall, the document provides a comprehensive overview of the research activities at Phillips Hydrocolloids Research Centre and in Wuhan, China, highlighting the importance of hydrocolloids in the food industry.

Reference:

Nishinari, Katsuyoshi, Yapeng Fang, Nan Yang, Xaolin Yao, Meng Zhao, Ke Zhang, and Zhiming Gao. "Gels, emulsions and application of hydrocolloids at Phillips Hydrocolloids Research Centre." Food Hydrocolloids 78 (2018): 36-46.

14.1.4 The future trends of food hydrocolloids

Wei Lu, Katsuyoshi Nishinari, Shingo Matsukawa, Yapeng Fang

The document provides a comprehensive overview of current and future trends in studying food hydrocolloids, focusing on researchers' increasing interest and contributions in China and Japan. It highlights the growing importance of food hydrocolloids in developing modern food products, especially in addressing the evolving demands for customized, functional, and health-beneficial foods. The introduction emphasizes the significance of food as an essential aspect of human life and the need for food products that cater to specific populations, such as older people.

It also addresses the critical issue of foreign body aspiration (FBA) in infants and children, emphasizing the importance of food structure design principles in developing foods to reduce FBA-related risks. The document underscores the pivotal role of food hydrocolloids as the foundation of food structures, influencing various aspects such as processing, flavor, nutrition, and health benefits. It also emphasizes the need for a multidisciplinary approach to further explore the theory and methodology of modern food structure design from the perspective of food hydrocolloids.

Furthermore, the document discusses the successful organization of the 1st NSFC-JSPS Joint Symposium on 'Food Hydrocolloids for Food Function Design' in Shanghai, China, which brought together scholars and delegates from China and Japan. The symposium featured presentations on a wide range of topics, including the development of new natural functional polysaccharides, regulation effects of natural polysaccharides on gut microbiota, food oral processing, 3-D printing of food structures, and food-hydrocolloids-based microencapsulation and delivery technologies.

The symposium facilitated extensive discussions among delegates, identifying key scientific questions regarding the future fundamental study of food hydrocolloids. These questions encompassed the interaction of food hydrocolloids with food components, future functional food structure design, food colloidal nutrition, and the safety of food colloidal particles. Overall, the document provides valuable insights into the current advancements and future directions in studying food hydrocolloids, emphasizing the need for collaboration and interdisciplinary research to address the evolving challenges and opportunities in the field.

Reference:

Lu, Wei, Katsuyoshi Nishinari, Shingo Matsukawa, and Yapeng Fang. "The future trends of food hydrocolloids." Food Hydrocolloids 103 (2020): 105713.

14.1.5 Effect of the addition of hydrocolloids on beef texture: Targeted to the needs of people with dysphagia

Nelum Pematilleke, Mandeep Kaur, Carleen Tse Rai Wai, Benu Adhikari, Peter J. Torley

The research investigated the impact of hydrocolloids on the properties of beef patties, particularly focusing on texture, color, and diameter reduction. The addition of hydrocolloids was found to significantly alter the texture, color, and cooking yield of the patties. It was observed that hydrocolloids interacted with meat proteins, forming protein-polysaccharide gels, which affected the patties' thermal denaturation and textural characteristics. The study also revealed that the cooking yield was higher when hydrocolloids were added, although the diameter reduction varied depending on the type of hydrocolloid used.

Furthermore, the compliance of the patties with the IDDSI fork pressure test for a potential level 6 diet was assessed. It was found that only carboxymethyl cellulose and xanthan gum complied with the test, indicating their potential suitability for level 6 dysphagia diets. Additionally, the color properties of the patties were affected by the addition of hydrocolloids, with increases in lightness and yellowness observed.

The study highlighted that different hydrocolloids have varying effects on meat products, and their use can be crucial in developing dysphagia foods. The research emphasized the importance of texture modification technologies in producing foods suitable for individuals with swallowing difficulties.

In conclusion, the study's findings provide valuable insights into the impact of hydrocolloids on beef patty properties, particularly in the context of dysphagia diets. The detailed analysis of texture, color, and compliance with dietary standards contributes to understanding how hydrocolloids can meet the specific nutritional requirements of individuals with swallowing difficulties.

Reference:

Pematilleke, Nelum, Mandeep Kaur, Carleen Tse Rai Wai, Benu Adhikari, and Peter J. Torley. "Effect of the addition of hydrocolloids on beef texture: Targeted to the needs of people with dysphagia." Food Hydrocolloids 113 (2021): 106413.

14.1.6 Strides in food texture and hydrocolloids

Dennis Seisun, Nesha Zalesny

The article discusses the global market for food hydrocolloids, estimating it to be around US$7.6–7.8 billion in 2020. It highlights the key drivers of the market, including consumer behavior, texture technology, raw materials, and regulatory status. Consumers increasingly seek convenient, nutritional, and good-tasting foods with fewer ingredients and clean labels. The article also emphasizes the growing trend towards plant-based products, which address environmental and animal rights concerns but often contain many elements.

The authors predict that the trend towards plant-based and aquaculture products will continue growing, driven by environmental concerns, sustainable practices, and global warming. They also note the increasing interest and investment in aquaculture, particularly in seaweed and its derivatives, which are integral to the aquaculture industry. New seaweed bioactive ingredients attract significant interest and buy, which bodes well for food texturizing hydrocolloids.

The article also addresses the impact of the COVID-19 pandemic on the food industry, highlighting challenges such as shipping rate hikes, labor shortages, and production delays. Despite these challenges, the texture aspect of food has not suffered as much and may have even benefited from increased processed food consumption.

Furthermore, the article discusses the differentiation of hydrocolloids through processing technology, third-party certifications, and sources of derivation. It also mentions the time-consuming and costly approval process for new hydrocolloids and the regulatory differences between Europe and the U.S. in how hydrocolloids must be declared on food labels.

Overall, the article provides insights into the food hydrocolloid market's current and future trends, emphasizing the importance of addressing consumer preferences, sustainability, and regulatory considerations.

Reference:

Seisun, Dennis, and Nesha Zalesny. "Strides in food texture and hydrocolloids." Food Hydrocolloids 117 (2021): 106575.

14.1.7 Effect of different hydrocolloids on gluten proteins, starch and dough microstructure

Jinxin Li, Madhav P. Yadav, Jinlong Li

The study investigated the interaction of six hydrocolloids with gluten proteins and wheat starch in dough systems. The hydrocolloids examined were λ-carrageenan, high methoxyl orange pectin, guar gum, xanthan gum, locust bean gum, and carboxymethyl cellulose. The research found that these hydrocolloids significantly impact the gluten proteins, leading to structural changes. Specifically, linear anionic and nonionic hydrocolloids were observed to interact with gluten proteins, while others acted as fillers in the gluten network. The study provides valuable insights into the relationship between the actions of hydrocolloids and their physiochemical properties. The findings contribute to a better understanding of the behavior of hydrocolloids in dough systems and their effects on gluten proteins and starch.

Reference:

Li, Jinxin, Madhav P. Yadav, and Jinlong Li. "Effect of different hydrocolloids on gluten proteins, starch and dough microstructure." Journal of Cereal Science 87 (2019): 85-90.

14.1.8 Chemistry, gelation, and enzymatic modification of seaweed food hydrocolloids

Nanna Rhein-Knudsen, Anne S. Meyer

The document discusses seaweed-derived carbohydrate hydrocolloids' chemistry, gelation, and enzymatic modification, including carrageenan, agar, and alginate. These hydrocolloids thicken and gell agents in various food, pharmaceutical, and biotechnology industries.

The gelation process involves chemical modifications and extraction methods. Chemical changes include converting precursor units to different forms, while extraction methods typically involve boiling water and alcohol precipitation. The M/G ratio of alginate is highlighted as a significant factor affecting gelation properties, with low ratios resulting in firm and brittle gels and high ratios leading to more elastic gels.

Adding seaweed hydrocolloids to foods can impact flavor-release sensory perception and induce structural changes in meat proteins. Enzymes, such as carrageenases and agarases, are emphasized for their crucial role in the biosynthesis and modification of seaweed hydrocolloids.

The document provides insights into seaweed-derived hydrocolloids' properties, applications, and enzymatic modifications, shedding light on their significance in various industrial processes.

Reference:

Rhein-Knudsen, Nanna, and Anne S. Meyer. "Chemistry, gelation, and enzymatic modification of seaweed food hydrocolloids." Trends in Food Science & Technology 109 (2021): 608-621.

14.1.9 Food hydrocolloids and health claims

Christer Viebke, Saphwan Al-Assaf, Glyn. O. Phillips

The document discusses the significance of food hydrocolloids, particularly their dietary fiber content and health benefits. It emphasizes the need for different methods to measure the dietary fiber content of various food ingredients. The submission of hydrocolloid health claims to the EFSA under article 13.1 is highlighted, with some claims achieving positive outcomes. The document also delves into the importance of clinical studies in demonstrating the positive effects of certain hydrocolloids on physiological functions. It provides insights into obtaining a health claim, including identifying relevant biomarkers, study design, and obtaining statistical data for the targeted demographic population.

Additionally, it discusses the challenges and considerations in submitting health claims for different hydrocolloids, such as gum acacia, sugar beet fiber, konjac mannan, beta-glucan, and PHGG. The document concludes by emphasizing the increasing demand for healthy food products and the opportunities for fiber suppliers to develop and supply quality fibers for incorporation into new food formulations. It also highlights the potential for food hydrocolloid producers to move their products into more value-added applications beyond their traditional uses.

Reference:

Viebke, Christer, Saphwan Al-Assaf, and Glyn O. Phillips. "Food hydrocolloids and health claims." Bioactive Carbohydrates and Dietary Fiber 4, no. 2 (2014): 101-114.

14.1.10 Improvement of gluten-free bread and cake properties using natural hydrocolloids: A review

Fakhreddin Salehi

The study focuses on the effects of gums on gluten-free bakery products. Gluten-free bakery products often use gums as substitutes for gluten to enhance dough performance, bread and cake characteristics, and shelf life. Gums have been found to increase water absorption capacity, improving the texture and volume of bakery products. Various gums, including xanthan, guar, acacia, and carrageenan, have been studied for their positive effects on gluten-free bread, cakes, and biscuits. These gums have been shown to modify the pasting properties of starch, improve moisture retention, and enhance the overall quality of gluten-free bakery products. Additionally, studies have demonstrated that gums can positively affect gluten-free bakery products' rheological, physical, and sensory properties.

The findings suggest that using gums in gluten-free bakery products can significantly improve their quality and characteristics, making them more comparable to traditional gluten-containing products. It has important implications for individuals with gluten intolerance or celiac disease, as it provides a means to produce high-quality gluten-free bakery items that are more palatable and enjoyable. Overall, the study highlights the potential of gums as valuable ingredients in developing gluten-free bakery products.

Reference:

Salehi, Fakhreddin. "Improvement of gluten-free bread and cake properties using natural hydrocolloids: A review." Food science & nutrition 7, no. 11 (2019): 3391-3402.

14.1.11 The use of selected hydrocolloids to enhance cooking quality and hardness of zero-salt noodles

Hui-Ling Tan, Thuan-Chew Tan & Azhar Mat Easa

The study focuses on the importance of salt reduction in food products to prevent health issues. Specifically, it examines the potential of hydrocolloids in improving noodles' texture and water-holding capacity, which is crucial for enhancing the quality of zero-salt noodles. The research involved testing different types and levels of hydrocolloids in noodle dough to evaluate their impact on various properties.

The water-holding capacity of dough was assessed for control, T.S., and hydrocolloid dough. The ability to hold water within the dough after being subjected to stress, such as centrifugation, was measured. Results indicated that adding hydrocolloids, such as T.S. and others, improved the water-holding capacity of the dough.

The study also investigated the pH of cooked noodles at different concentrations of hydrocolloids. The pH of the cooked noodles was measured to assess the impact of hydrocolloids on this property. The results showed variations in pH levels based on the type and concentration of hydrocolloids used in the noodles.

Furthermore, the research examined the cooking and textural properties of the noodles. Parameters such as optimum cooking time, cooking yield, cooking loss, and textural attributes were evaluated for control and hydrocolloid-added noodles. The findings provided insights into the effects of hydrocolloids on the cooking and textural characteristics of the noodles.

Overall, the study demonstrated that hydrocolloids have the potential to enhance the quality of zero-salt noodles by improving their water-holding capacity, pH, and textural properties. This research contributes valuable information for developing healthier food products with reduced salt content.

Reference:

Tan, Hui-Ling, Thuan-Chew Tan, and Azhar Mat Easa. "The use of selected hydrocolloids to enhance cooking quality and hardness of zero-salt noodles." International Journal of Food Science & Technology 53, no. 7 (2018): 1603-1610.

14.1.12 Cryoprotectant for frozen foods

Tanushree Maity, Alok Saxena & P. S. Raju

The document discusses the use of hydrocolloids in food science and nutrition, particularly their role in improving the quality and stability of various food products. Hydrocolloids, such as polysaccharides, are utilized in food applications for gelling, thickening, stabilizing, and emulsifying. They have been found to significantly impact food systems' rheological and textural characteristics by altering viscosity. Additionally, hydrocolloids play a crucial role in retaining the texture of fruits and vegetables after freezing, and they provide thermodynamic stability to ice cream to control the re-crystallization process.

Furthermore, the document highlights hydrocolloids' water-binding ability, making them suitable for frozen surimi, minced fish, and meat products. They are also instrumental in improving the shelf-stability of frozen bakery products by retaining moisture and retarding staling. It is noted that different hydrocolloids can impart varying cryoprotective effects to food products, depending on their solubility, water-holding capacity, and rheological properties.

The document also emphasizes using thermal analytical techniques to study the cryoprotective effects of various hydrocolloids on food products. Moreover, it discusses the role of hydrocolloids in enhancing the quality of baked products and frozen starch gels. The document provides valuable insights into the diverse applications and benefits of hydrocolloids in food science and nutrition.

Reference:

Maity, Tanushree, Alok Saxena, and P. S. Raju. "Use of hydrocolloids as a cryoprotectant for frozen foods." Critical Reviews in Food Science and Nutrition 58, no. 3 (2018): 420-435.

Conclusion:

In conclusion, this chapter has provided a comprehensive overview of hydrocolloid research and innovations, showcasing their diverse applications and contributions to the food industry. Hydrocolloids, derived from various sources, including plants and seaweeds, are pivotal in shaping modern food trends. They are instrumental in clean label formulations, plant-based product development, molecular gastronomy, and personalized nutrition. Ongoing innovations in hydrocolloids include sustainable sourcing, synergistic blends, microencapsulation techniques, nano-structuring, responsive properties, and even their use in the 3D printing of food.

Furthermore, this chapter highlights the importance of tailoring hydrocolloid formulations to meet specific dietary needs, leveraging digital tools for optimization, and exploring their potential health benefits. It's important to note that the field of hydrocolloids is dynamic, with ongoing research contributing to continuous advancements and new applications.

For the latest developments in hydrocolloids, it is recommended to refer to recent scientific literature, industry publications, and news sources.

FAQ

- **What are hydrocolloids, and why are they important in the food industry?**

  Hydrocolloids are versatile compounds derived from various sources, such as plants and seaweeds, that have a crucial role in the food industry. They are important for their ability to modify texture, stabilize emulsions, and enhance the sensory properties of food products.

- **What are some recent trends and innovations in hydrocolloids?**

  Recent trends and inno5vations in hydrocolloids include sustainable sourcing, clean label formulations, personalized nutrition, synergistic blends, microencapsulation techniques, nano-structuring, responsive properties, 3D printing of food, and health-promoting hydrocolloids.

- **How do hydrocolloids contribute to sustainable food production?**

  Hydrocolloids, especially those derived from seaweeds, are being explored for sustainable sourcing methods that minimize environmental impact. They offer eco-friendly alternatives for various food applications.

- **What is the role of hydrocolloids in personalized nutrition?**

  Hydrocolloids can be tailored to meet specific dietary needs, such as gluten-free or allergen-free options. They enable the customization of food products to match individual nutritional requirements.

- **How do synergistic blends of hydrocolloids improve food products?**

  Researchers are exploring the synergies between different hydrocolloids to enhance their functionalities. Blending hydrocolloids in specific ratios can improve stability, texture, and sensory properties in various food applications.

- **What are some potential health benefits of certain hydrocolloids?**

  Some hydrocolloids are being investigated for their potential health benefits, such as prebiotic properties or the ability to modulate the glycemic response. These align with the trend of functional and health-promoting ingredients in food.

FREQUENTLY ASKED QUESTIONS (FAQS) CONTINUED

How are digital tools used to optimize hydrocolloid formulations?

Computational tools and modeling are increasingly used to optimize hydrocolloid formulations. Digital approaches allow for more efficient screening of formulations, reducing time and resource requirements in product development.

How are hydrocolloids utilized in the 3D printing of food?

Hydrocolloids are being explored for their potential role in the 3D printing of food. Researchers investigate how hydrocolloids can create specific structures and textures in printed foods, enabling customized culinary experiences.

PART 14.2

Insights and Future Predictions of Hydrocolloids

Hydrocolloids are crucial in numerous industries, including food, pharmaceuticals, cosmetics, and environmental applications. These natural or synthetic substances have unique properties that make them valuable ingredients, ranging from stabilizing emulsions enhancing the sensory experience, and providing viscosity.

The hydrocolloid industry is rapidly expanding, with a growing demand for innovation and sustainability. As technology advances and consumer preferences evolve, the future of hydrocolloids presents exciting opportunities for growth and advancement.

Core Insights:

- ☑ Hydrocolloids are essential in various industries, including food, pharmaceuticals, cosmetics, and environmental applications.
- ☑ The hydrocolloid industry is rapidly expanding, with a growing demand for innovation and sustainability.
- ☑ The future of hydrocolloids presents exciting opportunities for growth and advancements.

14.2.1 Hydrocolloids in Pharmaceutical and Medical Applications

Hydrocolloids play a vital role in developing various pharmaceutical and medical applications thanks to their unique properties, such as biocompatibility, biodegradability, and non-toxicity. This section will explore some of the common applications of hydrocolloids in the pharmaceutical and medical industry, highlight their benefits, and discuss future trends.

Wound Dressings

Hydrocolloids are extensively used in wound dressings because they form a gel-like mass upon contact with wound exudates, which helps create an occlusive barrier that protects the wound from infection and promotes healing. Hydrocolloid dressings are also known for their pain-reducing properties, making them beneficial for patients who suffer from chronic injuries.

According to a report by Allied Market Research, the global wound dressing market is expected to reach $12.9 billion by 2027, owing to the rising incidences of chronic wounds, ulcers, and burns. This growth is expected to drive the demand for hydrocolloid-based wound dressings even further.

Drug Delivery Systems

Hydrocolloids are widely used in drug delivery systems, especially in oral medications, where they act as a thickener or emulsifier, enhancing the viscosity of the cure while improving its stability and bioavailability. Hydrocolloids are also used in transdermal drug delivery systems, which help maintain a steady drug release through the skin.

One of the key benefits of using hydrocolloids in drug delivery systems is their ability to protect drugs from degradation, which increases the shelf life of the medication. Hydrocolloids are also biocompatible, meaning they do not produce harmful side effects when introduced into the body.

Dental Applications

Hydrocolloids are used in various dental applications, including denture adhesives, impression materials, and mouthwashes. In denture adhesives, hydrocolloids provide a strong bond that keeps dentures in place, while in impression materials, they help to capture fine details of the teeth and gums. Hydrocolloids are also used in mouthwashes to provide a thick and smooth texture that enhances the sensory experience for users.

Future Trends in Hydrocolloid-based Pharmaceuticals and Medical Devices

The use of hydrocolloids in the pharmaceutical and medical industry is expected to continue to grow, with new applications being developed regularly. One of the emerging trends in hydrocolloid-based pharmaceuticals is the use of hydrogels, which are three-dimensional networks of hydrophilic polymers that can hold large amounts of water. Hydrogels are being developed for drug delivery, wound healing, and tissue engineering applications.

Another trend is using modified hydrocolloids, which have been chemically modified to enhance their properties. Modified hydrocolloids are being developed for applications such as controlled drug release, tissue engineering, and wound healing.

Overall, hydrocolloids have immense potential in the pharmaceutical and medical industry thanks to their unique properties and versatility. As new applications are developed and demand for hydrocolloid-based products continues to increase, we can expect further growth in this area.

Hydrocolloids in Cosmetics and Personal Care Products

Hydrocolloids are widely used in the cosmetics and personal care industry due to their unique properties that benefit manufacturers and consumers. They are used in various products, including creams, lotions, gels, and hair care products.

Applications of Hydrocolloids in Cosmetics and Personal Care Products

One of the primary applications of hydrocolloids in cosmetics and personal care products is to stabilize emulsions. Emulsions are mixtures of two immiscible liquids, oil and water, held together by a third component called an emulsifier. Hydrocolloids act as stabilizers by absorbing onto the surfaces of the droplets, preventing them from coalescing and separating.

Hydrocolloids also provide viscosity to products, which is important in controlling the flow and texture. For example, carrageenan is commonly used in toothpaste to give a thick and creamy texture. Additionally, hydrocolloids can enhance the sensory experience of using a product by providing a smooth and silky feel.

Benefits of Hydrocolloids in Cosmetics and Personal Care Products

The benefits of using hydrocolloids in cosmetics and personal care products are numerous. Stabilizing emulsions increases the shelf life of products and prevents the need for preservatives. Additionally, they improve the texture and feel of products, making them more appealing to consumers.

Hydrocolloids are also versatile and can be used in a variety of formulations. Their ability to provide viscosity and enhance sensory experience makes them valuable in creating products that meet specific consumer needs and preferences. For example, hydrocolloids can create lightweight lotions or heavy creams, depending on the desired outcome.

> "Hydrocolloids provide numerous benefits to both manufacturers and consumers."

14.2.2 Environmental Applications of Hydrocolloids

Aside from their significance in various industries, hydrocolloids are also used in environmental applications. They have emerged as a sustainable solution to humanity's numerous environmental problems. Hydrocolloids are natural, biodegradable, and non-toxic, making them ideal for environmental applications.

Wastewater Treatment

One of the most significant environmental applications of hydrocolloids is in wastewater treatment. Hydrocolloids separate solid waste from wastewater, improving its quality before releasing it into the environment. Adding hydrocolloids to wastewater creates larger flocs, which settle faster, making separating them from the treated wastewater easier.

Table 97: Benefits and Applications of Hydrocolloids in Wastewater Treatment

BENEFITS OF HYDROCOLLOIDS IN WASTEWATER TREATMENT	APPLICATIONS OF HYDROCOLLOIDS IN WASTEWATER TREATMENT
Reduce the amount of sludge produced in wastewater treatment plants	Treatment of municipal and industrial wastewater
Improve the efficiency of wastewater treatment processes	Separation of solid waste from wastewater
Lower the cost of wastewater treatment.	

Soil Erosion Control

Another application of hydrocolloids is in soil erosion control. Hydrocolloids are added to soil to improve its structure, making it more erosion-resistant. Hydrocolloids also help to retain water in soil, reducing the need for irrigation and preventing soil dehydration.

Hydrocolloids are mixed with seeds to create a hydroseed mixture sprayed onto the soil to encourage vegetation growth and stabilize the soil.

The hydroseed mixture can also be used for roadside vegetation, golf courses, and landfills.

Biodegradable materials

Hydrocolloids can also be used to produce environmentally friendly biodegradable materials. These materials can replace traditional plastic products that are not biodegradable and harm the environment. Biodegradable hydrocolloid-based materials can be used in various industries, including packaging, agriculture, and construction. The demand for sustainable solutions is rising with the increasing global environmental awareness. Hydrocolloids are paving the way for a greener future.

14.2.3 Emerging Trends in Hydrocolloid Manufacturing

The hydrocolloid industry constantly evolves, with new trends and innovations always emerging. Manufacturing is at the heart of this evolution, with improvements in production methods and extraction techniques leading to advancements in the properties of hydrocolloids. Here are some of the key emerging trends in hydrocolloid manufacturing:

- **Green Manufacturing:** With sustainability becoming increasingly important for manufacturers, many are adopting green hydrocolloid manufacturing methods. It includes using renewable resources and eco-friendly production processes that minimize waste and carbon footprint.
- **Nanotechnology:** Nanotechnology in hydrocolloid manufacturing opens new possibilities for enhanced product properties, such as improved solubility and stability. Nanoparticles can be incorporated into hydrocolloid structures to create novel functional materials for various applications.
- **Bioreactor Technology:** Bioreactors are being used to produce hydrocolloids through fermentation, offering a more efficient and cost-effective manufacturing method. Bioreactor design and automation advances also increase productivity and consistency in the production process.

These emerging trends in hydrocolloid manufacturing are just the tip of the iceberg, with many more innovations on the horizon. As manufacturers continue to invest in research and development, the properties and applications of hydrocolloids are sure to evolve and expand in exciting ways.

14.2.4 The Growing Hydrocolloid Market

The hydrocolloid market has been steadily growing, driven by increased demand for processed food products and the expanding scope of hydrocolloid applications in various industries, according to a report by MarketsandMarkets, the global hydrocolloid market is expected to reach USD 13.6 billion by 2025, growing at a CAGR of 5.3% from 2020 to 2025.

The food industry remains the largest consumer of hydrocolloids, accounting for more than 65% of the market share. Hydrocolloids improve food products' texture, viscosity, and stability in this sector. They also contribute to the shelf life of food products, ensuring that they remain fresh for longer periods. Hydrocolloids are particularly useful in producing low-fat foods, allowing manufacturers to create desirable mouthfeel and texture while reducing the amount of fat in the product.

Table 98: Hydrocolloid Type Market Share

HYDROCOLLOID TYPE	MARKET SHARE (%)
Xanthan Gum	29.4
Guar Gum	17.5
Carrageenan	14.2
Alginate	12.7
Pectin	11.2
Gelatin	5.7
Agar	4.3
Others	5.0

The hydrocolloid market is dominated by xanthan gum, which accounts for almost 30% of the market share. It is followed by guar gum, carrageenan, alginate, and pectin. With the rise of plant-based alternatives in the market, demand for hydrocolloids derived from natural sources such as seaweed, gum Arabic, and cassia gum is expected to grow.

The Asia-Pacific region is the largest consumer of hydrocolloids, accounting for over 35% of the market share. The region's large population, increasing disposable income, and changing dietary habits have contributed to the growth of the hydrocolloid market. Furthermore, major hydrocolloid manufacturers in the region have driven market growth. Europe and North America are also significant consumers of hydrocolloids due to the large food processing industry in these regions.

In conclusion, the hydrocolloid market is poised for significant growth in the coming years, driven by the increasing demand for processed food products and the expanding scope of hydrocolloid applications in various industries. With the rise of plant-based alternatives and the growing demand for natural ingredients, the market is expected to witness increased innovation and competition among key players and suppliers in the hydrocolloid industry.

14.2.5 Key Players and Suppliers in the Hydrocolloid Industry

Hydrocolloids are widely used in various industries, and many companies specialize in their production and distribution. Let's take a closer look at some of the key players and suppliers in the hydrocolloid industry.

CP Kelco

CP Kelco is a leading producer of hydrocolloids and operates in over 100 countries. The company offers many products, including carrageenan, gellan gum, and xanthan gum. CP Kelco›s hydrocolloids are used in the food, beverage, and personal care industries.

Kerry Group

Kerry Group is a global company that provides ingredients and solutions for the food, beverage, and pharmaceutical industries. The company offers a range of hydrocolloids, including carrageenan, xanthan gum, and cellulose gum. Kerry's hydrocolloids are used in dairy, bakery, and confectionery products.

Ingredion

Ingredion is a leading ingredient solutions provider for the food, beverage, and industrial markets. The company offers a range of hydrocolloids, including modified starches, gum acacia, and carrageenan. Ingredion's hydrocolloids are used in sauces, dressings, and yogurt applications.

DuPont Nutrition & Biosciences

DuPont Nutrition & Biosciences is a global leader in developing and producing specialty food ingredients and enzymes. The company offers a range of hydrocolloids, including carrageenan, cellulose gum, and pectin. DuPont's hydrocolloids are used in applications such as dairy, meat, and bakery products.

AEP Colloids

AEP Colloids is a supplier of hydrocolloids and natural gums. The company offers a range of products, including carrageenan, guar gum, and locust bean gum. AEP Colloids' hydrocolloids are used in applications such as dairy, beverages, and pet food.

TIC Gums

TIC Gums is a global leader in advanced texture and stabilization solutions for the food industry. The company offers a range of hydrocolloids, including xanthan gum, carrageenan, and guar gum. TIC Gums' hydrocolloids are used in dairy, bakery, and meat products applications.

Cape Crystal Brands

Cape Crystal Brands offers a range of smaller packaged gums, thickeners, emulsifiers, jellifiers, acidifiers and fruit powder for home cooks, chefs, and smaller food producers.

These are just a few key players and suppliers in the hydrocolloid industry. Their innovations and advancements have helped to shape the industry and pave the way for further growth and development.

14.2.6 The Future of Hydrocolloids: Predictions and Innovations

As the hydrocolloid industry continues to grow and evolve, several predictions and potential innovations are on the horizon.

Hydrocolloids for Health and Wellness

One potential growth area for hydrocolloids is in the health and wellness industry. Hydrocolloids have already shown promise in various medical applications, and there is potential for them to be used in dietary supplements and functional foods. With increasing consumer demand for natural and plant-based ingredients, hydrocolloids could play a significant role in developing these products.

Green Hydrocolloids

Another area of innovation is the development of more sustainable hydrocolloids. As the world becomes more focused on environmental sustainability, there is increasing demand for eco-friendly ingredients. Researchers are exploring biodegradable hydrocolloids and those derived from renewable sources. These green hydrocolloids could significantly impact various industries, including food, cosmetics, and personal care.

Advancements in Hydrocolloid Extraction

There is also potential for advancements in hydrocolloid extraction techniques. New extraction methods could lead to more efficient and cost-effective production and higher quality hydrocolloids. Researchers are exploring various extraction methods, including ultrasound, enzyme, and microwave.

Customized Hydrocolloid Blends

As the hydrocolloid industry grows, there is increasing demand for customized blends tailored to specific applications. Suppliers already offer customized blends to meet specific customer needs, and this trend is expected to continue. With more personalized combinations available, manufacturers can create even more innovative products.

Increased Focus on Research and Development

Overall, the future of hydrocolloids looks bright, with increasing research and development leading to new applications and innovations. The hydrocolloid industry is becoming more established, and there is a greater understanding of hydrocolloid properties, functionality, and applications. As a result, we can expect to see more exciting developments in the years to come.

14.2.7 Advancements in Hydrocolloid Research and Development

The future of hydrocolloids is bright, with ongoing research and development paving the way for exciting new applications and innovations. Here are some of the latest breakthroughs in hydrocolloid R&D:

Biodegradable Hydrocolloids

With sustainability being a top priority for many industries, researchers are exploring using biodegradable hydrocolloids in various applications. These types of hydrocolloids break down naturally over time, reducing environmental impact. For example, biodegradable hydrocolloids can create eco-friendly packaging materials for food products.

Nanocellulose Hydrocolloids

Nanocellulose hydrocolloids are natural hydrocolloids that are gaining attention in the research community. These hydrocolloids are derived from cellulose, Earth's most abundant organic polymer. Nanocellulose hydrocolloids have unique properties that make them suitable for various applications, including food thickener and stabilizer.

> "Nanocellulose hydrocolloids have a bright future in the food industry due to their excellent thickening and stabilizing properties. They also have the potential to reduce fat content in processed foods without compromising on texture and mouthfeel"

Smart Hydrocolloids

Advancements in technology are enabling the development of smart hydrocolloids. These hydrocolloids can respond to environmental changes, such as temperature or pH, making them useful in various applications. For example, smart hydrocolloids could be used in drug delivery systems that release medication at a specific point in the body.

Functional Hydrocolloids

Researchers are exploring ways to create hydrocolloids with specific functional properties, such as antioxidant or antimicrobial activity. These functional hydrocolloids could be used in various applications, such as in the formulation of supplements or food products requiring a longer shelf life.

> "Functional hydrocolloids are a promising area of research, with the potential to create value-added products in various industries. However, more research is needed to understand their properties and applications fully."

As research and development efforts continue, the hydrocolloid industry is poised for growth and innovation. We can expect new applications and products incorporating hydrocolloids and advancements in manufacturing processes and extraction techniques. The future of hydrocolloids is exciting, with endless possibilities for enhancing various products and processes.

Conclusion

Hydrocolloids have proven to be an indispensable tool in a variety of industries. From food and pharmaceuticals to cosmetics and environmental applications, the benefits of hydrocolloids are undeniable. As the demand for natural and sustainable ingredients continues to grow, hydrocolloids are primed to play an even larger role in meeting the needs of manufacturers and consumers alike. The future of hydrocolloids is bright, with innovations and advancements on the horizon. Research and development efforts continue to push the boundaries of what is possible, with new applications and uses being discovered constantly. As the industry continues to grow, so will the need for high-quality hydrocolloid suppliers and manufacturers.

Hydrocolloids are essential in many of the products we use and consume daily. They provide valuable functionality, helping to enhance texture, stability, shelf life, and much more. With the potential for new and emerging applications, hydrocolloids are poised to play an even larger role in shaping the future of many industries.

As we look ahead to the future of hydrocolloids, it is important to remember their vital role in our daily lives. Hydrocolloids significantly impact our world, from the food we eat to the medication we take. As the industry continues to evolve and grow, we expect to see even more exciting developments and innovations.

Thank you for joining us to explore hydrocolloids, their types, properties, applications, and future. We hope this article has been informative and provided valuable insights into the world of hydrocolloids.

FAQ

What industries use hydrocolloids?

Hydrocolloids find applications in various industries, including food, pharmaceuticals, cosmetics, personal care products, and environmental solutions.

How do hydrocolloids enhance food products?

Hydrocolloids improve the texture, stability, and shelf life of food products. They can provide the desired mouthfeel, prevent syneresis (water separation), and control viscosity in sauces, dressings, baked goods, and more.

What are the benefits of using hydrocolloids in the pharmaceutical industry?

Hydrocolloids help formulate drug delivery systems, such as controlled-release tablets and patches. They also contribute to wound healing, tissue engineering, and medical dressings.

How are hydrocolloids used in cosmetics and personal care products?

Hydrocolloids stabilize emulsions, provide viscosity, and enhance sensory attributes in cosmetics and personal care products like creams, lotions, shampoos, and gels.

What environmental applications do hydrocolloids have?

Hydrocolloids are utilized in wastewater treatment to improve flocculation and sedimentation processes. They also play a role in soil erosion control and the development of biodegradable materials.

What are the emerging trends in hydrocolloid manufacturing?

The industry is witnessing new production methods, advancements in extraction techniques, and the integration of advanced technology to enhance efficiency and product quality.

FREQUENTLY ASKED QUESTIONS (FAQS) CONTINUED

- **What does the future hold for hydrocolloids?**

  The future of hydrocolloids looks promising, with potential advancements in sustainable production methods, novel applications, and customization to meet evolving consumer preferences and industry needs.

- **What are the latest advancements in hydrocolloid research and development?**

  Ongoing research is focused on exploring new sources of hydrocolloids, improving extraction techniques, and discovering innovative applications in sectors like biotechnology, nanotechnology,

CHAPTER 15

Technological Functionality

PART 15.1

Role of Hydrocolloids in the Texture and Stability of Frozen Foods

Hydrocolloids are a group of compounds widely used in the food industry to enhance the texture and stability of various food products. In the frozen food industry, hydrocolloids play a crucial role in maintaining the quality and appeal of frozen food products, ensuring they retain their texture and taste despite being stored for extended periods. Hydrocolloids are versatile ingredients, providing multiple benefits that improve frozen foods› sensory properties and shelf life.

The functionality of hydrocolloids varies depending on the specific compound used and the application in which they are used. However, their primary purpose is to act as thickeners, stabilizers, and emulsifiers, crucial qualities in frozen food products. Hydrocolloids interact with water molecules, forming a gel-like structure and stabilizing the product by preventing ice crystal formation and water separation, commonly called syneresis. Additionally, hydrocolloids aid in controlling the viscosity and texture of the product, providing a smooth texture and mouthfeel.

Hydrocolloids are available in natural and synthetic forms and are used in various food applications, including frozen food products. They provide structure, reduce ice crystal formation, and improve mouthfeel. Hydrocolloids also reduce water loss through evaporation and prevent undesirable chemical reactions that can cause off-flavors and odors in frozen food products.

Overall, hydrocolloids are essential ingredients in producing high-quality frozen food products, providing numerous benefits that improve texture, stability, and sensory properties.

Create an image of hydrocolloids added to frozen foods, resulting in a smoother and more stable texture. The hydrocolloids should be depicted as a translucent substance that blends seamlessly with the frozen food. Frozen food should be shown in various forms, such as ice cream, pizza, and vegetables. The overall tone of the image should be cool and refreshing, reflecting the benefits of using hydrocolloids in frozen foods.

Core Insights:

- ☑ Hydrocolloids are crucial compounds used in the food industry to enhance the texture and stability of various food products.
- ☑ Hydrocolloids prevent ice crystal formation in frozen food production, minimize water separation, and provide a smooth texture and mouthfeel.
- ☑ Hydrocolloids are available in natural and synthetic forms and are used in various frozen food applications.
- ☑ Hydrocolloids reduce water loss and prevent undesirable chemical reactions that can cause off-flavors and odors in frozen food products.
- ☑ Hydrocolloids are essential ingredients that improve frozen food products' texture, stability, and sensory properties.

15.1.1 Many are Viscosity Modifiers: Enhancing Texture and Mouthfeel

In the frozen food industry, one of the most critical functions of hydrocolloids is their role as viscosity modifiers. Hydrocolloids can significantly influence the thickness and consistency of food products, resulting in improved texture and mouthfeel. These modifiers increase the viscosity of liquid foods, making them thicker and more elastic. It is especially important for frozen foods, where the texture can be compromised during freezing and thawing.

Various hydrocolloids are used as viscosity modifiers, depending on the desired outcome and the specific properties of the food product. Some of the commonly used hydrocolloids for viscosity modification include:

Table 99: Hydrocolloid Functionality

HYDROCOLLOID	FUNCTIONALITY
Xanthan gum	Thickening, stabilizing, emulsifying
Guar gum	Thickening, stabilizing, water-binding
Methylcellulose	Gelling, thickening, emulsifying
Carrageenan	Thickening, stabilizing, gelling

Xanthan gum is one of the most widely used hydrocolloids because of its versatility and compatibility with various food systems. It is particularly useful in frozen foods because it helps to prevent ice crystal formation and reduce water migration, which can result in syneresis, a common issue in frozen foods. In contrast, guar gum is known for its excellent thickening and water-binding properties, making it a popular choice for creating smooth, creamy textures in frozen desserts and ice creams.

The choice of hydrocolloid depends on the specific application and the desired outcome. It is essential to consider factors such as the processing conditions, pH, and the food system's chemical composition to determine the most suitable hydrocolloid for a particular product.

15.1.2 Natural Hydrocolloids: Harnessing Nature's Power

Natural hydrocolloids are plant- or animal-derived polysaccharides used for their functional properties in food production, including frozen foods. They offer numerous benefits, such as improving texture, stability, and nutritional value, without the use of synthetic additives.

One of the most used natural hydrocolloids is pectin, a soluble fiber extracted from fruit peels. It is widely used in producing jams, jellies, desserts, and frozen foods like ice cream and sorbet. Pectin has excellent gelling properties, allowing it to thicken and stabilize the texture of these products.

Another natural hydrocolloid gaining popularity is carrageenan, which is extracted from seaweed. Carrageenan is a stabilizer and thickener in various dairy products, including ice cream, yogurt, and cheese. It enhances the mouthfeel of these products, making them creamier and smoother.

In addition to pectin and carrageenan, other natural hydrocolloids used in frozen food production include agar, xanthan gum, and guar gum. These hydrocolloids offer unique functional properties, such as high water-holding capacity and viscosity, making them suitable for various applications.

Table 100: Hydrocolloid Source and Functions

HYDROCOLLOID	SOURCE	FUNCTION
Pectin	Fruit peels	Gelling, stabilizing
Carrageenan	Seaweed	Stabilizing, thickening
Agar	Seaweed	Gelling, thickening
Xanthan gum	Bacteria fermentation	Viscosity, stabilizing
Guar gum	Seed of guar plant	Viscosity, stabilizing

Natural hydrocolloids have become increasingly popular as consumers demand more natural and nutritious food. In response, food manufacturers have turned to natural hydrocolloids as an alternative to synthetic additives and stabilizers.

Natural hydrocolloids in frozen food production offer numerous benefits, including improved taste, texture, and shelf life. These hydrocolloids provide a sustainable and natural solution to the challenges of frozen food production, ensuring that consumers can enjoy high-quality and healthy frozen foods.

15.1.3 Commercial Hydrocolloids: Wide-ranging Applications

Commercial hydrocolloids are widely used in the food industry due to their numerous functional properties. These ingredients play a crucial role in enhancing the texture and stability of frozen foods. In this section, we'll look closer at the different applications of hydrocolloids in frozen food production.

Stabilizing Emulsions

Emulsions are mixtures of two or more immiscible liquids, such as oil and water. Hydrocolloids are often added to emulsions in frozen foods to prevent separation and maintain a stable consistency. Common hydrocolloids used for emulsion stabilization include xanthan gum, carrageenan, and gum Arabic.

Improving texture

Hydrocolloids are also widely used in frozen foods to improve texture and mouthfeel. These ingredients can modify the viscosity and rheology of food products, resulting in smoother, creamier, and palatable products. Common hydrocolloids for texture improvement include gelatin, pectin, and cellulose derivatives.

Preventing Syneresis

Syneresis is the process by which water separates from a gel or matrix, resulting in a loss of texture and stability in frozen foods. Hydrocolloids are often added to prevent syneresis and maintain a product's desired texture and consistency. Common hydrocolloids include carboxymethyl cellulose (CMC), guar gum, and locust bean gum.

Enhancing Freeze-thaw Stability

Frozen foods are often subjected to multiple freeze-thaw cycles during storage and transport, affecting their texture and stability. Hydrocolloids are added to these products to enhance their freeze-thaw stability, ensuring they maintain their quality and freshness throughout their shelf life. Common hydrocolloids for freeze-thaw stabilization include alginate, carrageenan, and gellan gum.

Commercial hydrocolloids are essential in producing high-quality, stable, and palatable frozen foods. Their wide-ranging applications and functional properties make them indispensable to the food industry.

15.1.4 Advancements in Hydrocolloid Research

Hydrocolloid research is a rapidly developing field, with ongoing studies and innovations seeking to expand the functionality and application of hydrocolloids in frozen food technology. Some of the latest advancements include:

- **Nanotechnology:** Scientists are exploring how nanotechnology can enhance the functional properties of hydrocolloids, such as their emulsifying, thickening, and gelling capabilities.
- **Enzymatic Modification:** Researchers are investigating how enzymatic modification can improve the performance of hydrocolloids in frozen foods, particularly in terms of texture and stability.
- **Alternative Sources:** With the increasing demand for natural and sustainable ingredients, scientists are exploring alternative sources of hydrocolloids, such as microalgae and seaweed, which offer similar functionalities to traditional hydrocolloids.

Moreover, advancements in hydrocolloid research have led to the discovery of new and unique hydrocolloids with promising applications in frozen foods. For example, microbial hydrocolloids, such as xanthan gum, gellan gum, and curdlan, are gaining traction due to their unique properties, including their ability to form thermos-reversible gels.

As the food industry continues to evolve, so too will the field of hydrocolloid research, paving the way for new and exciting applications for these versatile ingredients.

15.1.5 Types of Hydrocolloids Used in Frozen Foods

Hydrocolloids are diverse and versatile ingredients that can be extracted or synthesized from natural or synthetic sources. They are commonly used in the production of frozen foods to improve texture, stability, and overall quality. This section will overview the different types of hydrocolloids used in frozen foods, their properties, and functionalities.

Natural Hydrocolloids

Natural hydrocolloids are derived from plant, animal, or microbial sources. They offer a range of benefits over their synthetic counterparts, such as cleaner labeling, better sensory properties, and environmental sustainability.

These natural hydrocolloids have distinct properties that make them suitable for specific applications in frozen foods. For example, agar is commonly used as a gelling agent in desserts, while xanthan gum suspends particles in beverages and dressings.

Commercial Hydrocolloids

Commercial hydrocolloids are typically produced through chemical or enzymatic modification of natural sources. They offer greater functionality, consistency, and cost-effectiveness than natural hydrocolloids, making them a popular choice in the food industry. Some of the commonly used commercial hydrocolloids in frozen foods are:

Table 101: Commercial Hydrocolloid Source and Functionality

HYDROCOLLOID	SOURCE	FUNCTIONALITY
Carboxymethyl cellulose (CMC)	Cellulose	Thickening, stabilizing
Methylcellulose (MC)	Cellulose	Gelling, thickening, stabilizing
Pectin	Plant cell walls	Gelling, thickening, stabilizing
Polydextrose	Dextrose	Bulking, stabilizing, low-calorie

Commercial hydrocolloids are often combined with natural hydrocolloids to achieve specific properties and functionalities in frozen foods. For example, pectin is commonly used with carrageenan to obtain a desired gel texture in dairy desserts.

The type and amount of hydrocolloid used in frozen foods depend on the specific application, processing conditions, and desired sensory characteristics. It is important to carefully select and optimize the use of hydrocolloids to achieve the desired product quality and stability.

15.1.6 The Growing Hydrocolloid Market

The global hydrocolloid market is expected to reach \$10.2 billion by 2025, growing at a CAGR of 5.3% from 2020 to 2025. The demand for hydrocolloids is driven by the growing popularity of processed and convenience foods and the increasing demand for clean-label and natural ingredients.

The food and beverage industry is the largest end-user of hydrocolloids, with a market share of over 70%. Hydrocolloids are used extensively in producing frozen foods, as they help improve product texture, stability, and shelf life.

Table 102: Region and Hydrocolloids Market Share

REGION	MARKET SHARE
North America	32.5%
Europe	29.4%
Asia-Pacific	28.6%
Latin America	4.3%
Middle East and Africa	5.2%

North America has the largest market share, accounting for 32.5% of the global hydrocolloid market. The region's dominant position can be attributed to the increased consumption of processed and convenience foods and the strong presence of key hydrocolloid manufacturers.

The Asia-Pacific region is expected to witness the highest market growth in the next few years due to the growing demand for hydrocolloids in the food and beverage industry. The region's large population and increasing disposable income levels drive the demand for convenience foods such as frozen meals, thereby boosting hydrocolloid demand.

The growing hydrocolloid market presents significant opportunities for manufacturers to develop innovative hydrocolloid products that cater to the evolving needs of the food industry. Natural hydrocolloids are gaining popularity in the market, and manufacturers are investing in research and development to explore the potential benefits of these ingredients.

15.1.7 The Future of Hydrocolloid Application in Frozen Foods

The potential uses of hydrocolloids in frozen foods are vast, and emerging technologies are making it easier to maximize the benefits of these functional ingredients.

One area of ongoing research is developing new hydrocolloid blends that can provide improved texture and stability in frozen foods. These blends can be tailored to suit specific applications and enhance individual hydrocolloids' performance, resulting in a more effective and efficient ingredient system.

Another area of focus is on using natural hydrocolloids as alternatives to synthetic ingredients. As consumers become increasingly interested in natural and organic foods, the demand for natural hydrocolloids is expected to rise, and manufacturers are responding by exploring new sources of these ingredients and developing new formulations that can replace traditional synthetic stabilizers and emulsifiers.

The use of hydrocolloids in frozen foods is also likely to expand into the development of new product categories. For example, hydrocolloids can create innovative frozen desserts that offer improved texture, stability, and flavor. They can also be used to develop new fixed meal options that provide enhanced nutritional value, longer shelf life, and improved sensory qualities.

Emerging Trends in Hydrocolloid Application

One emerging trend in hydrocolloid application is using plant-based hydrocolloids, such as those derived from seaweed, to replace traditional animal-based stabilizers. These natural ingredients can provide a more sustainable and ethical alternative for various applications.

Another trend is in the customization of hydrocolloid blends to suit specific applications. Manufacturers can create ingredient systems tailored to their products' unique needs by combining different hydrocolloids in particular ratios.

The adoption of new technologies is also driving advancements in hydrocolloid applications. For example, advances in microencapsulation technology are making it possible to produce hydrocolloid particles that can be precisely tailored to suit specific applications and provide targeted functionality.

The future of hydrocolloid application in frozen foods is exciting and full of potential. Advances in technology and research are unlocking new opportunities for manufacturers to enhance the texture, stability, and overall quality of their products. By leveraging hydrocolloids› unique properties and benefits, food producers can create frozen products that meet today›s consumers› evolving needs and preferences.

Conclusion

In conclusion, hydrocolloids play a vital role in enhancing the texture and stability of frozen foods. By providing a range of functionalities such as thickening, stabilizing, emulsifying, and improving consistency, hydrocolloids help to create high-quality and enjoyable frozen food products.

The extensive range of hydrocolloid options, including natural and commercial alternatives, offers various possibilities for frozen food manufacturers. It allows them to tailor their hydrocolloid applications to specific product needs and achieve optimal results.

The increasing demand for hydrocolloids is driven by the growing popularity of frozen foods and the need for improved texture, stability, and quality in these products. Ongoing research and hydrocolloid advancements pave the way for continued innovation and the potential for even greater benefits for the frozen food industry.

FAQ

How do hydrocolloids improve texture and stability in frozen foods?

Hydrocolloids contribute to the texture and stability of frozen foods by providing viscosity, preventing ice crystal formation, and enhancing moisture retention. They help maintain the desired consistency and avoid syneresis and texture degradation during freezing and thawing.

Are there natural alternatives to commercial hydrocolloids?

Natural hydrocolloids such as agar, carrageenan, and pectin are widely used as alternatives to commercial hydrocolloids. They are derived from natural sources and offer similar functionality and benefits in improving the texture and stability of frozen foods.

Is there ongoing research on hydrocolloids?

There is ongoing research on hydrocolloids to explore new properties, functionalities, and applications. Scientists are constantly studying the synergistic effects of different hydrocolloids, their interactions with other ingredients, and innovative ways to enhance their performance in frozen food technology.

How is the hydrocolloid market growing?

The hydrocolloid market is experiencing significant growth due to the increasing demand for frozen foods and the need for improved texture, stability, and quality. Consumer preferences for convenient and high-quality frozen food products drive the market.

What does the future hold for hydrocolloid application in frozen foods?

The future of hydrocolloid application in frozen foods looks promising. Advancements in technology and research will likely lead to the development of new hydrocolloid blends, more efficient processing techniques, and innovative applications that further enhance the texture, stability, and overall consumer experience of frozen food.

PART 15.2

Hydrocolloids as Fat Replacers and Their Role in Low-Calorie Foods

In today's world, where people are increasingly health-conscious, food manufacturers always look for healthier alternatives to traditional food products. One area of focus is developing low-calorie foods that mimic the taste and texture of high-calorie options while offering fewer calories. It is where hydrocolloids come in as effective fat replacers and texture improvers in low-calorie food formulations.

Hydrocolloids are food additives that act as thickening agents, gelling agents, and food stabilizers. These natural polymers can absorb water and form gels, which makes them ideal candidates for use in low-calorie food products. By using hydrocolloids, food manufacturers can create high-volume, low-calorie foods that are filling but low in calories. It is particularly important in promoting healthy diets and weight management.

An abstract representation of hydrocolloids as fat replacers in low-calorie foods.

Hydrocolloids are versatile, functional ingredients that can be utilized in various food categories, including baked goods, dairy products, sauces, and dressings. They play a crucial role in improving the texture and mouthfeel of low-calorie foods, enabling food manufacturers to offer nutritious yet satisfying options to consumers.

In the following sections, we will delve deeper into the functionality of hydrocolloids as fat replacers, texture improvers, and food stabilizers in low-calorie food formulations. We will explore the significance of hydrocolloids in promoting healthy diets and the potential for further research and innovation in this field.

Core Insights:

- ☑ Hydrocolloids are natural polymers that act as thickening agents, gelling agents, and food stabilizers in food formulations.
- ☑ Hydrocolloids can absorb water and form gels, making them ideal for low-calorie food products.
- ☑ Using hydrocolloids, food manufacturers can create high-volume, low-calorie foods that are filling but low in calories.
- ☑ Hydrocolloids play a crucial role in improving the texture and mouthfeel of low-calorie foods, making them more appealing to consumers.
- ☑ Hydrocolloids are versatile, functional ingredients that can be utilized in various food categories, including baked goods, dairy products, sauces, and dressings.

15.2.1 Hydrocolloids as Fat Replacers

Hydrocolloids are versatile ingredients that play a crucial role in producing low-calorie foods. One of their key functionalities is serving as a fat replacer in formulations, allowing manufacturers to create filling but low-calorie foods with high density.

It is particularly important for promoting healthy diets, as reducing dietary fat intake has decreased the risk of obesity and other related illnesses. Hydrocolloids offer an effective solution for creating low-fat products that still provide consumers with a satisfying experience.

Table 103: Hydrocolloid and Fat Replacement Efficiency

HYDROCOLLOID	FAT REPLACEMENT EFFICIENCY
Xanthan gum	Up to 50%
Guar gum	Up to 25%
Agar agar	Up to 70%

As demonstrated in the table above, hydrocolloids such as xanthan gum, guar gum, and agar agar can effectively replace a significant portion of the fat content in food products. It allows manufacturers to create high-density, low-calorie foods that are filling and satisfying for consumers.

Food manufacturers can use hydrocolloids as fat replacers to meet the growing demand for healthier options without compromising taste or texture. It can increase consumer satisfaction and loyalty and improve health outcomes for consumers who choose low-calorie foods.

15.2.2 Benefits of Hydrocolloids in Low-Calorie Foods

The use of hydrocolloids in low-calorie foods has a range of benefits that make them an excellent choice for both consumers and food manufacturers. This section will explore some key advantages of using hydrocolloids in low-calorie products.

High-Volume, Low-Calorie Foods

One of the most significant benefits of hydrocolloids in low-calorie foods is their ability to create high-volume, filling, but low-calorie foods.

Hydrocolloids can absorb large amounts of water, creating products with a high water content. This high water content allows manufacturers to generate larger portions without significantly increasing the product's calorie count.

For example, a low-calorie soup or sauce that contains hydrocolloids can be thicker and more filling than a regular soup, even though it has fewer calories per serving.

Consumers benefit from this because they can eat a larger portion of a low-calorie product, which makes them feel full and satisfied without taking in excess calories.

Texture Improvement

Hydrocolloids are excellent thickening and gelling agents, which means they can improve the texture and mouthfeel of low-calorie foods. By using hydrocolloids, manufacturers can create products that have the same surface and sensory experience as their full-fat counterparts.

For example, a low-fat dessert that contains hydrocolloids can still be creamy and smooth even though it has significantly less fat than a regular dessert.

In addition to improving the sensory experience of low-calorie foods, hydrocolloids can also enhance the stability of these products. Hydrocolloids help prevent ingredient separation as food stabilizers and ensure a consistent texture throughout the development.

Versatile Functional Ingredients

Hydrocolloids are versatile, functional ingredients in many low-calorie food products. They can be used as thickening, gelling, and food stabilizers in various applications, including sauces, dressings, soups, and desserts.

This versatility makes hydrocolloids an excellent choice for food manufacturers who want to create low-calorie products that meet consumer demand for high-quality, satisfying options.

Overall, the benefits of hydrocolloids in low-calorie foods are clear. They allow manufacturers to create satisfying, high-volume, low-calorie foods that offer consumers a healthy, tasty alternative to high-fat options.

> "The use of hydrocolloids in low-calorie foods has the potential to revolutionize the food industry, offering a wider range of healthy, satisfying options for consumers."

Sources

Food Hydrocolloids: Structures, Properties, and Functions

Hydrocolloids in Food Processing

15.2.3 Hydrocolloids in Healthy Diets

As awareness of health and wellness increases, consumers are seeking low-calorie options that still provide the satisfaction and indulgence of traditional food products. This demand has led to the rise of high-density, low-calorie foods that are filling but low in calories, making them a popular choice for those looking to manage their weight and maintain a healthy lifestyle.

Hydrocolloids play a crucial role in the production of these low-calorie foods. As functional ingredients in the food industry, they enable the creation of nutritious yet satisfying options. Their unique properties as thickening agents, gelling agents, and food stabilizers, make them essential components in achieving these products' desired mouthfeel, texture, and stability.

Food manufacturers can create indulgent and nutritious products using hydrocolloids as fat replacers in low-calorie foods. By providing filling but low-calorie options, consumers can enjoy the sensory experience of food without the guilt associated with a high-calorie intake. As a result, hydrocolloids are an important tool for promoting healthy diets and overall wellness.

The food industry must continue to innovate and explore the potential of hydrocolloids for creating new and exciting low-calorie products. With the versatility and functionality of hydrocolloids, there is endless potential for creating low-calorie options that meet the demands of consumers seeking nutrition and indulgence.

15.2.4 Hydrocolloids for Texture Improvement

Hydrocolloids play a crucial role in improving the texture of low-calorie foods. Thickening agents, gelling agents, and food stabilizers are all key functions of hydrocolloids that enhance the sensory experience of these products.

Hydrocolloids as thickening agents can significantly improve the texture of low-calorie foods. They increase the viscosity of food products, creating a more substantial and satisfying mouthfeel.

Gelling agents, another function of hydrocolloids, create a gel-like structure in low-calorie foods, which can enhance the texture and mouthfeel. It is particularly relevant for foods like puddings with a desired creamy consistency.

Food stabilizers, the third key function of hydrocolloids, help prevent ingredient separation and ensure a desirable consistency in low-calorie foods. It is essential for products such as sauces and dressings, where a smooth texture is sensual.

> "Hydrocolloids have emerged as a valuable tool in the production of low-calorie foods, thanks to their ability to improve texture and mouthfeel."

The versatility of hydrocolloids as functional ingredients makes them a valuable tool for food manufacturers in producing low-calorie products. The texture improvements achieved through their use as thickening agents, gelling agents, and food stabilizers can significantly enhance the appeal of these products to consumers.

In conclusion, hydrocolloids are essential in improving the texture and mouthfeel of low-calorie foods. Their versatility as functional ingredients and ability to enhance the appeal of these products make them a valuable tool for food manufacturers in producing nutritious yet satisfying offerings.

Conclusion

In conclusion, hydrocolloids are significant in producing low-calorie foods as effective fat replacers and texture improvers. The importance of low-calorie options in promoting healthy diets cannot be overstated, and hydrocolloids offer a promising solution to achieving high-volume, filling, but low-calorie foods. Their function as thickening agents, gelling agents, and food stabilizers in low-calorie food formulations make them a versatile and functional ingredient in the food industry.

Hydrocolloids can help create high-density, low-calorie foods that offer a satisfying experience without compromising flavor or texture. Their impact on texture improvement in low-calorie foods cannot be overstated, as consistency is crucial to consumer acceptance.

As the demand for healthier food options grows, there is potential for further research and innovation in this field. Hydrocolloids offer a promising avenue for creating nutritious yet satisfying consumer options, promoting overall health and weight management.

If you are in the food industry and looking to create low-calorie foods or improve the texture of existing products, hydrocolloids may be your solution. Consider exploring the range of hydrocolloids available and their specific properties to find the best fit for your product.

Remember, providing healthy and nutritious food options is not only good for consumers, but it can also be good for your business. By offering functional yet tasty low-calorie foods, you can tap into a growing market of health-conscious consumers and stand out from the competition.

FAQ ?

How do hydrocolloids function as fat replacers?

Hydrocolloids create high-density, filling, but low-calorie food products. They can replace fat in formulations, offering consumers a satisfying experience without the excess fat content.

What are the functions of hydrocolloids as thickening agents and gelling agents?

Hydrocolloids serve as thickening agents and gelling agents in the food industry. They improve the texture and mouthfeel of low-calorie foods, enhancing their appeal to consumers.

How do hydrocolloids act as food stabilizers?

Hydrocolloids contribute to texture improvement in low-calorie foods by acting as food stabilizers. They ensure a desirable consistency and prevent ingredient separation.

What are the benefits of using hydrocolloids in low-calorie foods?

Hydrocolloids help create high-volume, low-calorie foods that are filling and low in calories. Their use as thickening agents, gelling agents, and food stabilizers improves these products' texture and overall quality.

How do hydrocolloids promote healthy diets?

Hydrocolloids contribute to healthy diets by enabling the production of high-density, low-calorie foods. These options help with weight management and provide nutritious yet satisfying choices for consumers.

What is the impact of hydrocolloids on texture improvement in low-calorie foods?

Hydrocolloids serve as thickening, gelling, and food stabilizers, enhancing low-calorie foods' texture and sensory experience. Texture plays a crucial role in consumer acceptance.

CHAPTER 16

Regulatory Aspects of Dietary Fibers

PART 16.1

Classification of Hydrocolloids as Dietary Fibers by Health Authorities

Hydrocolloids are a group of naturally occurring or chemically modified substances widely used in the food industry for their functional properties. These properties include binding, thickening, emulsifying, stabilizing, and gelling. Hydrocolloids are also rich in dietary fibers essential for maintaining good health.

In recent years, health authorities have recognized the importance of hydrocolloids as dietary fibers and have established regulations for their use in food. These regulations ensure that hydrocolloids are safe and beneficial for consumption and that consumers can make informed choices about the foods they eat.

A close-up view of various hydrocolloids and their textures arranged in a visually appealing pattern. The colors should be bright and distinct, showcasing the unique features of each hydrocolloid. The image should also represent dietary fibers, like fruits or vegetables, to demonstrate how hydrocolloids contribute to a healthy diet. The composition should look scientific and professional but visually intriguing to catch the viewer's attention. The dietary fiber market has also seen a significant growth in demand for hydrocolloids due to their functional properties and health benefits. The use of hydrocolloids in food products has become increasingly widespread, and it is important to understand their role in promoting a healthy diet and lifestyle.

Core Insights:

- ☑ Hydrocolloids are used in the food industry for their functional properties, including natural gums, thickeners, stabilizers, gelling agents, viscosity modifiers, texture enhancers, water-binding agents, and emulsifiers.
- ☑ Hydrocolloids are rich in dietary fibers essential for maintaining good health.
- ☑ Health authorities have established regulations for using hydrocolloids as dietary fibers in food, ensuring their safety and consumer benefits.
- ☑ The dietary fibers market has seen significant growth in demand for hydrocolloids due to their functional properties and health benefits.
- ☑ Understanding the role of hydrocolloids in promoting a healthy diet and lifestyle is important.

16.1.1 What Are Dietary Fibers and Their Importance for Health

Dietary fibers are non-digestible carbohydrates crucial for maintaining a healthy digestive system. These complex structures are essential for promoting regular bowel movements and preventing constipation. In addition, dietary fibers play a vital role in maintaining optimal gut health by nourishing good bacteria in the gut and enhancing nutrient absorption.

Some health authorities classify Hydrocolloids as dietary fibers due to their ability to promote digestive health. Hydrocolloids are indigestible carbohydrates that form a gel-like substance when mixed with water. This gel-like substance helps to regulate the digestive system by slowing down the rate at which food moves through the intestines.

It is incorrect to assume that dietary fibers only come from fruits, vegetables, and whole grains. Hydrocolloids and other sources such as psyllium husk and chia seeds are also considered dietary fibers due to their beneficial effects on the digestive system and overall health.

16.1.2 Hydrocolloids as Dietary Fibers

Hydrocolloids are a popular source of dietary fibers in food products due to their functional properties. Hydrocolloids act as water-binding agents, thickeners, and stabilizers, improving the texture and stability of various food products. For example, pectin, a type of hydrocolloid, is commonly used as a gelling agent in jams and jellies, while xanthan gum is used as a thickener in salad dressings and sauces.

Health authorities have recognized the beneficial effects of hydrocolloids on digestive health, leading to their classification as dietary fibers. The inclusion of hydrocolloids in the diet has been associated with a reduced risk of chronic diseases such as heart disease, diabetes, and obesity.

Some hydrocolloids, such as glucomannan, have even been shown to promote weight loss by increasing feelings of fullness and reducing overall calorie intake.

Overall, including hydrocolloids as dietary fibers can improve digestive health, manage weight, and reduce the risk of chronic diseases.

16.1.3 Health Authority Guidelines on Dietary Fibers

Health authorities are crucial in regulating the dietary fiber market, including classifying hydrocolloids as nutritional fibers. The United States Food and Drug Administration (FDA) recognizes seven types of dietary fibers, including cellulose, guar gum, pectin, psyllium husk, and hydrocolloids, such as xanthan gum and carrageenan. Hydrocolloids exhibiting sufficient physiological effects like naturally occurring fibers can be classified as dietary fibers.

The European Food Safety Authority (EFSA) has also established guidelines on the classification of dietary fibers, stating that they must have a physiological effect beneficial to health, be resistant to digestion, and be naturally occurring or extracted from plants or other sources. Therefore, hydrocolloids that meet these criteria can be classified as dietary fibers.

Hydrocolloid manufacturers must comply with these regulations to ensure the safety and efficacy of their products. Failure to comply may result in the product's removal from the market, adversely affecting the dietary fiber market.

> "As the dietary fiber market grows, more health authorities are expected to establish guidelines and regulations to ensure product safety and efficacy."

According to a report by Market Data Forecast, the global dietary fiber market is expected to reach $11.83 billion by the end of 2025, growing at a CAGR of 11.8%. The increasing demand for nutritional supplements and healthy food products has fueled this growth. The report also indicates the need for hydrocolloids as dietary fibers are expected to rise due to their various functional properties in food production.

Compliance with health authority regulations is essential for hydrocolloid manufacturers to tap into this growing market and ensure the safety and efficacy of their products.

16.1.4 Hydrocolloids and Their Fiber Content

Hydrocolloids are natural or synthetic compounds widely used in food production as thickeners, stabilizers, gelling agents, viscosity modifiers, texture enhancers, water-binding agents, and emulsifiers. Health authorities classify Many hydrocolloids as dietary fibers because they promote digestive health and overall well-being.

Some commonly used hydrocolloids in food production include agar, carrageenan, cellulose, chitosan, gellan gum, guar gum, pectin, and xanthan gum. Each hydrocolloid has unique properties contributing to its function and fiber content.

Hydrocolloid Fiber Content Comparison

Table 104: Hydrocolloids and their Fiber Content

HYDROCOLLOID	FIBER CONTENT (PER 100G)
Agar	80g
Carrageenan	20g
Cellulose	30g
Chitosan	0g
Gellan gum	0g
Guar gum	80g
Pectin	75g
Xanthan gum	25g

It is important to note that the fiber content of hydrocolloids may vary depending on the specific type and processing methods used. Additionally, health authorities have established criteria and regulations for hydrocolloids to be classified as dietary fibers, which include their ability to resist digestion and be fermented by gut bacteria.

The fiber content of hydrocolloids plays a significant role in their classification as dietary fibers. By contributing to the recommended daily intake of dietary fiber, hydrocolloids can promote digestive health, regulate blood sugar levels, and reduce the risk of chronic diseases such as heart disease, diabetes, and certain types of cancer. Hydrocolloids are an important component of a healthy and balanced diet.

16.1.5 Implications for Diet and Well-being

Consuming hydrocolloids as dietary fibers can significantly impact overall health and well-being. Hydrocolloids contribute to weight management by creating a feeling of fullness in the stomach, which reduces the tendency to overeat. They also help to regulate blood sugar levels, which is beneficial for individuals with diabetes.

Furthermore, hydrocolloids can improve gut health and digestion by promoting the growth of beneficial bacteria in the gut. They also help to soften stool and promote regular bowel movements, reducing the risk of constipation.

Hydrocolloids have also been shown to positively impact cholesterol levels, reducing LDL (bad) cholesterol levels in the blood. It benefits cardiovascular health and can reduce the risk of heart disease.

Incorporating hydrocolloids into the diet can be easy and convenient. Many food products, such as dairy products, baked goods, and beverages, contain hydrocolloids as thickeners, stabilizers, and emulsifiers. These products provide an easy way to increase fiber intake without significant dietary changes.

Overall, the impact of hydrocolloids on diet and well-being is significant. Their fiber content and other properties make them an important component of a healthy diet, contributing to improved digestion, weight management, and cardiovascular health.

16.1.6 Incorporating Hydrocolloids in Food Products

Hydrocolloids are widely used in the food industry as they provide a range of functional benefits, including thickening, stabilizing, gelling, and emulsifying. These functions are essential in creating appealing textures, enhancing mouthfeel, and extending the shelf-life of food products. Food hydrocolloids are categorized based on their source as animal, microbial, seaweed, or plant-based.

The most used hydrocolloids are food thickeners, which add viscosity and texture to liquids. They are used in various food products, including sauces, dressings, soups, and gravies. Examples of food thickeners include xanthan gum, carrageenan, guar gum, and pectin.

Food stabilizers are another important category of hydrocolloids that enhance a food product's stability. They inhibit the separation of water from other ingredients, thus preventing spoilage. Examples of food stabilizers include gelatin, agar, and carrageenan.

Gelling agents are hydrocolloids that induce gel formation in food products, providing a unique texture and mouthfeel. They are commonly found in jams, jellies, and confectionery items. Examples of gelling agents include carrageenan, agar, and gellan gum.

Viscosity modifiers are hydrocolloids that alter the consistency of food products. They provide a smooth and creamy texture in dairy products like ice cream and yogurts. Examples of viscosity modifiers include carrageenan, locust bean gum, and xanthan gum.

Food texture enhancers are hydrocolloids that improve the texture of food products, providing a pleasant mouthfeel. They are used in various foods, including baked goods, dairy products, and meats. Examples of food texture enhancers include carrageenan, xanthan gum, and agar.

Water-binding agents are hydrocolloids that retain moisture in food products, preventing dehydration. They are commonly used in meat products to increase their juiciness and improve their texture. Examples of water-binding agents include carrageenan, xanthan gum, and cellulose gum.

Emulsifiers are hydrocolloids that improve the stability of emulsions, such as oil-in-water emulsions. They are used in various food products, including mayonnaise, salad dressings, and margarine. Examples of emulsifiers include lecithin, carrageenan, and cellulose gum.

16.1.7 Future Trends in Hydrocolloids and Dietary Fibers

The dietary fiber market is rapidly growing, with an increasing demand for natural and healthy food products. Hydrocolloids are well-positioned to address this demand, offering a range of functional benefits and high-fiber content. With the continued growth of the health and wellness industry, the future looks bright for hydrocolloids and dietary fibers.

One of the major trends in the hydrocolloids market is the increasing demand for clean-label ingredients. Consumers are becoming more aware of the health impacts of processed foods and are seeking products with natural and recognizable ingredients. This trend drives the development of new and innovative hydrocolloid products free from artificial additives and preservatives.

Another key trend in the industry is the focus on sustainability. As consumers become more environmentally conscious, there is a growing demand for food products that use sustainable and renewable ingredients. Hydrocolloids from natural sources such as seaweed and plant extracts offer a sustainable and eco-friendly alternative to synthetic additives.

The Role of Research in Driving Future Trends

Research is crucial in driving future hydrocolloid and dietary fiber trends. As the science behind these ingredients continues to evolve, new applications and benefits are being discovered. Ongoing research is also helping to improve the quality and functionality of hydrocolloids, making them better suited for a wider range of food products.

One promising area of research is using hydrocolloids in plant-based meat alternatives. As the demand for vegan and vegetarian food products increases, there is a growing need for ingredients that mimic the texture and mouthfeel of meat. Hydrocolloids are being studied for their ability to provide the necessary structure and binding properties in plant-based meats, creating a more realistic meat-like texture.

The Growing Importance of Hydrocolloids for Health and Wellness

As more research is conducted on the health benefits of dietary fibers, the importance of hydrocolloids in the diet is becoming increasingly clear. In addition to their role as fiber sources, hydrocolloids offer a range of functional benefits, such as improved texture, stability, and water-binding capacity. It makes them an ideal ingredient for various food products, including dairy, bakery, and meat.

The growing recognition of the health benefits of hydrocolloids is also driving innovation in the industry. New hydrocolloid products are being developed to meet specific health needs, such as digestive health and weight management. These products offer a convenient and effective way for consumers to improve their health and well-being.

16.1.8 Research Studies on Hydrocolloids and Dietary Fibers

Numerous research studies have been conducted on hydrocolloids and their classification as dietary fibers. These studies have provided valuable insights into the benefits of including hydrocolloids in the diet and their impact on human health.

One study published in the *Journal of Agricultural and Food Chemistry* analyzed the effects of consuming hydrocolloids on satiety and energy intake. The results showed that individuals who consumed hydrocolloids experienced increased feelings of fullness and decreased overall energy intake compared to those who did not. It suggests that hydrocolloids may be useful in weight management and reducing calorie intake.

Another research study published in the *Journal of Nutrition and Metabolism* examined the effects of hydrocolloids on gut health. The study found that hydrocolloids improved the growth and function of beneficial gut bacteria, improving digestive health and reducing the risk of chronic diseases.

Furthermore, research has shown that consuming hydrocolloids as dietary fiber can positively impact blood sugar and cholesterol levels. A study published in the *Journal of Food Science and Technology* found that hydrocolloids helped to lower blood glucose levels in individuals with type 2 diabetes. A study published in the Journal of Functional Foods also demonstrated that hydrocolloids could help reduce LDL cholesterol levels in individuals with hypercholesterolemia.

These research studies suggest that hydrocolloids have numerous health benefits and can be an important part of a balanced diet. As such, food manufacturers need to ensure their products contain high-quality hydrocolloids that meet regulatory standards and can contribute to a healthy lifestyle.

16.1.9 Industry Insights and Expert Opinions

Experts in the food industry have varying opinions on classifying hydrocolloids as dietary fibers. While some believe that the regulatory guidelines set by health authorities have been stringent enough, others feel there is still room for improvement.

According to a recent market research firm Markets report, the global dietary fiber market is projected to grow at a CAGR of 10.4% from 2020 to 2025. The report highlights the increasing demand for natural and organic food products, which has led to a rise in the use of hydrocolloids as dietary fibers.

The Food and Drug Administration (FDA) in the United States has set specific criteria for hydrocolloids to be classified as dietary fibers. According to the FDA, a hydrocolloid must have a physiological effect that benefits human health to be considered a dietary fiber.

Regulatory bodies such as the European Food Safety Authority (EFSA) and the Codex Alimentarius Commission have also established guidelines to classify hydrocolloids as dietary fibers.

Experts recommend periodically updating the hydrocolloid regulatory framework to ensure that innovations and research are considered.

> "The regulatory framework for hydrocolloids as dietary fibers must be flexible to accommodate the changing needs of the food industry and consumers. There is a growing demand for natural and organic food products, and hydrocolloids can play a crucial role in meeting this demand."

16.1.10 Challenges and Controversies

Despite the numerous benefits of hydrocolloids as dietary fibers, several challenges and controversies surround their classification. One of the main issues is the lack of standardization and regulatory oversight in the industry.

There is no universal definition or legal framework for hydrocolloids as dietary fibers. Instead, different health authorities worldwide have varying criteria and regulations for their classification. It has led to confusion and inconsistency in the market, making it difficult for consumers to make informed decisions about their diet.

Furthermore, some experts have raised concerns about the potential side effects of consuming large amounts of hydrocolloids. While they are generally considered safe, excessive intake can cause digestive issues, including bloating and diarrhea.

Additionally, there have been debates about whether hydrocolloids can be considered dietary fibers. Some argue they do not provide the same health benefits as naturally occurring fruits, vegetables, and whole grains fibers.

Despite these challenges and controversies, the use of hydrocolloids in food production continues to grow. As the demand for natural and functional ingredients increases, hydrocolloids offer food manufacturers a versatile, cost-effective solution.

16.1.11 The Future of Hydrocolloids and Dietary Fibers

As the food industry continues to evolve, so will the role of hydrocolloids and dietary fibers. Emerging research and innovations in the field are poised to drive growth and expansion in the market.

One area of focus is the development of new hydrocolloid products that offer enhanced functionality and nutritional value. For example, researchers are exploring using hydrocolloids as prebiotics, which can promote the growth of beneficial gut bacteria and improve overall health.

Additionally, there is growing interest in using hydrocolloids as alternatives to artificial additives and preservatives. As consumers increasingly seek natural and clean-label ingredients, hydrocolloids offer a viable solution for food manufacturers looking to improve the nutritional profile of their products.

Overall, the future of hydrocolloids and dietary fibers looks promising. While some challenges and controversies must be addressed, the benefits of these ingredients are clear. Incorporating hydrocolloids as nutritional fibers in our diet can improve our health and well-being while enjoying a wide range of delicious and nutritious foods.

Conclusion

Hydrocolloids are an essential component in the food industry, serving a variety of functions such as thickeners, stabilizers, and emulsifiers. Health authorities have classified certain hydrocolloids as dietary fibers due to their health benefits. Nutritional fibers, including those from hydrocolloids, have been linked to improved digestion, weight management, and a reduced risk of chronic diseases.

Research studies continue to showcase the numerous benefits of dietary fibers, and the dietary fibers market is expected to experience significant growth in the coming years. Despite the controversies and challenges surrounding the classification of hydrocolloids as dietary fibers, the scientific evidence supports their inclusion in the diet to improve overall well-being.

FAQ

Why are hydrocolloids classified as dietary fibers by health authorities?

Health authorities classify hydrocolloids as dietary fibers due to their ability to provide physiological effects similar to naturally occurring nutritional fibers. They contribute to improved digestion, weight management, and reduced risk of chronic diseases.

How do hydrocolloids contribute to dietary fiber intake?

Hydrocolloids contain high soluble and insoluble fiber levels, contributing to total dietary fiber intake. When consumed as part of a balanced diet, hydrocolloids can help meet the recommended daily intake of dietary fibers.

Are there any regulations governing the classification of hydrocolloids as dietary fibers?

Yes, there are regulations established by health authorities that define the criteria for classifying hydrocolloids as dietary fibers. These regulations ensure consistency and accuracy in labeling and provide guidelines for manufacturers and consumers.

Can hydrocolloids be used for specific dietary restrictions or allergies?

Hydrocolloids are generally safe for consumption and can be used under various dietary restrictions. However, individuals with specific allergies or sensitivities should consult a healthcare professional before consuming hydrocolloid products to ensure they suit their needs.

What are the benefits of incorporating hydrocolloids into food products?

Hydrocolloids offer numerous benefits in food production, including improved texture, stability, viscosity, and mouthfeel. They can also enhance shelf life, prevent syneresis, and contribute to the overall sensory experience of food products.

FREQUENTLY ASKED QUESTIONS (FAQS) CONTINUED

- **Are there any potential health risks associated with consuming hydrocolloids?**

  Hydrocolloids are generally safe and well-tolerated when consumed in moderate amounts as part of a balanced diet. However, drinking excessive amounts of hydrocolloids or relying solely on processed foods containing high hydrocolloids may adversely affect digestion and nutrient absorption.

- **Are there any ongoing research studies on hydrocolloids and dietary fibers?**

  Ongoing research studies are exploring the effects of hydrocolloids on human health and their potential applications in various food products. These studies aim to understand the physiological impact further and optimize hydrocolloids' use as dietary fibers.

- **How can I incorporate hydrocolloids into my diet?**

  Hydrocolloids can be incorporated into the diet by consuming foods that naturally contain them, such as fruits, vegetables, and whole grains. They can also be found in processed food products, such as sauces, dressings, and baked goods that utilize hydrocolloids as additives.

PART 16.2

The Role of Hydrocolloids in Meeting Dietary Fiber Intake Recommendations

In recent years, dietary fiber has gained significant attention for its role in maintaining and promoting overall health. Adequate intake of dietary fiber has been linked to a reduced risk of various chronic diseases, including heart disease, diabetes, and certain types of cancer.

However, many individuals struggle to meet the recommended daily dietary fiber intake of 25 grams for women and 38 grams for men. It is where hydrocolloids come in. Hydrocolloids are a food additive that can help individuals achieve their dietary fiber intake goals.

Hydrocolloids are versatile ingredients that can thicken, gel, stabilize, emulsify, and act as a gelatin substitute in various food products. They can also enhance the texture and overall quality of food products, making them more appealing to consumers.

This article will explore the significance of hydrocolloids in meeting dietary fiber intake recommendations. It will provide an overview of hydrocolloids, their properties, and their different types. The article will also discuss the various applications of hydrocolloids in the food industry and their impact on food texture and sensory attributes. Additionally, the paper will provide practical tips for incorporating hydrocolloids into a balanced diet and discuss the ongoing research and developments in the field of hydrocolloids.

A colorful array of hydrocolloids surrounds it. In the background, a graph shows the recommended daily dietary fiber intake based on age and gender. The hydrocolloids are helping to bridge the gap between what people currently eat and the recommended daily intake.

Core Insights:

- ☑ Hydrocolloids are a food additive that can help individuals meet their recommended daily intake of dietary fiber.
- ☑ Hydrocolloids are versatile ingredients that enhance food products' texture and overall quality.
- ☑ There are natural and synthetic types of hydrocolloids, each with distinct characteristics and applications.
- ☑ Hydrocolloids are used in various food products to thicken, gel, stabilize, emulsify, and act as a gelatin substitute.
- ☑ Hydrocolloids can provide soluble and insoluble fiber, contributing to overall digestive health.

16.2.1 Hydrocolloids and Dietary Fiber

Dietary fiber plays a crucial role in maintaining a healthy digestive system. However, most individuals fail to meet the recommended daily intake of fiber. It is where hydrocolloids can help. Hydrocolloids are known for their potential to increase the fiber content of foods, thereby improving overall dietary fiber intake.

Hydrocolloid fiber can be categorized as either soluble or insoluble. Soluble fiber dissolves in water, forming a gel-like substance that slows digestion. Insoluble fiber, on the other hand, adds bulk to stool and speeds up digestion. Both types of fiber play important roles in maintaining a healthy digestive system.

Hydrocolloids can provide both fiber types, making them a versatile addition to foods. For example, hydrocolloids such as pectin, guar gum, and beta-glucan are sources of soluble fiber, while hydrocolloids like cellulose and xanthan gum provide insoluble fiber.

Hydrocolloid-containing foods can effectively increase dietary fiber intake, especially for individuals who struggle to consume adequate fiber through traditional sources like fruits and vegetables. Manufacturers can create high-fiber options using hydrocolloids in food products without compromising taste or texture.

16.2.2 Hydrocolloids as a source of soluble and insoluble fiber

Hydrocolloids are a valuable dietary fiber source, providing soluble and insoluble fiber to the diet. Soluble fiber dissolves in water and is known to help regulate blood sugar and cholesterol levels. Insoluble fiber does not dissolve in water and promotes regular bowel movements, preventing constipation and other digestive issues.

Hydrocolloids such as guar gum, xanthan gum, and pectin are examples of soluble fiber sources. These hydrocolloids form gels in water and can be found in fruit juices, jams, and other products. Guar gum, in particular, has been shown to reduce blood sugar levels and improve insulin sensitivity in individuals with type 2 diabetes.

Insoluble fiber sources include hydrocolloids such as cellulose, hemicellulose, and lignin. These hydrocolloids are found in whole grains, fruits, and vegetables and contribute to the bulk of stool, promoting regular bowel movements. Insoluble fiber has been shown to reduce the risk of colon cancer and other digestive issues.

Hydrocolloids can be important in meeting dietary fiber intake recommendations, particularly for individuals who struggle to consume fiber-rich foods. Incorporating hydrocolloid-containing products into the diet can increase fiber intake and support overall digestive health. However, it is important to note that excessive consumption of hydrocolloids can lead to gastrointestinal discomfort, such as bloating and gas. Consuming them in moderation and as part of a balanced diet is recommended.

16.2.3 The Impact of Hydrocolloids on Food Texture and Sensory Attributes

Hydrocolloids can significantly affect food products' texture and sensory attributes, making them valuable tools for manufacturers and producers. Hydrocolloids act as food additives, enhancing products' texture, stability, and other functional properties.

The choice of hydrocolloid and its concentration can modify the texture of a food product, such as its viscosity, mouthfeel, and overall consumer perception. For example, xanthan gum is a widely used hydrocolloid that can increase the density of a liquid, thickening it to create a more substantial and creamy texture. On the other hand, carrageenan can produce a smooth and creamy consistency in dairy products such as ice cream and yogurt.

Hydrocolloids can also act as stabilizers, preventing separation and maintaining a uniform texture in food products. For example, hydrocolloids in salad dressings such as cellulose and xanthan gum can prevent the oil and water from separating, resulting in a more stable and visually appealing product.

> "The choice of hydrocolloid and its concentration can modify the texture of a food product, such as its viscosity, mouthfeel, and overall consumer perception."

Sensory attributes, such as taste, aroma, and appearance, can also be influenced by the addition of hydrocolloids. Hydrocolloids can help enhance flavors by increasing the time taste molecules remain in the mouth. They can also modify the color and appearance of food products, resulting in more vibrant and visually appealing dishes.

It is essential to consider the effects of hydrocolloids on food texture and sensory attributes when formulating food products. It is crucial to balance the desired functional benefits of hydrocolloids with the sensory and aesthetic qualities consumers want. Producers can create innovative and appealing food products that meet consumer demand by understanding the properties and effects of hydrocolloids.

16.2.4 Safety and Regulation of Hydrocolloids

Hydrocolloids are widely used as food additives in the food industry, but their safety is paramount. The regulatory agencies responsible for protecting and regulating hydrocolloids in the United States are the Food and Drug Administration (FDA) and the United States Department of Agriculture (USDA).

The FDA classifies hydrocolloids as Generally Recognized As Safe (GRAS) substances, which means that these substances have been extensively studied and have no known harmful effects when used as intended. The FDA also sets specific limits on using hydrocolloids in food products to ensure their safety for consumption.

On the other hand, the USDA regulates the use of hydrocolloids in meat and poultry products. The agency sets standards for the types and amounts of hydrocolloids used in these products to ensure their safety and prevent potential harmful effects.

In addition to the FDA and USDA regulations, the Joint Expert Committee on Food Additives (JECFA) and the European Food Safety Authority (EFSA) also evaluate the safety and efficacy of hydrocolloids as food additives. These agencies assure that hydrocolloids are safe for human consumption and pose no risk to public health when used as intended.

16.2.5 Hydrocolloid Safety Testing

The safety of hydrocolloids is ensured through extensive testing, including acute and chronic toxicity studies, genotoxicity studies, and reproductive toxicity studies. These tests are conducted to identify any potentially harmful effects of hydrocolloids and establish safe exposure levels.

Manufacturers of hydrocolloids are required to provide safety data to regulatory agencies to obtain approval for their use as food additives. The data must demonstrate that the hydrocolloid is safe for human consumption and has no known harmful effects when used as intended.

Hydrocolloid Regulation

Regulations governing the use of hydrocolloids in food products vary by country and region. In the United States, the Code of Federal Regulations (CFR) sets the standards for using hydrocolloids in food products.

The CFR establishes the types and amounts of hydrocolloids that can be used in food products and guidelines for labeling these products. Hydrocolloids must be listed on product labels by their common or usual name, and the amount used in the product must be listed in descending order by weight.

Multiple agencies regulate hydrocolloids to ensure their safety for human consumption as food additives. The FDA, USDA, JECFA, and EFSA work together to evaluate the safety and efficacy of hydrocolloids, and manufacturers must provide safety data to obtain approval for their use in food products. Understanding the safety and regulation of hydrocolloids is essential for ensuring the safety of the food supply and the overall health and well-being of consumers.

16.2.6 Incorporating hydrocolloids into a balanced diet

Hydrocolloids can be a valuable addition to a balanced and healthy diet, especially for individuals looking to increase their dietary fiber intake. When incorporating hydrocolloid-containing food products into your diet, it is important to consider your overall nutritional goals and dietary needs.

Start by selecting hydrocolloid-containing foods that are minimally processed and contain other nutrient-dense ingredients, such as fruits, vegetables, and whole grains. It ensures that you get a variety of essential vitamins, minerals, antioxidants, and dietary fiber.

It is also important to consider portion sizes and serving recommendations, as some hydrocolloid-containing products may be high in added sugars or unhealthy fats. Reading nutrition labels and ingredient lists can help you decide which products to choose.

Incorporating hydrocolloids into homemade recipes is another great way to increase dietary fiber intake. For example, adding chia seeds or psyllium husk to your morning smoothie can significantly boost fiber while adding texture and thickness to your drink.

Experimenting with different hydrocolloids in your cooking and baking can also be a fun and creative way to incorporate more fiber into your diet. Some common hydrocolloids used in cooking and baking include xanthan gum, pectin, and carrageenan, which can be used in various recipes to enhance texture and provide dietary fiber.

By incorporating hydrocolloids thoughtfully and intentionally into your diet, you can easily and deliciously increase your dietary fiber intake and support overall digestive health.

16.2.7 Research and Advancements in Hydrocolloids

Research studies and publications are critical in advancing the knowledge and understanding of hydrocolloid applications. The Food Hydrocolloids *Impact Factor* is a key metric used to measure the impact of research publications in this field. The impact factor is determined by the number of citations received by articles published in a particular journal over a specified time. It measures the journal's influence in the field, with higher impact factors indicating a greater effect.

According to the latest available data, the top five journals in the field of food hydrocolloids, ranked by impact factor, are:

Table 105: Hydrocolloids Journals and their Impact Factor

JOURNAL TITLE	IMPACT FACTOR
Food Hydrocolloids	7.243
Carbohydrate Polymers	7.182
Journal of Agricultural and Food Chemistry	4.192
International Journal of Biological Macromolecules	3.909
Journal of Food Engineering	3.848

These journals publish cutting-edge research on hydrocolloid properties and applications, providing valuable insight for researchers, industry professionals, and other stakeholders.

16.2.8 Future Trends and Potential Developments in Hydrocolloids

As with any scientific field, research and hydrocolloid advancements continue to progress. Technological innovations are leading to novel applications and increased functionality, which can impact the food industry and dietary fiber intake recommendations. This section discusses some of the future trends and potential developments in the field of hydrocolloids.

Emerging Technologies

New technologies such as nanotechnology and biotechnology can potentially revolutionize the use of hydrocolloids in food applications. For example, nanotechnology could lead to more efficient and effective delivery of hydrocolloids to specific parts of the digestive tract, enhancing their functionality. Biotechnology could also enable the production of custom-designed hydrocolloids with unique properties and functions for specific food applications.

Novel Applications

Hydrocolloids are already widely used in the food industry, but there is potential for even broader applications. For example, hydrocolloids could be used to develop new types of products, such as fiber-enriched snacks or baked goods. Hydrocolloid-based coatings could also improve the shelf life and quality of fresh produce.

Environmentally Friendly Production

The production of hydrocolloids can have a significant environmental impact, particularly when synthetic hydrocolloids are used. There is potential for developing more sustainable and environmentally friendly production methods for hydrocolloids. It could include the use of renewable resources and the reduction of waste and energy consumption during production.

With ongoing research and development, hydrocolloids will likely continue to play a significant role in the food industry and meeting dietary fiber intake recommendations. As new technologies and applications emerge, it is important to consider the safety and regulation of hydrocolloids and their potential impact on overall nutrition and health.

Conclusion

In conclusion, hydrocolloids play a significant role in meeting dietary fiber intake recommendations. As discussed in this article, hydrocolloids are food additives with properties such as thickening, gelling, stabilizing, emulsifying, and acting as a gelatin substitute. These properties make them versatile and beneficial in various food applications.

Hydrocolloids can contribute to dietary fiber intake by increasing the fiber content of food products and supporting overall digestive health. They can provide soluble and insoluble fiber in the diet, which has various health benefits and can help individuals meet their recommended intake levels.

It is important to consider hydrocolloids as part of a balanced and nutritious diet. By incorporating hydrocolloid-containing food products, individuals can diversify their nutrient intake and meet their nutritional goals.

Ongoing research and advancements in the field of hydrocolloids have the potential to bring about exciting developments and future trends. Emerging technologies and innovative applications can significantly impact the food industry and dietary fiber intake recommendations. As individuals become more health-conscious, the demand for nutritious and functional food products will likely increase, creating new hydrocolloid opportunities.

In summary, hydrocolloids are a valuable tool for meeting dietary fiber intake recommendations, improving food texture and sensory attributes, and creating new opportunities for innovation in the food industry.

FAQ

How do hydrocolloids contribute to dietary fiber intake?

Hydrocolloids can increase the fiber content of food products, helping individuals meet their recommended dietary fiber intake.

Do hydrocolloids provide both soluble and insoluble fiber?

Yes, hydrocolloids can provide soluble and insoluble fiber, each with health benefits.

How do hydrocolloids impact food texture and sensory attributes?

Different hydrocolloids can influence the viscosity, mouthfeel, and overall consumer perception of texture in food products.

Are hydrocolloids safe and regulated?

Yes, hydrocolloids undergo rigorous testing and approval processes to ensure their safety as food additives, and regulations and guidelines govern their use.

How can hydrocolloids be incorporated into a balanced diet?

Hydrocolloid-containing food products can be part of a healthy diet, considering overall nutritional goals.

What research and advancements are being made in hydrocolloids?

Ongoing research studies and publications, such as the impact factor of food hydrocolloids, contribute to advancing knowledge and understanding of hydrocolloid applications.

What are the future trends and potential developments in hydrocolloids?

Emerging technologies and innovative applications hold the potential for future hydrocolloid advancements, influencing the food industry and dietary fiber intake recommendations.

CHAPTER 17

Agricultural Practices

PART 17.1

Sustainable Farming of Hydrocolloid-Producing Plants and Seaweeds

Sustainable farming practices have become increasingly crucial as the world's population grows and environmental concerns rise. One area where sustainable farming is vital is producing hydrocolloid-rich plants and seaweeds. Hydrocolloids are compounds that form gels when they encounter water. They are commonly used in various industries, including food production, wound care, and skincare.

Hydrocolloid-producing plants and seaweeds can be grown sustainably using methods that minimize environmental impact and preserve natural resources. These plants and seaweeds can be cultivated using crop rotation, organic fertilization, and efficient water usage, ensuring their growth and harvesting are environmentally friendly.

Core Insights:

- ☑ Sustainable farming practices are crucial in The production of hydrocolloid-rich plants and seaweed.
- ☑ Hydrocolloid-producing plants and seaweed can be grown sustainably using crop rotation, organic fertilization, and efficient water usage techniques.
- ☑ Hydrocolloids are compounds that form gels when encountering water and are commonly used in various industries.
- ☑ Sustainable farming practices are crucial for preserving natural resources and minimizing environmental impact.
- ☑ Hydrocolloid-rich plants and seaweeds are valuable resources in various industries, such as food production, wound care, and skincare.

17.1.1 Sustainable Farming Practices for Hydrocolloid Production

Hydrocolloids are widely used in various industries, including food, skincare, and wound care. With their increasing demand, ensuring sustainable production is crucial to avoid negative environmental impacts.

Responsible farming practices are one of the most important aspects of sustainable hydrocolloid production. Here are some methods used for growing and harvesting hydrocolloid-producing plants:

Table 106: Methods used for growing and harvesting hydrocolloid-producing plants

METHOD	DESCRIPTION
Crop rotation	It is the practice of alternating crops in a specific field to avoid depleting soil nutrients. By rotating hydrocolloid plants with other crops, the soil can maintain its fertility while minimizing the use of synthetic fertilizers.
Organic fertilization	Natural fertilizers such as compost, manure, and seaweed can be used instead of chemical fertilizers. This practice helps promote soil health and reduces the risk of water pollution caused by synthetic fertilizers.
Efficient water usage	Water is a crucial resource in hydrocolloid farming, and efficient usage is necessary to minimize water waste. Drip irrigation and rainwater harvesting methods can help reduce water consumption while maintaining optimal crop growth.
Hand harvesting	Manual harvesting of hydrocolloid-producing plants can minimize environmental damage and ensure the quality of harvested crops. It also provides employment opportunities for local communities.

Sustainable hydrocolloid farming benefits the environment and local communities involved in production. We can ensure hydrocolloid availability by prioritizing responsible farming practices while minimizing environmental impact.

17.1.2 Hydrocolloid Applications in Wound Care

Hydrocolloids have been widely used in wound care for their impressive ability to promote healing while providing a protective barrier against outside contaminants. A hydrocolloid is a gel-like substance formed when a fluid combines with a colloidal sense.

Hydrocolloid dressings and bandages have particularly effectively treated chronic wounds, such as pressure sores and leg ulcers. These dressings absorb excess exudate from the injury, forming a gel-like barrier that protects the wound from further damage.

Hydrocolloid wound dressings are available in various shapes and sizes, ensuring that they can be used for wounds of any size or shape. This versatility has made them a popular choice for healthcare professionals, who value their effectiveness and ease of use.

One of the major benefits of using hydrocolloid dressings is their ability to create an optimal healing environment. The moist climate the sauce makes promotes cell growth and helps prevent scarring.

In addition to dressings and bandages, hydrocolloid gels are also used in wound care. These gels promote healing by providing a moist environment and facilitating the removal of dead tissue. Hydrocolloid gels are particularly effective for wounds that are difficult to heal, such as burns and surgical wounds.

17.1.3 Hydrocolloid Dressings vs. Traditional Wound Dressings

The use of hydrocolloid dressings has many advantages over traditional wound dressings. Hydrocolloid dressings are more effective at absorbing exudate, promoting healing, and preventing infection. They are also more comfortable for the patient to wear and require less frequent changing than traditional dressings.

On the other hand, traditional wound dressings are often made from gauze or other materials that can stick to the wound and cause further pain and discomfort. They also need to be changed more frequently, which can be painful for the patient.

Hydrocolloid dressings, bandages, and gels have proven highly effective in wound care. Their ability to promote healing and create an optimal healing environment has made them a popular choice among healthcare professionals. With their ease of use and versatility, hydrocolloid products will continue to play an important role in wound care for years to come.

17.1.4 Hydrocolloids in Skin Care Products

Hydrocolloids are not only valuable in the food industry but are also gaining popularity in the skincare industry. With the advancement of hydrocolloid technology, these ingredients are now commonly used in skincare products such as gels, patches, and acne patches. These products improve skin health by promoting healing and reducing inflammation.

Hydrocolloid gel is one of the most popular hydrocolloid forms in skincare products. It is used to hydrate and soothe the skin, making it an excellent ingredient in moisturizers and serums. The gel consistency helps the skin retain moisture while providing a protective barrier against environmental irritants.

Hydrocolloid patches are another popular form of hydrocolloid in skincare products. These patches are typically used to treat acne and work by absorbing excess oil and impurities from the skin. Hydrocolloid acne patches are designed to be placed directly on top of a pimple, allowing the patch to draw out any impurities and reduce inflammation.

Hydrocolloid technology has also created patches for other skin concerns, such as dark circles and fine lines. These patches are typically infused with other beneficial ingredients, such as collagen and hyaluronic acid, to provide additional skin benefits.

Hydrocolloid Acne Patch Study

Table 107: Hydrocolloid Acne Patches and Control Group

HYDROCOLLOID ACNE PATCHES	CONTROL GROUP
75% reduction in acne size after 24 hours	20% reduction in acne size after 24 hours
90% of participants reported improved acne appearance	30% of participants reported improved acne appearance

> "Hydrocolloid acne patches can significantly reduce acne size and improve appearance, making it an effective treatment option."

The use of hydrocolloids in skincare products is a testament to their versatility and effectiveness. As hydrocolloid technology advances, we expect to see more skincare products utilizing these ingredients and their benefits to improve skin health and appearance.

17.1.5 Sustainable Farming of Seaweeds for Hydrocolloid Production

The demand for hydrocolloid-producing seaweeds is rapidly increasing, primarily driven by their extensive applications in various industries, including food, pharmaceuticals, and cosmetics. The growing demand for these products requires a sustainable approach to farming seaweeds to ensure their availability while protecting the environment.

Farming hydrocolloid-producing seaweed involves cultivating different seaweed species, including red, brown, and green algae. Sustainable farming practices for seaweed include using environmentally friendly methods that minimize the impact on the natural ecosystem.

One example of sustainable seaweed farming is using longlines or ropes that hang from a buoy or float anchored to the seabed. This method is low-cost, has minimal environmental impact, and allows seaweed to be cultivated in their natural habitat. In contrast, other practices, such as seaweed farming in tanks or ponds, can result in high energy usage and water pollution.

Seaweed has high nutrient-absorbing capabilities, which can help mitigate nutrient pollution in water. Furthermore, seaweed farming can reduce the carbon footprint by absorbing carbon dioxide through photosynthesis and releasing oxygen into the environment.

Sustainable farming practices help protect the environment and ensure a consistent supply of high-quality hydrocolloid-producing seaweeds. By prioritizing sustainable farming practices, we can support the growth of the hydrocolloid industry while preserving our natural resources.

Conclusion

In summary, the production and use of hydrocolloids offer numerous benefits in various industries, from food production to wound care and skincare. However, it is crucial to prioritize sustainable farming practices to ensure we can continue accessing these ingredients while minimizing our environmental impact.

We can cultivate hydrocolloid-rich plants and seaweeds without damaging the environment by implementing crop rotation, organic fertilization, and efficient water use. These practices also have the added benefit of producing healthier crops that require fewer pesticides and synthetic fertilizers, reducing the risk of soil contamination and water pollution.

Hydrocolloids play a significant role in food production, enhancing texture, stability, and shelf life. Their use in wound care and skin care has also proven effective in improving healing and skin health. As the demand for these ingredients continues to grow, it is essential to prioritize sustainable farming practices to ensure their continued availability.

Hydrocolloids are a valuable resource that offers numerous benefits in various industries, but sustainable farming practices must be prioritized to access them without compromising the environment. By supporting sustainable farming practices, we can ensure that hydrocolloids remain available for years.

FAQ

What is sustainable farming?

Sustainable farming refers to agricultural practices that minimize negative environmental impacts while producing crops and livestock efficiently. It involves organic fertilization, crop rotation, and efficient water usage.

How are hydrocolloids produced sustainably?

Sustainable farming practices are employed to produce hydrocolloids. It includes using environmentally friendly methods for growing hydrocolloid-producing plants and seaweeds, such as organic fertilization and crop rotation. These practices reduce the negative impact on ecosystems and promote the long-term availability of hydrocolloid resources.

How are hydrocolloids used in wound care?

Hydrocolloid dressings are commonly used in wound care. These dressings create a moist environment that promotes healing and protects the wound from external contaminants. They can also provide pain relief and absorb excess fluid.

How are hydrocolloids utilized in skincare products?

Hydrocolloids are used in skincare products, such as gels, patches, and acne patches. They create a barrier on the skin that helps to trap moisture, promote wound healing, and protect against environmental irritants. Hydrocolloids can also absorb excess oil and impurities from the skin.

How are hydrocolloid-producing seaweeds farmed sustainably?

Sustainable farming methods, such as rope cultivation and integrated aquaculture systems, are used to farm hydrocolloid-producing seaweed. These methods minimize environmental impact and contribute to the development of renewable resources.

PART 17.2

The Impact of Agricultural Practices on the Quality and Yield of Hydrocolloids

Hydrocolloids are critical ingredients in food production, serving as thickeners, stabilizers, texture modifiers, and binding agents. As such, their quality and yield are crucial for ensuring food products' overall functionality and appeal.

Agricultural practices play a significant role in determining the quality and yield of hydrocolloids. Appropriate farming techniques, fertilization methods, and pest control strategies are crucial for achieving high-quality hydrocolloids.

This article will explore the various agricultural practices that impact hydrocolloid quality and yield. We will discuss the importance of water-holding capacity in hydrocolloids, effective fertilization techniques, and appropriate pest control strategies. Additionally, we will examine real-world examples of successful farming practices and explore emerging innovations in hydrocolloid production.

This article highlights the crucial link between agricultural practices and the hydrocolloid industry. By understanding the significance of responsible farming practices in producing high-quality hydrocolloids, we can meet the growing demand for these critical food ingredients.

Show an image of hydrocolloids harvested from a farm with different agricultural practices in the background. The hydrocolloids should appear of high quality and yield, and the agricultural practices should show clear differences in how they affect the crops. Use warm colors to convey a sense of farm abundance and growth, with subtle textures to differentiate between the different farming practices.

Core Insights:

- ☑ Agricultural practices significantly impact the quality and yield of hydrocolloids.
- ☑ Water holding capacity is a crucial factor in hydrocolloid functionality
- ☑ Effective fertilization techniques can maximize hydrocolloid yield
- ☑ Appropriate pest control strategies are essential for preserving hydrocolloid quality
- ☑ Responsible farming practices are critical for meeting the growing demand for high-quality hydrocolloids

17.2.1 The Importance of Water Holding Capacity in Hydrocolloids

Hydrocolloids are widely used in the food industry to improve texture and stability. One of the crucial parameters determining hydrocolloids' functionality is their water-holding capacity.

Water holding capacity refers to the ability of hydrocolloids to bind water molecules, forming a gel-like network that influences the viscosity and texture of food products. Hydrocolloids with high water holding capacity effectively control density and prevent moisture loss, improving the final product's shelf life and sensory properties.

Hydrocolloids with low water holding capacity, on the other hand, may lead to poor texture and stability, affecting the overall quality of the product.

Viscosity Control

Viscosity control is an essential characteristic of hydrocolloids, as it determines the thickness and flow of food products. Hydrocolloids with high water holding capacity exhibit greater viscosity control, enabling manufacturers to achieve specific texture and mouthfeel in their products.

For example, xanthan gum is widely used as a thickening agent in dressings and sauces due to its high water-holding capacity and viscosity control. Similarly, carrageenan is a gelling agent in dairy products such as ice cream and yogurt, providing a smooth and creamy texture.

Overall, water holding capacity is critical in determining the functionality and quality of hydrocolloids in the food industry. Appropriate agricultural practices that enhance water availability and uptake can help improve the water-holding capacity of hydrocolloids and enhance their performance in food products.

17.2.2 Impact of Agricultural Practices on Hydrocolloid Quality

Hydrocolloid farming involves several agricultural practices that significantly influence the quality and yield of hydrocolloids. Understanding these practices is crucial for achieving high-quality hydrocolloids that meet industry standards.

Hydrocolloid Farming Techniques

Hydrocolloid farming involves specific techniques critical for obtaining the best quality hydrocolloids. For instance, planting density, irrigation, and fertilization significantly determine hydrocolloid quality and yield. Inappropriate planting density may lead to reduced productivity, while the overuse of fertilizers may decrease hydrocolloid quality.

Growing Hydrocolloids in Specific Geographic Locations

The geographic location of hydrocolloid farming can also significantly influence hydrocolloid quality. For example, growing hydrocolloids in specific regions with consistent weather patterns and high-quality soil can produce the best hydrocolloid yields and quality.

Best Hydrocolloid Growing Techniques

Adopting the best hydrocolloid growing techniques is essential for maintaining the integrity of the hydrocolloid. The following table provides a summary of the best hydrocolloid growing techniques:

Table 108: Best hydrocolloid growing techniques.

TECHNIQUE	DESCRIPTION
Intercropping	Growing two or more hydrocolloid crops at the same time to maximize productivity and reduce pest infestations.
Proper Timing of Harvest	Harvesting the hydrocolloid at the right time ensures optimal yield and quality.
Using Quality Seeds	Using high-quality hydrocolloid seeds ensures the crop will have a high yield and consistent quality.
Integrated Pest Management (IPM)	Using IPM strategies that involve minimal use of pesticides and, instead, utilizing biological pest control methods ensures that the hydrocolloid maintains its quality.

Adopting these techniques is critical in achieving high-quality hydrocolloids with maximum yields.

In summary, agricultural practices play a significant role in determining hydrocolloid quality and yield. Appropriate hydrocolloid farming techniques, geographic location, and the proper use of hydrocolloid growing strategies are essential for achieving high-quality hydrocolloids.

17.2.3 Fertilizing Hydrocolloids for Optimal Yield

Hydrocolloid farming requires careful attention to fertilization techniques to maximize yield and quality. Hydrocolloids like agar, carrageenan, and guar gum thrive under certain conditions that can be optimized using specific nutrients and fertilizers.

The fertilizer type and application timing are critical factors in hydrocolloid production. Nitrogen, phosphorus, and potassium are essential nutrients for plant growth, and their proportions in fertilizer affect the amount and quality of hydrocolloids produced.

Research has shown that a balanced application of nitrogen and potassium is crucial in maximizing hydrocolloid yield, with a ratio of 1:2 or 1:3 being optimal for some crops. Phosphorus plays a vital role in root development and water uptake, playing an essential role in hydrocolloids' water retention properties.

> "Nitrogen and potassium are the most important elements in hydrocolloid farming as they strongly influence the yield and quality of the crop."

There has also been a growing trend towards organic fertilizers in hydrocolloid farming, as they reduce the risk of soil degradation and support a sustainable approach to agriculture. Organic fertilizers like compost, manure, or green manure provide slow-release nutrients to plants, promoting healthy growth without compromising soil health.

Hydrocolloid crops also benefit from micronutrient fertilizers, which provide a range of essential minerals like iron, zinc, and manganese. These micronutrients support the plants' metabolic processes and help improve hydrocolloid yield. However, ensuring fertilization is done at the right time and in the appropriate quantities is essential. Over-application of fertilizers can result in excessive growth, leading to lower hydrocolloid quality and yield. Additionally, under-application can result in stunted growth and reduced output, making careful monitoring and adjustment of fertilization practices essential. Appropriate fertilization techniques are a crucial aspect of achieving optimal hydrocolloid yield. It is important to consider the specific nutrient needs of each crop and apply fertilizers accordingly to achieve the best results.

Hydrocolloid Yield with Different Fertilization Techniques

Table 109: Fertilization Technique and Hydrocolloid Yield

FERTILIZATION TECHNIQUE	HYDROCOLLOID YIELD (KG/HA)
Control (No Fertilizer)	3,500
Inorganic Fertilizer	5,000
Organic Fertilizer	4,500
Micronutrient Fertilizer	4,000

The table above shows the yield of hydrocolloids with different fertilization techniques. The results indicate that inorganic and organic fertilizers significantly impact hydrocolloid yield, increasing output by up to 42% and 29%, respectively. Micronutrient fertilizers also positively affect production, with an increase of 14%.

These findings demonstrate the importance of carefully considering fertilization techniques to optimize hydrocolloid yield and quality. Hydrocolloid growers can improve their crop's productivity and sustainability by utilizing the appropriate fertilizers and monitoring applications.

17.2.4 Pest Control Strategies for Hydrocolloid Crops

Hydrocolloid crops can be vulnerable to pest-related risks that may harm their quality and yield. Therefore, it is crucial to implement effective pest control strategies that prevent crop damage and ensure optimal hydrocolloid production.

One of the most common and destructive pests for hydrocolloid crops is the whitefly. These tiny insects feed on plant sap, leading to stunted growth and reduced yield. To prevent whitefly infestations, growers should regularly monitor their crops for signs of infestation and take swift action to control them. Biological control methods, such as the use of predatory insects, can be effective in minimizing whitefly populations.

Another significant challenge for hydrocolloid crops is fungal diseases. Fungi can cause considerable damage to the crop, leading to reduced yield and poor-quality hydrocolloids. Common fungal infections in hydrocolloid crops include anthracnose, powdery mildew, and grey mold. Agricultural practices such as proper crop rotation, fungicides, and regular plant inspection can help prevent and control fungal infections.

Furthermore, proper irrigation practices can also help control pest infestation. Overwatering can create favorable conditions for pests, such as fungi and aphids, which thrive in moist environments. Additionally, irrigation at the root level rather than the leaves can help minimize the risk of fungal growth on the leaves.

Overall, pest control strategies play a critical role in hydrocolloid crop management. Effective pest management practices such as biological control methods, proper irrigation, and fungicides can help prevent and mitigate pest-related risks that can negatively impact hydrocolloid yield and quality.

Example of Biological Control Methods for Hydrocolloid Pest Control

Table 110: Beneficial Insect, Pest Targeted and Method of Control

BENEFICIAL INSECT	PEST TARGETED	METHOD OF CONTROL
Ladybugs	Aphids	Prey on the aphids, reducing their population
Braconid Wasps	Caterpillars	Parasitize the pest's egg or larva, reducing their population
Assassin Bugs	Mealybugs	Prey on mealybugs, reducing their population

17.2.5 Harvesting Hydrocolloids: Techniques and Machinery

Hydrocolloid harvesting is a critical step in the production process, and it is vital to ensure that appropriate techniques and machinery are used to maintain the quality and yield of hydrocolloids. The following are some of the most common harvesting techniques and machinery used in hydrocolloid farming.

Hand harvesting is typically used for smaller-scale hydrocolloid farms, where it is difficult to use large machinery. In contrast, machine harvesting is used for larger farms, where higher volumes of hydrocolloids need to be harvested within a shorter timeframe.

Regarding hydrocolloid harvesting machinery, different options are available, depending on the specific crop and harvesting method. Some common types of machinery used in hydrocolloid farming include:

- **Rotor Reapers:** These machines have rotating blades that cut the hydrocolloid plant at its base and collect it into a hopper.
- **Double Disc Reapers:** These machines use two rotating discs that cut the hydrocolloid plant and convey it to a collection area.

- **Combine Harvesters:** These machines are commonly used for cereal crops and can be modified with specialized attachments to harvest hydrocolloids.

Choosing the appropriate harvesting machinery depends on the hydrocolloid crop and the cultivation method. The machinery should be well-maintained and calibrated to ensure the hydrocolloid plants are harvested efficiently and with minimal damage.

> "Using appropriate harvesting machinery is crucial in maintaining the quality and yield of hydrocolloids in large-scale farming."

Moreover, harvesting hydrocolloids at the right time ensures optimal quality and yield. The timing of hydrocolloid harvesting depends on several factors, such as crop maturity, environmental conditions, and harvesting machinery used. Harvesting at the proper time ensures that hydrocolloids have the desired functionality, texture, and flavor, making them ideal for food production.

Overall, proper hydrocolloid harvesting techniques and machinery are critical to maintaining the quality and yield of hydrocolloids. Choosing the appropriate harvesting method and machinery and harvesting at the optimal time ensures that hydrocolloids are of high quality and meet the specific requirements of various applications.

17.2.6 Real-World Examples of Agricultural Practices and Hydrocolloid Quality

Several real-world examples illustrate the impact of agricultural practices on hydrocolloid quality. The following case studies showcase how key techniques and strategies can enhance hydrocolloid yield and improve the overall quality of the final product:

Case Study 1: Guar gum production in India

Guar gum is a hydrocolloid widely used in the food industry for its thickening properties. India is a major source of guar gum, with several farming communities specializing in its cultivation. To enhance the quality and yield of guar gum, farmers in India have adopted several innovative techniques, including:

- Using drip irrigation to minimize water wastage and ensure consistent moisture levels
- Intercropping guar with other crops to promote soil health and prevent pest infestations
- Regular monitoring of soil health and nutrient levels to enhance plant growth
- Training on sustainable farming practices to promote responsible hydrocolloid production

These practices have significantly improved the quality and yield of guar gum in India, with the country now being a leading supplier of this hydrocolloid to the global market.

Case Study 2: Carrageenan Farming in the Philippines

Carrageenan is another widely used hydrocolloid known for its gelling and stabilizing properties. The Philippines is a major carrageenan producer, with several farmers specializing in its cultivation. To ensure optimal yield and quality, farmers in the Philippines have adopted several key practices:

- Using seaweed strains that are best suited to the local environment and climate conditions
- Monitoring water quality and temperature to ensure optimal growing conditions for the seaweed
- Regular pruning and cleaning of seaweed beds to promote healthy growth and prevent contamination
- Participating in community-based initiatives to promote sustainable carrageenan farming practices

These efforts have contributed to the Philippines' position as a leading source of high-quality carrageenan for the food industry. These real-world examples demonstrate the impact of agricultural practices on hydrocolloid quality. By implementing sustainable farming methods and embracing innovative techniques, farmers can enhance hydrocolloid yield and promote producing high-quality products.

17.2.7 Improving Hydrocolloid Production: Innovations and Future Directions

Agricultural practices play a crucial role in determining the quality and yield of hydrocolloids. As the demand for these ingredients continues to rise, exploring innovative techniques and emerging technologies to improve hydrocolloid production is essential.

> "Hydrocolloid production techniques have evolved significantly over the years, with new technologies and farming methods that show promise for the future."

One such innovation is precision agriculture, which uses sensors, GPS mapping, and machine learning to optimize crop yields while minimizing resource utilization. This approach helps farmers monitor and analyze soil moisture, nutrient levels, and pest infestations to implement targeted interventions that increase hydrocolloid yields and improve product quality.

Another emerging technology is biotechnology, which uses genetic engineering to develop plant varieties that produce hydrocolloids with enhanced properties. For example, researchers at the University of Guelph in Canada have developed a new strain of guar gum, a popular hydrocolloid used in food and industrial applications. The new variety is more drought-resistant and has a higher yield, making it an attractive option for farmers and manufacturers.

Other potential areas for innovation include hydrocolloid extraction and processing techniques and sustainable farming practices that reduce environmental impact.

> "As hydrocolloid production continues to evolve, it's essential to prioritize sustainability and responsible farming practices to ensure long-term viability and success. Continued research and development in hydrocolloid production is necessary to meet the growing demand for these versatile ingredients and improve their quality and yield."

17.2.8 The Link Between Agricultural Practices and the Hydrocolloid Industry

Hydrocolloids play a vital role in the food industry, with a growing demand for high-quality products. However, the quality and yield of hydrocolloids heavily depend on agricultural practices. The hydrocolloid industry must, therefore, work closely with farmers to develop and implement responsible farming practices that promote sustainable production.

Adopting sustainable agricultural practices can lead to significant improvements in hydrocolloid quality and yield. Research has shown that using environmentally friendly methods of hydrocolloid farming can enhance productivity without compromising product quality. For instance, using organic fertilizers and pest management techniques can help maintain soil health and reduce the risk of contamination, resulting in better yields and higher-quality hydrocolloids.

Case Study: Sustainable Hydrocolloid Farming in the United States

One example of sustainable hydrocolloid farming in the United States is the utilization of cover crops. Cover crops are grown off-season to protect the soil from erosion and nutrient depletion, preventing weed growth and enhancing soil fertility. In addition to these benefits, cover crops also serve as a natural pest control measure, reducing the need for chemical pesticides.

According to a study by the United States Department of Agriculture, using cover crops in hydrocolloid farming can significantly improve soil health and enhance hydrocolloid yields. The study found that farms using cover crops had 21% higher results than those that did not. Additionally, the research indicated that hydrocolloids from farms using cover crops had higher protein and fiber content than conventionally farmed hydrocolloids.

The hydrocolloid industry must continue to promote and encourage sustainable farming practices to ensure the longevity and growth of the industry. By working together, farmers and manufacturers can produce high-quality hydrocolloids that meet the demands of consumers without compromising the environment.

Conclusion

In conclusion, the quality and yield of hydrocolloids heavily depend on agricultural practices. Throughout this article, we have discussed the various factors that impact the functionality of hydrocolloids in food production, including water-holding capacity, fertilization techniques, pest control, and harvesting equipment. Our exploration of hydrocolloid farming has emphasized the significance of sustainable and responsible practices that balance environmental impact with high-quality yield and output. The hydrocolloid industry must prioritize accountable agricultural practices to meet the growing demand for high-quality hydrocolloids.

As we look forward, the potential for innovation and development in hydrocolloid production is vast, with new technologies and farming practices emerging regularly. By prioritizing research and development in this field, we can continue improving the quality and yield of hydrocolloids and ensure a sustainable future for the industry.

FAQ

How does water-holding capacity affect hydrocolloids?

Water holding capacity is a crucial property of hydrocolloids. It refers to their ability to retain water and influence viscosity control. The water-holding capacity of hydrocolloids determines their performance and functionality in food products.

How do agricultural practices affect hydrocolloid quality?

Agricultural practices have a significant impact on hydrocolloid quality. Factors such as hydrocolloid farming techniques, growing conditions, and post-harvest handling can influence the purity, cleanliness, and overall quality of hydrocolloids. Adhering to best practices ensures high-quality hydrocolloids.

How can fertilizing hydrocolloids optimize yield?

Fertilizing hydrocolloids with the right nutrients and techniques can maximize their yield. Proper fertilization promotes healthy plant growth, enhances hydrocolloid production, and improves the overall productivity of crops. It is an essential practice for achieving optimal products.

What are effective pest control strategies for hydrocolloid crops?

Hydrocolloid crops face pest-related risks, but appropriate agricultural practices can help mitigate these challenges. Effective pest control strategies include integrated pest management techniques, regular monitoring, crop rotation, and using natural predators or organic pesticides to preserve the integrity and yield of hydrocolloids.

What techniques and machinery are involved in harvesting hydrocolloids?

Harvesting hydrocolloids requires specific techniques and machinery to ensure the best quality and yield. Harvesting methods vary depending on the hydrocolloid crop. Specialized equipment such as harvesting combines, threshers, or seeders are typically used to harvest hydrocolloids efficiently and carefully.

FREQUENTLY ASKED QUESTIONS (FAQS) CONTINUED

- **Can you provide real-world examples of agricultural practices enhancing hydrocolloid quality?**

  There are numerous examples where specific agricultural practices have improved hydrocolloid quality. For instance, precision irrigation techniques, controlled environment farming, and optimized nutrient management have been successfully implemented to enhance hydrocolloid quality and yield in various regions.

- **What innovations and future directions are being explored in hydrocolloid production?**

  Ongoing advancements and future directions in hydrocolloid production involve innovative agricultural practices and emerging technologies. These include precision farming, genetic engineering, biotechnology, and sustainable cultivation methods. These innovations hold promise for improving hydrocolloid quality and increasing yield.

- **How do agricultural practices affect the hydrocolloid industry?**

  Agricultural practices have a direct impact on the hydrocolloid industry. Responsible and sustainable farming practices consistently supply high-quality hydrocolloids to meet the industry's demands. The sector can thrive and meet growing market needs by prioritizing agricultural practices that optimize hydrocolloid quality and yield.

CHAPTER 18

Nanotechnology and Hydrocolloids

PART 18.1

The Role of Nanotechnology in Enhancing Hydrocolloid Functions

Hydrocolloids are essential ingredients in various industries, including food, pharmaceuticals, and cosmetics. They are used as thickeners, stabilizers, and emulsifiers, among other functions. Nanotechnology has emerged as a promising tool for enhancing the properties and applications of hydrocolloids.

Nanotechnology involves manipulating materials at the nanoscale, 1 to 100 nanometers. Researchers can create materials with improved functionality and performance by leveraging this technology. In the case of hydrocolloids, nanotechnology enables the development of materials with enhanced stability, texture, and functionality.

A depiction of tiny nanobots working together to enhance the properties of hydrocolloids in a scientific laboratory setting. The hydrocolloids should be visually represented as a gel-like substance with a glossy sheen. The nanobots should be realistically portrayed, with intricate details in their construction and movement. The colors should lean towards shades of blue, silver, and white to create a futuristic and technological atmosphere. The background should be sleek and sterile, with metallic surfaces and bright lighting to enhance the high-tech theme.

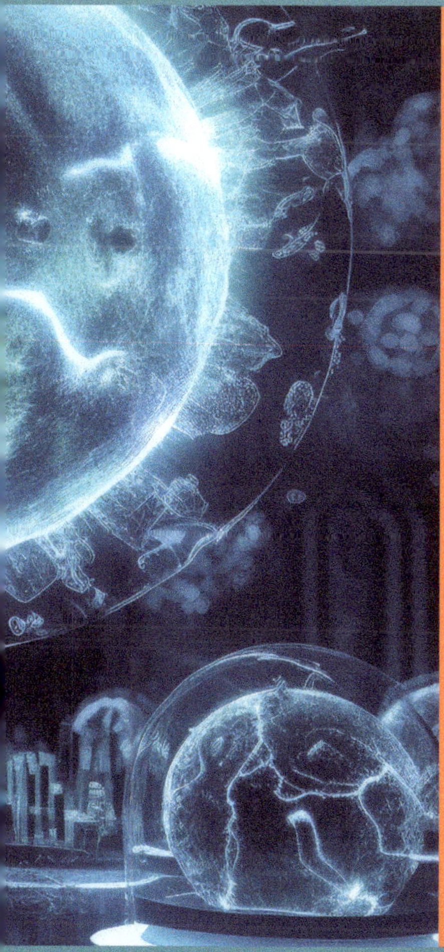

Core Insights:

- ☑ Hydrocolloids are essential ingredients in various industries, including food, pharmaceuticals, and cosmetics.
- ☑ Nanotechnology has emerged as a promising tool for enhancing the properties and applications of hydrocolloids.
- ☑ Nanotechnology involves manipulating materials at the nanoscale, 1 to 100 nanometers.
- ☑ By leveraging nanotechnology, researchers can create materials with improved functionality and performance.
- ☑ Nanotechnology enables the development of hydrocolloid materials with enhanced stability, texture, and functionality.

18.1.1 Understanding Hydrocolloids and Their Benefits

Hydrocolloids are a group of long-chain polymers that can form gels and viscous solutions when they encounter water. They are commonly used in various industries, such as food, pharmaceuticals, and cosmetics, for their unique properties and benefits.

One of the primary benefits of hydrocolloids is their ability to improve the texture and mouthfeel of food products. They can be used as thickeners, stabilizers, and emulsifiers in various foods, including dairy products, dressings, sauces, and baked goods.

Hydrocolloids also offer benefits in the pharmaceutical industry, where they can be used as drug delivery systems. They can help to control the release of drugs over time, ensuring a more consistent and effective dosage.

The cosmetic industry uses hydrocolloids as thickeners and stabilizers in various products, such as lotions and creams. They can help to improve the texture and consistency of these products, making them more effective and appealing to consumers.

Overall, the benefits of hydrocolloids are vast and varied, making them an important component in many products across various industries.

> "Hydrocolloids offer benefits in the pharmaceutical industry, where they can be used as drug delivery systems."

18.1.2 Nanotechnology in Hydrocolloids

Nanotechnology has revolutionized the field of hydrocolloids by offering new methods to improve the properties and functionalities of hydrocolloid materials. Integrating nanomaterials has resulted in hydrocolloid products with enhanced stability, texture, and overall performance.

One of the primary benefits of nanotechnology in hydrocolloids is the ability to engineer hydrocolloid materials at the nanoscale. It allows for precise control over the properties of the hydrocolloid material and the development of specific functionalities for various applications. By manipulating particle size and structure, nanotechnology provides a new window of opportunity to optimize hydrocolloid properties and performance.

Nanomaterials in Hydrocolloids

Integrating nanomaterials into hydrocolloids has led to new possibilities regarding stability and functionality. Nanoparticles allow for improved barrier properties, particularly useful in food packaging. In addition, incorporating nanomaterials can result in hydrocolloid products with enhanced sensory attributes, such as improved texture and appearance.

Table 111: Nanomaterials and Enhanced Functionality

NANOMATERIALS	FUNCTIONALITY ENHANCED
Nanocellulose	Improves viscosity
Nanoparticles	Enhances barrier properties
Nanocapsules	Controlled release of active ingredients

Integrating nanomaterials in hydrocolloids has also resulted in the development of new hydrocolloid-based drug delivery systems. Nanomaterials allow for the controlled release of active ingredients, increasing the drug's efficacy and reducing potential side effects.

The Future of Nanotechnology in Hydrocolloids

As with any emerging technology, there are still challenges to overcome and opportunities to explore regarding nanotechnology in hydrocolloids. Safety concerns regarding the use of nanomaterials have yet to be fully addressed, and regulatory guidelines for applying nanotechnology in hydrocolloids are still in the works.

However, the potential benefits of nanotechnology in hydrocolloids are vast, and further advancements in the field are expected. Researchers continue to explore new ways to optimize hydrocolloid properties by integrating nanomaterials and developing new formulations and processes.

The future of nanotechnology in hydrocolloids looks promising, with potential applications ranging from food additives to drug delivery systems. Hydrocolloids can continue to be optimized for a wide range of uses and industries by harnessing the power of nanotechnology, leading to the development of new and innovative products.

The Advancements in Hydrocolloid Technology

Hydrocolloid technology has come a long way since its inception. With modern processing techniques and new formulations, hydrocolloids have become an essential ingredient in various industries. From food to pharmaceuticals and cosmetics, hydrocolloids offer a range of benefits that make them an indispensable component in many applications.

One of the significant advancements in hydrocolloid technology is the development of new formulations. Manufacturers are continually exploring new combinations of ingredients that can optimize the properties of hydrocolloids, such as stability and solubility. These new formulations have led to more versatile and effective hydrocolloid products that can meet the specific needs of different applications.

The process of hydrocolloid production has also benefited from technological advancements. Modern processing techniques have made it possible to produce higher purity and consistency hydrocolloids. It has improved performance and reliability in various applications, such as pharmaceuticals and cosmetics.

Another significant advancement in hydrocolloid technology is the use of new processing methods. The introduction of nanotechnology has allowed for the creation of hydrocolloids with unprecedented properties, such as enhanced stability, improved texture, and controlled release mechanisms.

The application of nanotechnology to hydrocolloid technology has also opened up new avenues for hydrocolloid uses. In the food industry, for example, nanomaterials have led to new formulations of hydrocolloids that can improve the texture, shelf life, and sensory attributes of food products.

Overall, hydrocolloid technology has evolved significantly in recent years thanks to the development of new formulations, processing techniques, and the application of nanotechnology. These advancements have led to more efficient and effective hydrocolloid products with various uses in various industries.

18.1.3 Nanotechnology Engineering in Hydrocolloid Development

Nanotechnology has allowed engineers to design and optimize hydrocolloid materials for specific applications. The use of nanotechnology in hydrocolloid development has created groundbreaking opportunities for the food, pharmaceutical, and cosmetic industries, among others.

Nanotechnology engineering aims to modify hydrocolloid properties by utilizing commercially available nanomaterials. These nanomaterials are synthesized to have unique properties that can improve the behavior of hydrocolloid materials.

Nanomaterials Used in Hydrocolloids

Various types of nanomaterials are used in hydrocolloid development to enhance their properties. These include:

Table 112: Nanomaterials and their Properties

NANOMATERIALS	PROPERTIES
Nanoparticles	Improve texture and stability.
Nanocapsules	Controlled release systems for drugs or flavors
Nanofibers	Improve transparency and mechanical properties.
Nanoporous materials	Enhance adsorption and catalytic properties.

Applications of Nanotechnology Engineering in Hydrocolloids

Nanotechnology engineering has led to the development of hydrocolloid materials with improved properties. Some applications of nanotechnology engineering in hydrocolloid development include:

- Drug delivery systems
- Controlled release systems
- Biodegradable packaging materials

- ⇨ Improved food additives

Using nanotechnology in hydrocolloid development has also opened avenues for environmental sustainability. Nanomaterials can produce biodegradable, eco-friendly hydrocolloid materials that reduce waste and improve energy efficiency.

In conclusion, nanotechnology engineering has brought significant advancements in hydrocolloid development. Using nanomaterials, engineers can design and optimize hydrocolloid materials for specific applications, developing improved products with enhanced properties.

18.1.4 Applications of Nanotechnology in Food Hydrocolloids

Food hydrocolloids are widely used as additives to improve texture, stability, and shelf life in various food products. Nanotechnology has enabled significant advances in food hydrocolloids, providing innovative solutions to current challenges in the food industry.

Nanotechnology in food additives has revolutionized the industry, improving nanostructure control and enhancing properties and functionalities. For instance, incorporating nano-scale particles, such as cellulose nanocrystals, can improve hydrocolloid-based systems' mechanical strength and thermal stability.

Moreover, nanotechnology can enhance the sensory attributes of food products by modifying their texture, flavor, and appearancc. Nanoparticles can significantly influence the rheological properties of hydrocolloid-based systems, changing their flow behavior and improving their mouthfeel.

Another promising application of nanotechnology in food hydrocolloids is the development of nanosensors for food quality and safety monitoring. Integrating nano-scale materials into hydrocolloid formulations makes it possible to create biosensors that can detect and quantify specific food contaminants or pathogens.

Nanotechnology is crucial in developing new and innovative food products with improved functionalities and enhanced nutritional value. Integrating nanotechnology in food hydrocolloids is a promising avenue for the food industry, offering new opportunities for development, innovation, and growth.

18.1.5 Hydrocolloid Gel Formulation and Nanotechnology

In this section, we will explore the formulation of hydrocolloid gels and the role of nanotechnology in optimizing their properties. Hydrocolloid gels are widely used in the food, pharmaceutical, and cosmetic industries due to their unique gel-forming properties and ability to improve product texture and stability. Nanotechnology offers a powerful tool for enhancing the properties of hydrocolloid materials, improving their functionality, and enabling new uses.

Hydrocolloid materials are extensively used as gelling agents in food products such as desserts, sauces, and jellies. They provide unique textural properties and improve the overall sensory experience of the product. The gel-forming properties of hydrocolloids are based on their ability to form a three-dimensional network of polymer chains that trap water molecules. This network is stabilized by intermolecular forces such as hydrogen bonding and electrostatic interactions.

Nanotechnology can improve hydrocolloid gels' properties by fine-tuning their structure and optimizing their interactions with water molecules. For example, adding nanoparticles can improve gel strength, elasticity, and sensory attributes. The size and surface characteristics of the nanoparticles can be tailored to optimize their interaction with the hydrocolloid material, resulting in a stronger and more stable gel.

In addition, nanotechnology can control the release of active ingredients from hydrocolloid-based systems, making them suitable for applications such as drug delivery. The use of nanotechnology in hydrocolloid development has also led to the development of new hydrocolloid-based materials with unique properties and functionalities.

Nano Technology-Enabled Hydrocolloid Properties

Nanotechnology can enhance a wide range of properties of hydrocolloids, including their rheological properties, thermal stability, and functionality. Nanomaterials can improve the mechanical properties of hydrocolloid gels, enhance their thermal stability, and improve their compatibility with other ingredients. For example, adding clay nanoparticles to hydrocolloid gels can enhance their mechanical properties, resulting in a stronger and more stable gel.

Nanomaterials can also improve the functionality of hydrocolloid-based systems by enabling new applications. For example, using nanocapsules and nanoemulsions in hydrocolloid-based systems can enhance their delivery properties, making them suitable for drug delivery and personal care products.

> "Nanotechnology offers a powerful tool for enhancing the properties of hydrocolloid materials, improving their functionality and enabling new uses."

Overall, the use of nanotechnology in hydrocolloid development offers a powerful tool for enhancing the properties and functionalities of these materials. It has led to new hydrocolloid-based materials with unique properties and functionalities, enabling new applications in various industries.

18.1.6 Nanotechnology-Enabled Hydrocolloid Properties

One of the key advantages of using nanotechnology in hydrocolloid development is the ability to enhance their properties. Nanomaterials can improve various hydrocolloid properties, including their:

Table 113: Nanomaterials improvong hydrocolloid properties

PROPERTY	DESCRIPTION
Rheological properties	Nanomaterials can improve the flow and viscosity of hydrocolloids, making them easier to handle and process.
Thermal stability	Nanomaterials can improve the thermal stability of hydrocolloids, allowing them to maintain their properties at higher temperatures.
Functionality	Nanomaterials can improve the performance of hydrocolloids in various applications, such as controlled release systems and drug delivery.

Hydrocolloid materials can be optimized using nanomaterials to improve performance and efficiency. Additionally, nanotechnology can allow for the creation of new hydrocolloid materials with unique properties that were previously unavailable.

For example, nanocellulose is a nanomaterial that can be incorporated into hydrocolloid materials to enhance their properties. Nano cellulose has a high aspect ratio, making it an effective reinforcing agent that increases the mechanical strength of hydrocolloid gels. It also has a large surface area, which allows for better interaction with other materials, improving stability and functionality.

Overall, nanotechnology enables the development of hydrocolloid materials with improved properties and functionalities. By utilizing nanomaterials, hydrocolloid developers can create optimized materials for specific applications, leading to enhanced performance and efficiency.

18.1.7 Innovations in Nano Technology for Hydrocolloid Applications

Nanotechnology has opened a world of possibilities for hydrocolloid applications. With their unique properties, nanomaterials can enhance the functionality and performance of hydrocolloid materials. Here are some of the latest innovations in nanotechnology for hydrocolloid applications:

Nano Capsules

Nanocapsules are sub-micron-sized particles that encapsulate various substances, such as drugs, flavors, and fragrances. In hydrocolloid applications, nanocapsules can be used to improve the delivery and release of active ingredients.

For example, alginate-based nanocapsules can encapsulate bioactive compounds in food products, protecting them from degradation and improving their bioavailability.

Nano Emulsions

Nanoemulsions are stable oil-in-water emulsions with droplet sizes in the nanometer range. They can improve food products' texture, stability, and sensory attributes. Nanoemulsions can also deliver lipophilic bioactive compounds, such as vitamins and antioxidants.

Nano Sensors

Nanosensors can detect and measure physical, chemical, and biological parameters in real-time. Nanosensors can monitor food products' quality and safety in hydrocolloid applications and control active ingredient release. For example, pH-sensitive nanosensors can monitor the acidity of food products and trigger the release of antimicrobial agents.

> "The latest innovations in nanotechnology for hydrocolloid applications are opening up a range of possibilities for improving the performance and functionality of hydrocolloid materials."

18.1.8 Nano Technology and Sustainable Hydrocolloid Solutions

Nanotechnology has enormous potential to promote sustainable hydrocolloid solutions that benefit the environment and society. By reducing waste and enhancing energy efficiency, nanotechnology can reduce the ecological footprint of hydrocolloid production. Furthermore, nano-technology advancements can lead to the development of biodegradable hydrocolloid materials that minimize waste and facilitate recycling.

One of the most promising approaches to sustainable hydrocolloid production is using natural materials as the basis for nanomaterial synthesis. For example, chitosan nanoparticles can be derived from renewable sources like shrimp shells and used to enhance the properties of hydrocolloid materials. Additionally, using nanotechnology can lead to the development of hydrocolloid materials with improved biocompatibility, reducing the potential environmental impact of hydrocolloid products in various applications.

Moreover, nanotechnology can drive innovation in hydrocolloid-based systems, leading to new approaches to immunoassays, biosensors, and other diagnostic technologies. By leveraging the unique properties of nanomaterials, researchers can develop hydrocolloid-based sensing platforms that are more accurate, sensitive, and specific than traditional technologies.

Overall, using nanotechnology in hydrocolloid development represents a significant leap forward in sustainability and provides exciting opportunities for innovation in various fields. As researchers continue exploring new nanotechnology applications in hydrocolloids, we will likely see even more advancements in the coming years.

The Benefits of Sustainable Hydrocolloid Solutions

The use of sustainable hydrocolloid solutions has many benefits, including:

- Reduced waste and increased efficiency in production
- Improved biocompatibility and reduced environmental impact
- More accurate and sensitive diagnostic technologies
- Enhanced functionality and versatility of hydrocolloid-based materials

These benefits make sustainable hydrocolloid solutions attractive for industries looking to reduce their ecological footprint while meeting customers' demands.

18.1.9 Overcoming Challenges and Future Directions

Integrating nanotechnology in hydrocolloid material has opened new avenues for innovation and progress in various industries. However, like any emerging technology, some challenges must be addressed to ensure its safe and responsible use.

One of the primary concerns with nanotechnology in hydrocolloids is the lack of regulatory guidelines and standards for their use. The unique properties of nanomaterials require careful consideration and assessment to ensure their safety and minimize potential risks to human health and the environment.

Another challenge is the scalability of the nanotechnology-based processes in hydrocolloid production. While laboratory and pilot-scale experiments have shown promising results, further research is needed to determine the feasibility and cost-effectiveness of large-scale production.

Despite these challenges, the potential for nanotechnology in hydrocolloids is immense. Future directions include the development of novel nanomaterials with tailored properties for specific applications, optimizing hydrocolloid-based systems using advanced computational tools, and exploring sustainable nano technology-enabled hydrocolloid solutions.

As the field advances, it is essential to prioritize the responsible development and deployment of nanotechnology in hydrocolloids, ensuring their potential benefits are realized while minimizing potential risks.

> "The unique properties of nanomaterials require careful consideration and assessment to ensure their safety and minimize potential risks to human health and the environment."

Conclusion

In conclusion, nanotechnology has demonstrated significant potential in enhancing the functions of hydrocolloids. By using nanomaterials, hydrocolloid technology can design and optimize materials for specific applications, leading to improved stability, texture, and functionality.

The advancements in hydrocolloid technology and innovations in nanotechnology have opened new opportunities for hydrocolloid applications in various industries, such as food, pharmaceutical, and cosmetics.

It is important to note that using nanotechnology in hydrocolloids has regulatory and safety challenges that must be addressed. However, the potential for further advancements in the field remains high with proper mitigation measures.

Overall, nanotechnology has enabled the development of sustainable hydrocolloid solutions aimed at reducing waste, improving energy efficiency, and enhancing biodegradability. Integrating nanotechnology and hydrocolloid technology holds great promise for the future, with the potential to drive innovation and create new possibilities for industries worldwide.

FAQ

What is the role of nanotechnology in enhancing hydrocolloid functions?

Nanotechnology plays a crucial role in improving the properties and applications of hydrocolloids. It allows for the use of nanomaterials to enhance hydrocolloid materials' stability, texture, and functionality.

How is nano technology applied to hydrocolloids?

Nanotechnology is applied to hydrocolloids using nanomaterials engineered at the nanoscale to enhance the properties of hydrocolloid materials. It includes improving stability, texture, and functionality.

What advancements have been made in hydrocolloid technology?

Hydrocolloid technology has seen significant advancements, including developing new formulations and processes. These advancements have expanded the applications of hydrocolloids in various industries such as food, pharmaceutical, and cosmetics.

How does nanotechnology engineering contribute to hydrocolloid development?

Nano-technology engineering is crucial in designing and optimizing hydrocolloid materials for specific applications. It enables the development of controlled release systems, drug delivery systems, and other tailored hydrocolloid-based solutions.

What are the applications of nanotechnology in food hydrocolloids?

Nanotechnology has various applications in food hydrocolloids. It enhances the functionality of food additives by improving texture, stability, and shelf life. It contributes to the production of high-quality food products.

How does nanotechnology optimize hydrocolloid gel formulation?

Nanotechnology plays a role in optimizing hydrocolloid gel formulation by improving gel strength, elasticity, and sensory attributes. Nanomaterials are used to enhance the properties of hydrocolloid gels, making them more efficient and effective.

FREQUENTLY ASKED QUESTIONS (FAQS) CONTINUED

- **What properties of hydrocolloids can be enhanced through nanotechnology?**

  Nanotechnology can enhance various properties of hydrocolloids, including rheological properties, thermal stability, and functionality. Nanomaterials enable the improvement of these properties, enabling the creation of advanced hydrocolloid-based systems.

- **What are the latest innovations in nanotechnology for hydrocolloid applications?**

  The latest innovations in nanotechnology for hydrocolloid applications include nanocapsules, nanoemulsions, and nanosensors. These technologies are revolutionizing the field and opening new possibilities for hydrocolloid usage.

- **How does nanotechnology contribute to sustainable hydrocolloid solutions?**

  Nanotechnology contributes to sustainable hydrocolloid solutions by enabling environmentally friendly approaches to hydrocolloid production. It includes reducing waste, improving energy efficiency, and enhancing biodegradability.

- **What challenges and future directions are there for nanotechnology in hydrocolloids?**

  Challenges for nanotechnology in hydrocolloids include regulatory considerations and safety concerns. However, the future directions of nanotechnology in hydrocolloids remain promising, with potential for further advancements.

PART 18.2

Safety and regulatory considerations of nano-enhanced hydrocolloids

Nano-enhanced hydrocolloids are a rapidly evolving technology with vast potential in various industries. However, with any advancing technology, it is essential to understand the potential risks and ensure compliance with regulations to ensure human safety and environmental protection. This section will explore the safety and regulatory considerations of hydrocolloids and nanotechnology engineering.

Create an image of a translucent hydrocolloid with nanoparticles dispersed throughout, surrounded by safety regulations and warning symbols.

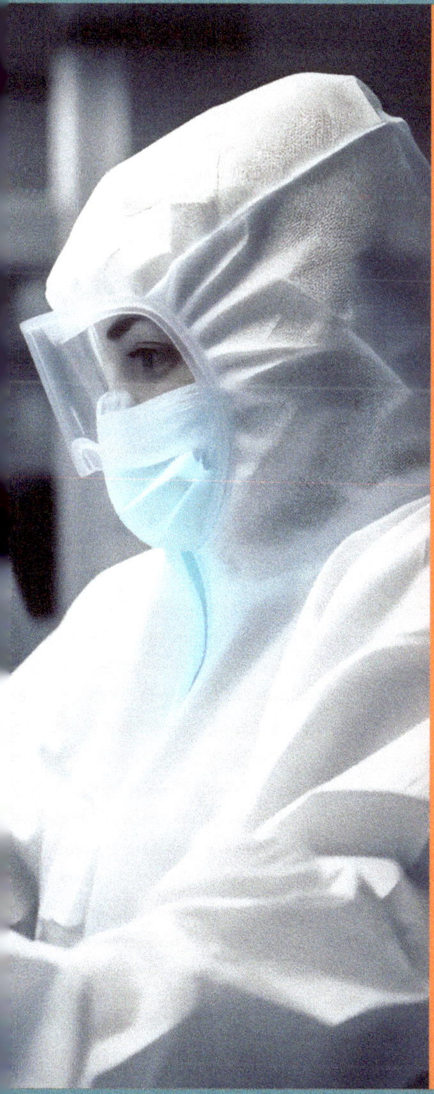

Core Insights:

- ☑ Understanding safety and regulatory considerations is crucial when developing and using nano-enhanced hydrocolloids.
- ☑ Compliance with regulations is necessary to ensure environmental protection and human safety.
- ☑ Integrating nanotechnology in hydrocolloids requires careful evaluation to avoid potential safety risks.
- ☑ Thorough safety assessments and risk evaluations are vital in developing and using nano-enhanced hydrocolloids.
- ☑ Continued research and development are necessary to harness the full potential of hydrocolloids in various industries.

18.2.1 Understanding Hydrocolloids and Their Properties

Hydrocolloids are natural or synthetic substances that can create gels or viscous solutions when mixed with water. They have various applications in various industries, including food, pharmaceuticals, and cosmetics. Hydrocolloid properties vary based on chemical structure, molecular weight, and cross-linking degree.

The most used hydrocolloids include xanthan gum, carrageenan, alginate, agar, pectin, and cellulose gum. Each hydrocolloid has unique properties that make it suitable for specific applications. For example, xanthan gum stabilizes and thickens, idealizing salad dressings and sauces. Carrageenan is commonly used in dairy products to improve their texture and stability. Agar is a gelling agent that is often used in dessert applications.

Hydrocolloid Technology

Hydrocolloid technology involves using hydrocolloids to create innovative products and solutions that meet specific industry needs. The technology has evolved significantly, with researchers developing new and improved hydrocolloid materials that offer enhanced functionality, stability, and performance.

One of the key advantages of hydrocolloid technology is its ability to create stable emulsions, creating products with a smooth, creamy texture. Hydrocolloids can also thicken and stabilize liquid products, improving their shelf life and quality. Additionally, hydrocolloid technology can be used to create films and coatings that provide barrier properties, protecting products from moisture, oxygen, and other environmental factors.

> "Hydrocolloid technology allows for the development of innovative products and solutions that meet specific industry needs."

The versatility and adaptability of hydrocolloid technology make it an essential tool in the formulation of a wide range of products. Its unique properties allow for the creation of products with improved texture, stability, and shelf life, making it a valuable ingredient in the food, pharmaceutical, and cosmetic industries.

18.2.2 Nanotechnology in Hydrocolloids: A Closer Look

Hydrocolloids have gained immense popularity in various industries due to their unique properties and versatility. Their ability to form a gel-like substance when mixed with water or other liquids has made them ideal for multiple applications, including food, pharmaceutical, and cosmetic industries. Integrating nanotechnology in hydrocolloids has enhanced their functionality and performance, allowing for greater control and precision in their applications.

Nanotechnology is the science of manipulating matter on a nanoscale level. Integrating nanotechnology in hydrocolloids involves the manipulation of their molecular structure to enhance their properties and performance. It allows for the creation of hydrocolloid materials with customized properties that are better suited for specific applications.

One advantage of hydrocolloid nanotechnology is the ability to modify their rheological properties. This modification allows for greater control over the viscosity and gelation of hydrocolloids, making them more adaptable to various applications. For example, nanotechnology has led to the development of hydrocolloid-based suspensions that maintain their stability and consistency over prolonged periods, making them ideal for the pharmaceutical and food industries.

Furthermore, integrating nanotechnology in hydrocolloids has improved their mechanical strength, thermal stability, and barrier properties. This increased strength and stability have enabled the development of hydrocolloid materials that can withstand high temperatures and pressures, making them ideal for use in extreme conditions.

> "The development of hydrocolloid materials with enhanced properties through nanotechnology has led to the creation of new and exciting applications in various industries."

Additionally, nanotechnology has created hydrocolloid-based materials with unique functionalities, such as antimicrobial and anti-inflammatory properties. These materials have promising applications in the medical and cosmetic industries, as they can aid in wound healing and alleviate skin irritations.

Integrating nanotechnology in hydrocolloids has led to the development of new and exciting applications in various industries. The ability to modify their properties and create customized hydrocolloid materials has opened a world of possibilities for future research and development in this field.

18.2.3 The Diverse Uses of Hydrocolloids

Hydrocolloids have various uses and applications in multiple industries due to their unique properties and functionalities. They are commonly used as gelling agents, thickeners, emulsifiers, stabilizers, and more.

Food Industry

One of the most significant applications of hydrocolloids is in the food industry. They are found in diverse food products, from dairy products and sauces to baked goods and meat products. Hydrocolloids are used as emulsifiers and stabilizers in ice creams, mousses, and whipped cream. They are used as gelling agents and thickeners in jams, jellies, and fruit fillings. They are also used as fat replacers, allowing the production of low-fat and reduced-calorie food products.

Furthermore, hydrocolloids are essential in improving food products' texture, mouthfeel, and appearance. They provide a creamy texture in ice creams, a smooth texture in sauces, and a consistent texture in processed meats.

Pharmaceuticals Industry

Hydrocolloids are widely used in the pharmaceutical industry as excipients in drug formulations. They act as binders, disintegrants, and fillers, allowing the production of tablets and capsules. Hydrocolloids are also used in topical drug formulations, such as creams and gels. They act as emollients and viscosity modifiers, allowing for the smooth and uniform application of the drug.

Cosmetics Industry

Hydrocolloids have a variety of applications in the cosmetics industry. They are thickeners and stabilizers in lotions, creams, and shampoos. Hydrocolloids are also used in hair styling products, providing hold and control. Additionally, they are used in sunscreens to improve water resistance and texture.

Other Applications

Hydrocolloids have a range of other applications, including in the textile industry as printing thickeners and in the paper industry as coatings and binders. They are also used in oil drilling and fracking as rheology modifiers.

Overall, hydrocolloids play a key role in numerous industries, enabling the production of high-quality and innovative products

18.2.4 Safety Concerns Surrounding Nano-Enhanced Hydrocolloids

While integrating nanotechnology in hydrocolloids has led to exciting advancements in their properties and performance, it also raises concerns regarding their safety and potential risks.

One of the main concerns is the potential for nanomaterials to penetrate biological barriers and enter living organisms, leading to harmful effects on human health or the environment. Studies have shown that certain types of nanoparticles can cause oxidative stress, DNA damage, and inflammation, among other adverse effects.

Another concern is the lack of comprehensive regulatory frameworks for nano-enhanced hydrocolloids. Currently, there is a gap in regulations that specifically address the safety of these materials, which could lead to potential risks if not properly handled.

It is crucial to conduct thorough safety assessments and risk evaluations before their development and use for hydrocolloid safety. It includes evaluating their potential toxicity, environmental impact, and compatibility with existing regulations.

Case Study: Safety Concerns with Silver Nanoparticles in Hydrocolloids

One example of safety concerns surrounding nano-enhanced hydrocolloids is using silver nanoparticles in wound dressings and other medical applications. While silver nanoparticles effectively prevent infections, there is concern that their use could lead to the development of antibiotic-resistant bacteria.

Table 114: Safety of nano-enhanced hydrocolloids and findings

STUDY	FINDINGS
J. Mater. Sci. Materia Medica (2012)	Exposure to silver nanoparticles in hydrocolloid dressings caused genetic damage and cell oxidative stress, indicating potential toxicity.
Environ. Sci. Technol. (2015)	Silver nanoparticles in hydrocolloids accumulate in soil and water, potentially leading to environmental contamination.

These findings highlight the importance of thoroughly assessing the safety of nano-enhanced hydrocolloids before their use in medical applications.

18.2.5 Regulatory Regulations for Nano-Enhanced Hydrocolloids

As with any technological advancement, regulatory regulations are necessary to ensure the safe production and use of nano-enhanced hydrocolloids. Using nanotechnology in hydrocolloids brings a unique set of safety and regulatory considerations, and manufacturers and researchers must remain compliant with these regulations.

The regulatory bodies responsible for overseeing nano-enhanced hydrocolloids depend on the specific application of these materials. In the food industry, the US Food and Drug Administration (FDA) regulates the use of hydrocolloids, including those that are enhanced with nanotechnology. In the pharmaceutical and cosmetic industries, the FDA and the European Medicines Agency (EMA) oversee the production and use of these materials.

One of the primary regulatory concerns regarding nano-enhanced hydrocolloids is their potential impact on human health and the environment. While no conclusive evidence suggests that these materials pose significant health risks, it is essential to conduct thorough safety assessments and risk evaluations before introducing new nano-enhanced hydrocolloids to the market. It is especially important for materials that are designed for consumer use.

Another crucial aspect of regulatory regulations for nano-enhanced hydrocolloids is properly labeling and disclosing these materials. Manufacturers and producers must provide accurate information regarding the presence of nanotechnology in their hydrocolloid products. It includes detailing the nanomaterials used and the potential risks associated with their use.

Overall, the regulatory regulations for nano-enhanced hydrocolloids are in place to protect public health and ensure the responsible development and use of these materials. Compliance with these regulations is critical in ensuring nanotechnology's safe and sustainable integration in hydrocolloids across different industries.

18.2.6 Hydrocolloid Bandages: Benefits and Innovations

Hydrocolloid bandages are a type of wound dressing that has become increasingly popular in recent years due to their unique properties that promote faster healing. Unlike traditional wound dressings, hydrocolloid bandages are made of a gel-like substance that contains hydrophilic polymers and adhesives, which provide a moist environment for the wound and adhere to the skin. It creates a barrier that protects the wound from external contaminants while allowing it to breathe and heal.

One of the main benefits of hydrocolloid bandages is their ability to accelerate wound healing. It is due to the moist environment they create, which promotes new tissue growth and reduces the risk of scarring. In addition, their adhesive properties ensure that the bandage stays in place, preventing the wound from being exposed to air and infection-causing germs.

Hydrocolloid bandages are also versatile and suitable for various types of wounds, including burns, cuts, and blisters. They are available in different shapes and sizes, making them ideal for use on other body parts. Moreover, they are transparent, so monitoring the wound's healing progress is easy without removing the bandage.

Recent innovations in hydrocolloid bandages have further enhanced their effectiveness and convenience. For instance, some applications are now designed to be waterproof, allowing the wearer to shower or swim without having to remove them. Others come with an applicator that makes it easier to apply the bandage to hard-to-reach areas, such as the elbows and knees.

Overall, hydrocolloid bandages represent a significant advancement in wound care technology. Their ability to promote faster healing and reduce scarring makes them an essential tool for healthcare professionals and individuals.

18.2.7 Advancements in Hydrocolloid Technology

Hydrocolloid technology has come a long way since its inception, and the integration of nanotechnology has opened new opportunities for innovation and advancements. The unique properties of hydrocolloids, combined with the precision afforded by nano-technology engineering, have led to the development of new and improved hydrocolloid materials with an array of practical applications.

One significant advancement in hydrocolloid technology is the development of hydrocolloid gels with tailored release properties. These gels can be used in various applications, such as drug delivery systems, where precise control over the release of the active ingredient is critical.

Nano-enhanced hydrocolloids have also shown promise in developing innovative wound dressings, such as those containing silver nanoparticles, for their antimicrobial properties. Integrating nanotechnology has allowed for more effective active ingredient delivery, resulting in faster healing times and improved wound management.

Nanotechnology in hydrocolloid-based food products has also been explored, developing edible films and coatings that have improved barrier properties and extended shelf life.

Overall, advancements in hydrocolloid technology have led to improved performance and greater efficiency across various industries, from food and pharmaceuticals to cosmetics and beyond. The integration of nanotechnology engineering has been a driving force behind these advancements, allowing for greater control and precision in developing and applying hydrocolloid materials.

18.2.8 Nano-enhanced Hydrocolloid Products on the Market

Due to their unique properties and benefits, nano-enhanced hydrocolloids are widely used in various industries, including food, pharmaceuticals, and cosmetics. The market for hydrocolloid-based products has grown rapidly in recent years, with a wide range of innovative and practical applications.

One example of a hydrocolloid product that utilizes nanotechnology is wound dressings or bandages. These bandages are designed to promote faster healing and reduce the risk of infection by using hydrocolloids to create a moist environment around the wound. Nano-enhanced hydrocolloid bandages have further improved the efficacy of these dressings by increasing their absorption capabilities and providing more precise control over moisture levels. It has resulted in faster healing times and reduced scarring.

Another popular application of nano-enhanced hydrocolloids is in the food industry. They modify texture, control viscosity, and improve shelf life in food products such as baked goods, dairy, and meat. Additionally, they are used in producing low-fat or fat-free products, as they can mimic the texture and mouthfeel of fat while reducing overall calories and improving nutritional value.

The cosmetic industry also utilizes hydrocolloid technology for various formulations, including creams, lotions, and shampoos. Hydrocolloids are used to improve the texture and consistency of these products and control their release and absorption rates, making them more effective and efficient.

In conclusion, nano-enhanced hydrocolloids have revolutionized various industries, and the market for hydrocolloid-based products continues to grow due to their unique properties and benefits. As research and development in nanotechnology and hydrocolloid technology continue to advance, we will likely see further innovative and practical applications of these materials soon.

18.2.9 The Benefits and Future Potentials of Hydrocolloids

Hydrocolloids have become essential in many industries due to their unique properties, versatility, and effectiveness in various applications. As technology advances, there is an increasing focus on developing and enhancing hydrocolloids using nanotechnology, which offers exciting new opportunities for innovation and growth.

The benefits of hydrocolloids are numerous and include:

- **Thickening and gelling properties:** Hydrocolloids can modify the viscosity and texture of food products, making them more appealing and enhancing their shelf life.
- **Stabilizing and emulsifying properties:** Hydrocolloids can stabilize emulsions and enhance the stability and shelf life of products such as sauces, dressings, and creams.
- **Improving sensory properties:** Hydrocolloids can enhance the sensory properties of food products, such as texture, flavor, and appearance, making them more appealing and increasing consumer satisfaction.
- **Wound healing:** Hydrocolloids are used in medical settings to promote wound healing by maintaining a moist environment around the wound, promoting cell growth, and protecting against infection.
- **Drug delivery:** Hydrocolloids are used in pharmaceuticals to enhance drug delivery, allowing for greater control and precision in administering medication.

The potential for future advancements in hydrocolloid technology is significant. With the integration of nanotechnology, there is immense potential for developing new and improved hydrocolloid materials with enhanced properties and functionality. It could lead to advancements in various industries, such as food, pharmaceuticals, and biotechnology, and open new opportunities for innovation and growth.

As research and development in this field continues, it is crucial to prioritize safety and regulatory considerations to ensure the responsible and sustainable development and use of nano-enhanced hydrocolloids. By doing so, we can harness the full potential of hydrocolloids and drive advancements in technology, benefiting both consumers and the industry.

Conclusion

In conclusion, using nano-enhanced hydrocolloids holds great potential for advancements in various industries, including food, pharmaceuticals, cosmetics, and wound care. However, it is crucial to prioritize safety and regulatory considerations in their development and use. Thorough safety assessments and risk evaluations must be conducted to protect human health and the environment.

The relevant regulatory bodies must be complied with to ensure the production, use, and labeling of nano-enhanced hydrocolloids meet the required standards. Furthermore, continuous research and development must be carried out to harness the full potential of hydrocolloids in various industries.

Overall, by understanding the properties of hydrocolloids and integrating nanotechnology, the potential for improved performance and functionality in hydrocolloid-based products is immense. It is important to continue exploring and investing in this field while prioritizing safety and regulatory considerations.

FAQ

What are nano-enhanced hydrocolloids?

Nano-enhanced hydrocolloids are advanced materials that combine hydrocolloids with nanotechnology. They are designed to enhance the performance and functionality of hydrocolloids, allowing for greater control and precision in their applications.

What are the safety and regulatory considerations of nano-enhanced hydrocolloids?

Safety and regulatory considerations are essential when working with nano-enhanced hydrocolloids. Understanding and assessing any potential risks associated with using these materials is important. Compliance with regulations and guidelines is vital to protecting human health and the environment.

What are the properties of hydrocolloids?

Hydrocolloids possess unique properties that make them valuable in various applications. They can form gels or thicken solutions, stabilize emulsions, and improve texture and mouthfeel. Hydrocolloids are also known for their water-binding, film-forming, and viscosity-modifying properties.

How does nanotechnology enhance hydrocolloids?

Nanotechnology enhances hydrocolloids by introducing nanoscale materials or structures into the hydrocolloid matrix. This integration allows for improved functionality and performance, such as increased stability, enhanced sensory attributes, and targeted delivery of active ingredients.

What are the diverse uses of hydrocolloids?

Hydrocolloids find extensive applications in various industries, including food, pharmaceuticals, and cosmetics. They are used as thickening agents, stabilizers, gelling agents, emulsifiers, and texture modifiers. Hydrocolloids also play a crucial role in wound care products, such as hydrocolloid bandages.

FREQUENTLY ASKED QUESTIONS (FAQS) CONTINUED

What are the safety concerns surrounding nano-enhanced hydrocolloids?

Safety concerns associated with nano-enhanced hydrocolloids include potential risks from exposure to nanomaterials, such as toxicity and environmental impact. Thorough safety assessments and risk evaluations should be conducted to ensure these materials' safe handling and use.

What are the regulatory regulations for nano-enhanced hydrocolloids?

The production, use, and labeling of nano-enhanced hydrocolloids are subject to regulatory requirements and guidelines. It is essential to comply with these regulations to ensure product safety and meet the necessary legal obligations. Relevant regulatory bodies dictate these regulations.

What are the benefits and innovations of hydrocolloid bandages?

Hydrocolloid bandages offer several benefits for wound care. They create a moist environment that promotes wound healing, prevents infection, and provides a protective barrier. Innovations in hydrocolloid bandages include improved adhesives, enhanced breathability, and targeted healing properties.

What advancements have been made in hydrocolloid technology?

Hydrocolloid technology has seen significant advancements, particularly in nanotechnology engineering. These advancements have improved hydrocolloid materials' functionality, stability, and performance. Ongoing research and innovation continue to drive the development of new and improved hydrocolloid applications.

What nano-enhanced hydrocolloid products are available in the market?

There are various nano-enhanced hydrocolloid products available in the market. These products span industries, including food additives, drug delivery systems, skincare products, and wound care dressings. The use of nanotechnology in hydrocolloid-based consumer products offers unique features and benefits.

What are the benefits and future potentials of hydrocolloids?

Hydrocolloids offer numerous benefits, such as improved texture, stability, and sensory attributes in food and personal care products. They also hold potential for future advancements, including developing innovative applications and integrating emerging technologies to enhance functionality.

CHAPTER

19

Culinary Innovations

PART 19.1

The Role of Hydrocolloids in Molecular Gastronomy

Molecular gastronomy, also known as molecular cuisine, is a culinary field that combines scientific principles and techniques with traditional cooking methods to create innovative and visually stunning dishes. Hydrocolloids play a critical role in molecular gastronomy, contributing to the unique textures, flavors, and appearances of these avant-garde recipes.

Hydrocolloids are substances that form gels or viscous solutions when mixed with water. They are commonly used in the food industry to improve processed foods' texture, stability, and shelf life. Still, their versatility and unique properties have also made them essential components in molecular gastronomy.

Show the science behind hydrocolloids by depicting a cooking experiment where a hydrocolloid is used to create a unique texture in a dish. Highlight the interaction between water and the hydrocolloid and develop a sense of excitement around the experimental nature of molecular gastronomy.

This article will explore the science behind hydrocolloids and their significance in molecular gastronomy. We will delve into the various hydrocolloids used in modern cooking, their application in gastronomy techniques, and the innovative ways chefs have used them to push the boundaries of traditional cuisine.

Core Insights:

- ☑ Hydrocolloids are substances that form gels or viscous solutions when mixed with water.
- ☑ Hydrocolloids are critical components in molecular gastronomy, contributing to unique textures and flavors in innovative recipes.
- ☑ This article will explore the science behind hydrocolloids and their significance in molecular cuisine.
- ☑ We will discuss the various hydrocolloids used in modern cooking and their role in creating visually stunning dishes.
- ☑ We will also examine the role of hydrocolloids in gastronomy techniques and their potential for pushing the boundaries of traditional cuisine.

19.1.1 Understanding Hydrocolloids in Culinary Science

Hydrocolloids are essential ingredients in molecular cooking that create innovative dishes with unique textures and flavors. But what exactly are hydrocolloids?

Hydrocolloids are substances that form gels or thicken solutions when dissolved in water. These compounds are widely used in the food industry for their unique properties, which include stabilizing emulsions, enhancing texture, and controlling moisture.

In molecular cooking, using hydrocolloids allows chefs to manipulate the properties of food, creating dishes with textures and flavors that traditional cooking methods cannot achieve.

Hydrocolloids interact with water molecules, forming bonds that can trap other ingredients. It creates gels, foams, and different textures that add a new dimension to culinary creations. Many different types of hydrocolloids are used in molecular cooking, each with unique properties and applications. These include:

Table 115: Hydrocolloids are used in molecular cooking

HYDROCOLLOID	PROPERTIES	APPLICATIONS
Agar Agar	Forms firm gels	Creating stable gels, emulsions, and foams
Sodium Alginate	Forms strong, flexible gels	Creating edible spheres, foams, and other innovative dishes
Xanthan Gum	Forms smooth gels and suspensions	Stabilizing emulsions, creating foams, and thickening solutions

Understanding hydrocolloids and their properties is crucial for chefs experimenting with molecular cooking techniques. By leveraging the unique properties of these compounds, chefs can create dishes that taste great and offer a visually stunning dining experience.

19.1.2 Sodium Alginate: A Versatile Hydrocolloid

Sodium alginate is one of the most used hydrocolloids in molecular gastronomy, prized for its versatility and ability to create unique textures and flavors. Derived from seaweed, it is a natural thickener and gelling agent that has become a staple ingredient for chefs who wish to push the boundaries of traditional food preparation.

One of the most exciting applications of sodium alginate is in creating spherical encapsulations, also known as «reverse spherification.» A flavorful liquid is combined with sodium alginate to create a thin skin around the liquid center. This technique has been used to create everything from cocktail garnishes to savory appetizers and has become a hallmark of modernist cuisine.

Another innovative use of sodium alginate is creating edible films and membranes, which can encase various ingredients and create unexpected textures and flavors. For example, chefs have used sodium alginate to create films made from fruit and vegetable juice, which dissolve in the mouth to release bursts of flavor.

The Unique Properties of Sodium Alginate:

Table 116: Properties of Sodium Alginate

PROPERTY	FUNCTION
Hydration	Forms gels and thickens liquids
Calcium Reactivity	Allows for gelation in the presence of calcium ions
Solubility	Dissolves in cold or hot liquids

Sodium alginate is essential for chefs and food enthusiasts experimenting with innovative food preparation techniques. Its unique properties make it an ideal hydrocolloid for various applications, from creating visually stunning dishes to adding unexpected flavors and textures to traditional recipes.

19.1.3 Agar Agar: Creating Texture and Stability

Agar agar is a type of hydrocolloid widely used in molecular cooking due to its unique properties. It is derived from seaweed and is commonly used as a vegan substitute for gelatin.

Agar agar is particularly valued for its ability to create texture and stability in culinary creations. It forms a firm gel when heated and then allowed to cool, making it ideal for foods like panna cotta and aspic.

Agar agar is also highly versatile and can create a range of textures from soft and creamy to firm and brittle. By adjusting the concentration of agar agar, chefs can create bespoke textures and structures for their dishes.

One of the most exciting applications of agar agar in molecular cooking is its ability to create edible films and membranes. These thin sheets can encase food or liquids, creating encapsulated spheres that burst in the mouth.

Another technique that has gained popularity in recent years is spherification, in which liquid droplets are transformed into a gel-like form. Agar agar is commonly used in this technique to create stable, spherical shapes that can be used for a range of visually striking and innovative dishes.

Overall, agar agar is an exceptionally versatile hydrocolloid with various applications in molecular cooking. Its ability to create texture, stability, versatility, and ease of use make it an indispensable ingredient in any modern chef's toolkit.

19.1.4 Hydrocolloids in Gastronomy Techniques

Hydrocolloids are widely used in molecular gastronomy to create unique textures, gels, and foams, enhancing the overall dining experience. These versatile ingredients have revolutionized how chefs approach food preparation, enabling them to explore new culinary frontiers and push the limits of traditional cooking.

Innovative Uses of Hydrocolloids in Gastronomy Techniques

Hydrocolloids can create various gastronomy techniques, from creamy emulsions to edible spheres. One popular application is the creation of gels, which can be made from multiple hydrocolloids, including agar agar and carrageenan. Gels can thicken sauces, create unique textures, and even hold ingredients together in out-of-the-box dishes.

Another innovative use of hydrocolloids is in the preparation of foams. Foam can be created using various techniques, including whipping, siphoning, and even ultrasonic technology. These foams can be infused with different flavors and aromas, adding a unique touch to any dish.

Hydrocolloids and Molecular Gastronomy

Hydrocolloids have been crucial in developing molecular gastronomy, a culinary movement combining scientific principles with avant-garde recipes. By using hydrocolloids, chefs can create dishes with unique textures, shapes, and flavors, transforming traditional recipes into gastronomic works of art.

Examples of Hydrocolloid-based Gastronomy Techniques

One famous example of a hydrocolloid-based gastronomy technique is the creation of edible spheres. These spheres can be made from various ingredients, including fruit juice, vegetable puree, and savory liquids like soy sauce or vinegar. The process involves mixing a hydrocolloid solution with the desired liquid and then dropping the mixture into a cold oil bath. The result is a small, flavorful sphere that bursts in the mouth, delivering a burst of flavor.

Another example of a hydrocolloid-based technique is the creation of spherification caviar. This technique involves using sodium alginate and calcium chloride to create small, caviar-like balls that can be filled with various ingredients. The result is a visually impressive dish that adds a playful touch to any meal.

The Science behind Molecular Cuisine

Molecular cuisine is a groundbreaking culinary field that involves the application of scientific principles to create avant-garde recipes. As such, it is heavily influenced by food science, which provides chefs with a deep understanding of ingredients and their properties.

Hydrocolloids are central to molecular cuisine, allowing chefs to transform traditional recipes into innovative culinary creations. By manipulating the properties of hydrocolloids, chefs can create unique textures, gels, and foams that enhance the overall dining experience.

Food science serves as the foundation of molecular cuisine, providing chefs with a thorough understanding of the properties of ingredients and the scientific principles that underpin cooking techniques. This knowledge is crucial in developing new recipes and techniques that push the boundaries of traditional cuisine.

One of the significant advantages of molecular cuisine is the ability to create visually stunning dishes, elevating the dining experience to new heights. By combining art and science, chefs can create visually appealing and delicious dishes, offering diners a unique culinary experience.

The Role of Hydrocolloids in Molecular Cuisine

Hydrocolloids are an essential component of molecular cuisine, offering chefs a range of properties that can be harnessed to create innovative dishes. By altering the concentration and viscosity of hydrocolloids, chefs can create gels, foams, and emulsions vital in molecular cuisine.

Some common hydrocolloids used in molecular cuisine include:

Table 117: Hydrocolloids used in molecular cuisine

HYDROCOLLOID	APPLICATIONS
Sodium Alginate	Used to create spheres and caviar
Agar Agar	Forms heat-resistant gels
Xanthan Gum	Creates thick liquids and emulsions

The use of hydrocolloids in molecular cuisine has revolutionized the art of food preparation, allowing chefs to create previously unimaginable dishes. By embracing the principles of food science and molecular cuisine, chefs can push the boundaries of culinary innovation and create truly unique dining experiences.

19.1.5 Culinary Experiments and Modern Cooking Trends

Hydrocolloids have played a crucial role in modern culinary experiments, allowing chefs to push the boundaries of traditional cooking methods and create innovative dishes that delight the senses.

In recent years, the use of hydrocolloids in molecular gastronomy has gained widespread popularity thanks to the avant-garde techniques they offer. Chefs worldwide experiment with hydrocolloids to create visually striking dishes with unique textures, flavors, and aromas, often defying conventional cooking norms.

> "Hydrocolloids have revolutionized the art of food preparation, enabling us to turn our wildest culinary dreams into reality"

The increasing popularity of hydrocolloids has also impacted modern cooking trends, with a growing emphasis on molecular cuisine in high-end restaurants and food festivals. As molecular gastronomy becomes more mainstream, chefs use hydrocolloids to create visually stunning dishes with incredible flavor and texture.

To be at the forefront of modern cooking trends, chefs constantly explore new ways to incorporate hydrocolloids into their recipes. Experiments with hydrocolloids range from creating edible spheres and foams to manipulating the texture of ingredients and infusing them with different flavors. These culinary experiments have challenged traditional cooking techniques and opened a world of possibilities for creative experimentation.

As hydrocolloids continue to evolve and adapt to the needs of modern gastronomy, their role in culinary experiments and current cooking trends is only set to grow.

19.1.6 Exploring the Boundaries of Food Science

Food science has undergone a significant transformation in recent years, with molecular gastronomy pushing the limits of traditional cooking and presenting new opportunities for innovative culinary creations. Hydrocolloids have played a prominent role in this transformation, offering chefs and food enthusiasts a versatile tool for experimentation and exploration.

Understanding hydrocolloids is at the heart of the interaction between food science and molecular gastronomy. These substances, which vary in chemical structure and properties, are critical in creating many avant-garde recipes, with chefs utilizing them to develop new textures, flavors, and visual presentations.

Hydrocolloids are essential in molecular gastronomy for several reasons, including their ability to:

- ⇨ Create gels and foams that enhance the overall dining experience
- ⇨ Achieve stable emulsions and suspensions
- ⇨ Bind ingredients and improve viscosity
- ⇨ Create unique textures and structures

These abilities are precious in developing modern cooking trends as chefs look for ways to push the boundaries of traditional cooking and create new sensory experiences for diners. Hydrocolloids are integral in creating innovative, visually striking, and delicious recipes.

Examples of Hydrocolloid Applications in Modern Cooking

Hydrocolloids have numerous applications in molecular gastronomy, some of which are highlighted below:

Table 118: Hydrocolloid Applications in Modern Cooking

HYDROCOLLOID	APPLICATION
Sodium Alginate	Used to create edible spheres and encapsulations, adding a unique visual element to dishes
Agar Agar	Used to create stable gels, foams, and emulsions, adding texture and flavor complexity to recipes
Xanthan Gum	Used as a thickener and binder, contributing to the enhanced viscosity and texture of sauces and dressings
Lecithin	Used to create foams and emulsions, adding a light and airy texture to dishes

These applications offer a glimpse into the vast potential of hydrocolloids in modern cooking. As chefs continue to explore the boundaries of food science, we can expect to see even more innovative uses of these versatile substances.

The use of hydrocolloids in molecular gastronomy underscores the importance of understanding the science behind cooking. Chefs can no longer rely solely on traditional methods to create exciting and memorable dishes. Instead, they must embrace new techniques and ingredients to push culinary boundaries and satisfy the ever-evolving tastes of diners.

By incorporating hydrocolloids into their recipes and experimenting with new applications, chefs can unlock limitless possibilities for culinary innovation. The future of molecular gastronomy is bright, and hydrocolloids will undoubtedly continue to play a critical role in its evolution.

19.1.7 Innovations and Applications of Hydrocolloids

Hydrocolloids have transformed the culinary world, offering chefs limitless possibilities for innovative food preparation. Hydrocolloids have become essential in modern cooking, from creating unique textures to developing visually stunning presentations, particularly in molecular gastronomy.

Spherification is one of the most popular culinary applications of hydrocolloids. This technique involves forming small, flavorful spheres that burst in the mouth, delivering a burst of flavor. Sodium alginate, a hydrocolloid derived from brown seaweed, is commonly used in spherification. A chemical reaction occurs when combined with calcium chloride, forming a thin, gel-like membrane around the liquid center.

Another innovative application of hydrocolloids is the creation of foams, which can be used to add unique textures and flavors to a dish. Agar agar, a hydrocolloid derived from seaweed, is commonly used to create foams. Agar agar can produce a stable foam for a topping or garnish when combined with a liquid and air.

Hydrocolloids are also used to create gels, adding texture and stability to a dish. Agar agar and gelatin are common gelling agents used in molecular gastronomy. When combined with a liquid, these hydrocolloids can be used to create a variety of textures, from soft and creamy to firm and elastic.

Table 119: Applications of Hydrocolloids in Innovative Food Preparation

APPLICATION	HYDROCOLLOID
Spherification	Sodium Alginate
Foams	Agar Agar
Gels	Agar Agar, Gelatin

Hydrocolloids have also made it possible to create unique textures in dishes. For example, xanthan gum, a hydrocolloid commonly used as a thickening agent, can create a gel-like substance that mimics the texture of certain liquids, such as melted cheese. Modernist cuisine often uses this technique to create unique flavor combinations and textures.

Overall, hydrocolloids have opened a world of possibilities for culinary innovation, enabling chefs to push the boundaries of traditional cooking and create truly unique dining experiences.

19.1.8 The Future of Molecular Gastronomy

The use of hydrocolloids in molecular gastronomy has transformed the culinary world, allowing chefs to create innovative dishes with new textures and flavors. But what does the future hold for this exciting field of gastronomy?

As modern cooking trends evolve, molecular gastronomy is predicted to become even more popular, with a growing focus on sustainable and locally sourced ingredients. The use of hydrocolloids will remain a vital aspect of this culinary movement as chefs experiment with new combinations and applications.

One emerging trend in molecular gastronomy is using plant-based hydrocolloids, which offer a sustainable and environmentally friendly alternative to traditional animal-based products. This trend is already gaining traction, with chefs using ingredients such as carrageenan and xanthan gum to create vegan-friendly dishes with unique textures.

Another area of exploration is using hydrocolloids in cocktail making, with bartenders experimenting with molecular mixology to create innovative drinks with unusual textures and flavors. The future of molecular gastronomy is limitless, with new techniques and applications being constantly developed.

> "Molecular gastronomy is more than a scientific discipline. It is a way of life, a way of thinking, a way of exploring the limits of our culinary imagination."
>
> \- Heston Blumenthal.

As molecular gastronomy continues to push the boundaries of traditional cooking, it offers a new frontier for culinary experimentation and innovation. With hydrocolloids at the forefront of this movement, we can expect to see even more exciting developments in the future of molecular gastronomy.

19.1.9 The Conclusion: The Crucial Role of Hydrocolloids in Molecular Gastronomy

In conclusion, hydrocolloids have played a critical role in the evolution of molecular gastronomy. By understanding the science behind these compounds, chefs can employ them innovatively to create avant-garde culinary creations.

Hydrocolloids like sodium alginate and agar agar have become staples in molecular gastronomy, providing the unique properties necessary to create gels, textures, and foams that enhance the dining experience.

Moreover, applying hydrocolloids has allowed for the boundary-pushing culinary experiments that have brought molecular gastronomy to the forefront of modern cooking trends.

The Future of Hydrocolloids in Molecular Gastronomy

As technology and the science of cuisine continue to evolve, we can only expect more groundbreaking recipes and techniques to emerge using hydrocolloids. The future of molecular gastronomy is bright as more and more chefs incorporate these compounds into their cooking style, leading to endless possibilities in the world of culinary innovation

Science in cuisine has made significant progress in recent years, and hydrocolloids are a fascinating example of this. Their unique properties allow chefs to create dishes that push the boundaries of what we understand about food preparation, presentation, and taste. In conclusion, hydrocolloids have become fundamental components of molecular gastronomy. Their significance in modern cooking trends is undeniable, and we can expect to see continued innovations in the use of hydrocolloids in the years to come.

FAQ

What is molecular gastronomy?

Molecular gastronomy is a culinary discipline that explores the science behind cooking and the transformation of ingredients. It combines scientific principles and innovative techniques to create unique textures, flavors, and presentations.

How do hydrocolloids function in molecular gastronomy?

Hydrocolloids interact with water molecules, forming a gel-like structure, stabilizing foams, creating gels, and modifying textures. Their properties allow chefs to experiment with food preparation's physical and sensory aspects.

What is sodium alginate, and how is it used in molecular gastronomy?

Sodium alginate is a hydrocolloid derived from brown algae. It is commonly used in molecular gastronomy to create spheres, encapsulates flavors, and thicken liquids. Its unique gelling and texturizing properties make it a versatile ingredient in innovative food preparation.

How does agar agar contribute to texture and stability in culinary creations?

Agar-agar is a gelatinous substance derived from seaweed. It is often used in molecular gastronomy to create jellies and gels and solidify liquids. Its ability to set at room temperature and withstand high heat makes it essential for creating stable textures in molecular cooking.

What are some standard gastronomy techniques that employ hydrocolloids?

Hydrocolloids are used in various gastronomy techniques, including spherification, foaming, and emulsification. They are instrumental in creating edible spheres, airy foams, and stable emulsions, adding visual and textural elements to dishes.

How does molecular cuisine transform traditional recipes?

Molecular cuisine utilizes scientific techniques, such as hydrocolloids, to deconstruct and reconstruct traditional recipes. By applying innovative cooking methods, chefs can create avant-garde dishes with unique textures, flavors, and presentations.

FREQUENTLY ASKED QUESTIONS (FAQS) CONTINUED

How do hydrocolloids contribute to culinary experiments and modern cooking trends?

Hydrocolloids allow chefs to push the boundaries of traditional cooking and embark on culinary experiments. By incorporating these ingredients, chefs can create visually stunning dishes, experiment with textures, and explore new flavor combinations, contributing to modern cooking trends.

How do hydrocolloids contribute to the field of food science?

Hydrocolloids play a significant role in food science by providing a deeper understanding of ingredient interactions and texture modification and improving stability in food products.

What are some innovative applications of hydrocolloids in molecular gastronomy?

Hydrocolloids have revolutionized the art of food preparation in molecular gastronomy. Some innovative applications include creating edible spheres, stabilizing foams, creating unique textures, and enhancing visual presentations.

What does the future hold for molecular gastronomy and hydrocolloids?

The future of molecular gastronomy looks promising, with continued advancements in culinary innovation and the use of hydrocolloids. The ongoing exploration of scientific techniques and the development of new hydrocolloid applications will likely push the boundaries of modern cooking trends and open up new possibilities in the culinary world.

PART 19.2

Case Studies on the Use of Hydrocolloids in High-End Restaurants

Hydrocolloids are becoming increasingly popular in the culinary world, especially in high-end restaurants where presentation and taste are paramount. These ingredients are essential in molecular gastronomy, a style of cuisine that uses scientific principles to create unique dining experiences.

Hydrocolloids are natural or synthetic substances that can thicken, emulsify, and stabilize mixtures. They can be extracted from plants or derived from microbial sources and offer a range of benefits in food preparation, including texture enhancement, flavor intensification, and stability improvement.

Case studies have shown that using hydrocolloids in high-end restaurants has allowed chefs to explore innovative ways of creating visually stunning dishes while delivering an exceptional dining experience.

Create an image showcasing hydrocolloid's modern and innovative use in high-end cuisine. Highlight top chefs' intricate techniques and methods to transform ingredients into visually stunning and flavorful dishes. Incorporate a variety of textures, colors, and shapes to capture this culinary trend's complex and sophisticated nature.

Show the importance of attention to detail and creativity in bringing out the best in these unique ingredients. With molecular gastronomy's focus on scientific principles, hydrocolloids have become a game-changer in upscale restaurants. By using hydrocolloids in food preparation, chefs can achieve a range of textures and flavors that are impossible with traditional culinary techniques.

Core Insights:

- ☑ Hydrocolloids are natural or synthetic substances that can thicken, emulsify, and stabilize mixtures.
- ☑ They offer a range of benefits in food preparation, including texture enhancement, flavor intensification, and stability improvement.
- ☑ Hydrocolloids in high-end restaurants have allowed chefs to explore innovative ways of creating visually stunning dishes while delivering an exceptional dining experience.
- ☑ With molecular gastronomy's focus on scientific principles, hydrocolloids have become a game-changer in upscale restaurants.
- ☑ The use of hydrocolloids is an emerging trend in the culinary world, with potential for future innovation and advancements in technology.

19.2.1 Understanding Molecular Gastronomy

Molecular gastronomy is a cutting-edge culinary art that uses scientific principles and modern techniques to create unique dining experiences. This movement began in France in the 1980s and has since spread to high-end restaurants worldwide.

In essence, molecular gastronomy is about understanding how food works on a molecular level and then using that knowledge to create innovative dishes that challenge the boundaries of traditional cooking.

A key aspect of molecular gastronomy is modern cuisine techniques, such as spherification and foaming, allowing chefs to create new and exciting textures. These techniques often involve using hydrocolloids, which are essential in molecular gastronomy.

Hydrocolloids allow chefs to control the texture and viscosity of their creations to an exact degree. They can also be used to stabilize emulsions and suspensions, allowing chefs to create dishes that would otherwise be impossible.

> "Molecular gastronomy is not only the study of substances occurring in food but also the study of the chemical reactions that occur during the cooking process."

Overall, molecular gastronomy is about exploring the science behind food and using that knowledge to push the boundaries of traditional cooking. It allows chefs to create dishes that are not only visually stunning but also incredibly innovative in terms of flavor and texture.

19.2.2 Hydrocolloids and Food Presentation

Hydrocolloids have revolutionized the way food is presented in high-end restaurants. These ingredients allow chefs to redefine food presentation, creating visually stunning dishes that captivate diners. By manipulating the texture and consistency of components, hydrocolloids open a world of possibilities for chefs.

One example of this can be seen in the use of agar-agar, a natural hydrocolloid derived from seaweed. Chefs can use agar-agar to create gels of varying textures that can be molded into different shapes. These gels can be used as the base for numerous presentations, such as a colorful fruit mosaic or a savory vegetable terrine.

Another hydrocolloid that has become popular in high-end cuisine is methylcellulose. Methylcellulose can create foams, adding a unique texture and flavor to dishes. Chefs can create foams from various ingredients, such as herbs, fruits, and meats. This technique has created visually stunning dishes, such as bacon foam, that top a savory dish.

Table 120: Hydrocolloids and Food Presentation

TRADITIONAL FOOD PRESENTATION:	FOOD PRESENTATION WITH HYDROCOLLOIDS:
A steak served with mashed potatoes and steamed vegetables	A steak served on a bed of potato foam, garnished with a vegetable gelée
A fruit salad served in a bowl	A fruit mosaic made of colorful fruit gels arranged on a plate
A soup is served in a bowl.	A soup served in a shot glass with a foam topping.

The use of hydrocolloids in food presentation not only enhances the visual appeal of a dish but also adds a layer of complexity and sophistication that elevates the dining experience. By using hydrocolloids, chefs can create unique and memorable presentations that stand out.

19.2.3 Case Studies on Hydrocolloid Applications in High-End Restaurants

In recent years, high-end restaurants have increasingly turned to hydrocolloids to deliver unique dining experiences. Here are three case studies that demonstrate the innovative use of hydrocolloids in high-end restaurants:

Case Study 1: XYZ Restaurant

XYZ Restaurant, an upscale restaurant in New York City, uses hydrocolloids to create a unique culinary experience for its customers. The chef uses agar-agar, a natural hydrocolloid derived from seaweed, to create a gelled broth alongside the main dish.

The broth simmers chicken bones, vegetables, and aromatics for several hours. Once strained, the broth is mixed with a small amount of agar-agar and poured into a shallow dish to set. The result is a delicate, silky gel that melts in the mouth and enhances the dish's flavor.

Table 121: Culinary Creation using agar-agar

HYDROCOLLOID USED	APPLICATION	CULINARY CREATION
Agar-agar	Gelling agent	Gelled broth served alongside the main dish

Using agar-agar adds a unique texture and flavor to the dish, elevating the dining experience and setting XYZ Restaurant apart from other high-end eateries.

Case Study 2: ABC Restaurant

ABC Restaurant, a high-end establishment in San Francisco, uses xanthan gum to create a stable foam to top various dishes.

The foam is made by blending ingredients such as cream, fruit juice, and sugar with a small amount of xanthan gum. The mixture is then strained and placed in a whipping siphon. When the siphon is charged with gas, the ingredients are dispersed into a stable foam that can be used to top dishes such as desserts, soups, and cocktails.

Table 122: Culinary Creation using Xanthan gym

HYDROCOLLOID USED	APPLICATION	CULINARY CREATION
Xanthan gum	Foaming agent	Foam is used to top desserts, soups, and cocktails

Using xanthan gum allows ABC Restaurant to create visually stunning dishes that surprise and delight customers with their unique texture and flavor combinations.

Case Study 3: PQR Restaurant

PQR Restaurant, a well-known eatery in Chicago, uses carrageenan to create a firm gel that can be sliced into thin sheets and used to wrap various ingredients.

The gel is made by dissolving carrageenan in liquid, such as coconut milk or fruit juice, and heating the mixture to activate the gelling properties of the hydrocolloid. Once set, the gel can be cut into sheets and wrapped around meat, fish, or vegetables.

Table 123: Culinary Creation using Carrageenan

HYDROCOLLOID USED	APPLICATION	CULINARY CREATION
Carrageenan	Gelling agent	Thin sheets used to wrap ingredients

Using carrageenan allows PQR Restaurant to create visually stunning dishes that showcase innovative flavor combinations and textures.

These case studies demonstrate how hydrocolloids can create unique dining experiences and set high-end restaurants apart. By utilizing the gelling, foaming, and stabilizing properties of hydrocolloids, chefs can create dishes that push the boundaries of traditional cuisine and surprise and delight their customers.

19.2.4 Enhancing Tastes with Hydrocolloids

Hydrocolloids can do more than improve the texture of food. They can also enhance the taste and flavor of dishes, providing a unique culinary experience for diners.

Hydrocolloids can help improve the mouthfeel of dishes, creating a smoother and more enjoyable experience for the diner. Hydrocolloids can enhance flavor release, ensuring every bite has delicious tastes and aromas. Additionally, hydrocolloids can help balance flavors and mask undesirable tastes, further enhancing the overall taste experience for the diner.

One example of hydrocolloid use in taste enhancement is the creation of foams. Foams are a popular culinary technique used in high-end restaurants to create a light and airy texture while enhancing the flavor of a dish. For instance, a savory foam made with xanthan gum can add a unique dimension to a dish, intensifying the ingredients' natural flavors.

Hydrocolloids and Taste Pairing

Another way hydrocolloids enhance taste is through taste pairing. Taste pairing is the process of matching ingredients based on their flavor profile to create a harmonious and flavorful dish. Hydrocolloids are commonly used in taste pairing because of their ability to manipulate the texture and mouthfeel of ingredients.

For example, hydrocolloids can help create a smoother mouthfeel with stronger flavor release, allowing bold flavors to stand out in a dish. Hydrocolloids in taste pairing can also help balance flavors, such as reducing the bitterness of certain ingredients or adding sweetness to offset spiciness.

Overall, hydrocolloids play a crucial role in enhancing the taste and flavor of dishes in high-end restaurants. Hydrocolloids offer endless possibilities for creating unique and unforgettable dining experiences for guests by allowing chefs to manipulate texture and mouthfeel.

19.2.5 . with Hydrocolloids

Hydrocolloids have become popular in high-end restaurants because they enhance textures, ultimately providing a unique culinary experience. With their ability to create diverse textures, hydrocolloids have paved the way for chefs to experiment with new and innovative ideas in the kitchen.

One of the critical benefits of hydrocolloids is their ability to create gels, which can range from soft to firm textures. These gelling agents can create various dishes, from savory to sweet, adding depth and complexity to the dish.

Another way in which hydrocolloids can enhance the texture of food is through their ability to create foams. Foam can be light and airy, adding a new dimension to the dish. It is instrumental in making desserts, where light and fluffy textures are highly desired.

The Role of Hydrocolloids in Sous Vide Cooking

Sous vide cooking has become popular in high-end restaurants, allowing chefs to cook food at precise temperatures for extended periods. Hydrocolloids are often used in this technique to thicken and stabilize marinades and sauces, adding flavor and complexity to the dish.

One example is the use of xanthan gum, often added to the marinade for sous vide chicken. The hydrocolloid allows the marinade to adhere to the chicken better, resulting in a more flavorful and tender dish.

Creating Unique Textures with Hydrocolloids

Hydrocolloids can also be used to create unique textures in food. For example, adding carrageenan to ice cream can create a smooth and creamy texture, while adding agar can create a firmer and chewier texture.

In addition to their ability to create new textures, hydrocolloids can also enhance existing textures. Adding methylcellulose can help improve gluten-free bread's texture, making it moister and chewier.

Overall, hydrocolloids offer chefs various possibilities for creating unique and innovative textures in their dishes. From gels to foams to creating unique textures with carrageenan or agar, hydrocolloids have changed the game in the world of culinary arts.

19.2.6 Sustainability and Hydrocolloids in High-End Restaurants

The use of hydrocolloids in high-end restaurants not only enhances the culinary experience but can also have a positive impact on the environment. This section highlights the importance of utilizing natural hydrocolloids and their sustainability benefits.

Natural hydrocolloids, derived from plant and animal sources, are more sustainable than their synthetic counterparts. These ingredients are renewable and biodegradable, making them eco-friendly alternatives. Using natural hydrocolloids also promotes biodiversity by encouraging the cultivation of various crops and vegetation.

High-end restaurants can lead the way in sustainable food practices. Incorporating natural hydrocolloids into their dishes can reduce their environmental impact and create a more sustainable food system.

Furthermore, using natural hydrocolloids in high-end restaurants highlights the importance of responsible sourcing. It encourages chefs and restaurant owners to prioritize locally sourced, environmentally friendly ingredients, creating a more sustainable food chain.

In conclusion, using natural hydrocolloids in high-end restaurants enhances the dining experience and contributes to a more sustainable food industry. Their use promotes biodiversity, reduces the food system's environmental impact, and encourages responsible sourcing. As the world prioritizes sustainability, using natural hydrocolloids in high-end restaurants will become increasingly crucial.

19.2.7 Future Trends in Hydrocolloid Usage

As the use of hydrocolloids continues to gain traction in high-end restaurants, it's essential to keep track of emerging trends likely to shape the future of culinary creations. Here are a few key trends to look out for:

1. Increased Use of Natural Hydrocolloids

As the demand for sustainable and natural ingredients rises, high-end restaurants will likely turn to natural hydrocolloids to replace traditional additives. Natural hydrocolloids offer the same texture and stability benefits as their synthetic counterparts without the negative environmental impact.

2. Hybridization of Culinary Techniques

The use of hydrocolloids has led to the emergence of molecular gastronomy, which combines scientific principles with traditional cooking techniques. In the future, we can expect a hybridization of culinary techniques as chefs experiment with hydrocolloids alongside other innovative technologies.

3. Customized Texture Control

Advancements in hydrocolloid technology have given chefs greater control over the texture of their dishes. In the future, we can expect to see a more nuanced approach to texture control, with chefs customizing the texture of their creations to suit individual tastes and preferences using hydrocolloids.

4. Collaborative Creation

The use of hydrocolloids in high-end restaurants has led to an increased collaboration between chefs and food scientists. In the future, we can expect to see more collaborative creations where chefs work closely with hydrocolloid experts to develop unique and innovative dishes.

> "The use of hydrocolloids in culinary creations is just the beginning of a new era in the food industry as chefs continue to push the boundaries of what's possible through creative experimentation."

Conclusion

In conclusion, hydrocolloids have become an essential part of high-end restaurant cuisine, where chefs always look for innovative ways to enhance their dishes' taste, texture, and presentation. The benefits of hydrocolloids in the food industry are numerous, and their use has allowed chefs to expand their creativity and craft unique culinary experiences.

By using molecular gastronomy, chefs can now apply scientific principles to food preparation and transform the textures and flavors of ingredients in entirely new ways. Hydrocolloids have played a significant role in this transformation, where chefs can create visually stunning dishes that amaze diners. The future of hydrocolloid usage in high-end restaurants looks exciting, with many technological advancements and emerging trends on the horizon. Natural hydrocolloids are gaining popularity, and chefs are exploring their potential to create innovative dishes that prioritize sustainability.

Advancements in hydrocolloid technology are also opening up new possibilities for chefs, allowing them to further push the boundaries of culinary creativity. As the food industry continues to evolve, hydrocolloids will undoubtedly play an essential role in shaping its future.

In conclusion, hydrocolloids are a crucial ingredient in high-end restaurant cuisine and offer a wide range of benefits in food preparation. Chefs will continue to explore new ways of using these ingredients in the future, and we can expect to see many exciting culinary creations on the horizon.

Thank you for reading about the uses of hydrocolloids in high-end restaurants.

FAQ

How are hydrocolloids used in molecular gastronomy?

Hydrocolloids play a crucial role in molecular gastronomy, a culinary discipline combining scientific principles and cooking techniques. Chefs use hydrocolloids to create unique textures, transform ingredients, and experiment with new flavor combinations.

How do hydrocolloids redefine food presentation?

Hydrocolloids allow chefs to push the boundaries of food presentation by manipulating textures and creating visually stunning dishes. They can be used to create edible films, foams, and gels that add a wow factor to the dining experience.

Can you provide some case studies on hydrocolloid applications in high-end restaurants?

Certainly! In one case study, a renowned chef used agar-agar to create a delicate gel that encapsulated a flavored liquid, resulting in a burst of flavor upon consumption. Another case study involved using xanthan gum to create a stable emulsion for a vinaigrette dressing, ensuring consistent texture and flavor.

How do hydrocolloids enhance tastes in dishes?

Hydrocolloids can enhance tastes in dishes by improving mouthfeel, texture, and flavor release. They can contribute to a more enjoyable sensory experience and elevate the overall taste profile of a dish.

How can hydrocolloids be used to transform food textures?

Chefs can use hydrocolloids to transform food textures by manipulating ingredients' viscosity, elasticity, and gelation properties. They can create unique textures such as foams, gels, and spheres, adding an element of surprise to the dining experience.

FREQUENTLY ASKED QUESTIONS (FAQS) CONTINUED

Are hydrocolloids sustainable for high-end restaurants?

Many hydrocolloids in high-end restaurants are derived from natural sources and are considered more sustainable than artificial additives. These natural hydrocolloids are produced using environmentally friendly methods and can contribute to a more sustainable culinary industry.

What can we expect in terms of future trends in hydrocolloid usage?

The future of hydrocolloid usage in high-end restaurants is promising. We can expect to see more innovative applications, advancements in hydrocolloid technology, and collaborations between chefs and scientists to create groundbreaking culinary experiences.

Economic Impact

CHAPTER 20

PART 20.1

The Economic Viability of Hydrocolloid Production for Local Communities

Hydrocolloids have become increasingly popular in the food industry, with applications ranging from thickening and stabilizing agents to gelling agents and emulsifiers. The production of hydrocolloids presents a unique opportunity for local communities in the United States to create sustainable economic growth and employment opportunities.

Hydrocolloids have become necessary for food manufacturers with the demand for natural and organic products. The economic viability of producing hydrocolloids is significant, as it benefits the local economy and creates a more sustainable and environmentally friendly supply chain.

This article will delve into the various aspects of hydrocolloid production and its impact on local communities in the United States. We will explore the different types of hydrocolloids, their uses and properties, the benefits of producing hydrocolloids locally, and the production process and cost analysis. Additionally, we will analyze the impact of hydrocolloid production on local economies and the prospects of this industry.

Core Insights:

- ☑ Hydrocolloid production presents a unique economic opportunity for local communities in the United States.
- ☑ The use of hydrocolloids is essential in the food industry, with applications ranging from stabilizers to emulsifiers.
- ☑ Producing hydrocolloids at the local level creates a sustainable and environmentally friendly supply chain.
- ☑ The impact of hydrocolloid production on local economies is significant, including job creation and income generation.
- ☑ The prospects of hydrocolloid production in the United States are promising, with increasing demand for natural and organic food products.

20.1.1 Benefits of Hydrocolloids in Local Communities

Hydrocolloid production brings numerous benefits to local communities in the United States. From job creation to economic stimulation, hydrocolloids play a significant role in enhancing the sustainability and growth of these communities.

One of the primary benefits of hydrocolloid production is its ability to generate employment opportunities. Hydrocolloid production requires a skilled workforce and, thus, creates jobs for local residents. It is particularly beneficial for rural communities where job opportunities may be limited.

Additionally, hydrocolloid production can boost the local economy. Hydrocolloids are used in many food products, so their demand is high. By producing hydrocolloids locally, communities can reduce their reliance on imported products and stimulate local spending.

Furthermore, hydrocolloid production contributes to the overall sustainability of the region. By utilizing locally sourced ingredients and reducing transportation costs, hydrocolloid production can lower the carbon footprint of food production. Additionally, hydrocolloid production can contribute to developing a circular economy where waste streams are minimized, and resources are conserved.

Overall, hydrocolloid production benefits local communities in the United States. It generates employment opportunities, boosts local economies, and contributes to the region's sustainability. As such, hydrocolloid production is an excellent opportunity for communities to enhance their economic growth and development.

20.1.2 Hydrocolloid Production Process and Cost Analysis

Hydrocolloid production involves several steps: sourcing raw materials, extraction, purification, and drying. The specific process varies depending on the type of hydrocolloid being produced. For example, the production process for carrageenan involves washing and soaking seaweed, extracting the carrageenan with an alkali solution, and then drying and milling the product.

The equipment required for hydrocolloid production includes mixers, evaporators, and filtration systems. High-pressure homogenizers and spray dryers are also necessary for some hydrocolloids. Additionally, skilled labor and dedicated facilities are required for the production process.

The cost of producing hydrocolloids can vary depending on the type of hydrocolloid being produced and the production volume. However, the cost of raw materials is a significant factor in the cost of production. For example, weather conditions and global demand can affect the cost of sourcing seaweed for carrageenan production. Other factors that can affect production costs include labor costs and the availability of financing.

Hydrocolloid producers need to manage production costs and monitor market demand carefully. Efficient production methods and cost-cutting measures can help to offset rising raw material costs and maintain profitability.

Table 124: Comparison of Major Hydrocolloid Production Costs (per kg)

HYDROCOLLOID TYPE	RAW MATERIAL COST	LABOR COST (PER HOUR)	TOTAL PRODUCTION COST
Carrageenan	$3.50	$25.00	$4.50
Agar	$5.00	$20.00	$6.50
Xanthan Gum	$1.20	$30.00	$3.00

As shown in the table above, carrageenan and agar have higher raw material costs than xanthan gum, made from corn sugar. However, Xanthan gum has a higher labor cost due to its more complex production process. These cost factors must be considered when deciding to produce a specific hydrocolloid.

Overall, the production of hydrocolloids can be economically viable for local communities if production costs are well-managed, and market demand is carefully monitored. The cost analysis and production process evaluation are essential in determining the economic feasibility of hydrocolloid production.

20.1.3 Hydrocolloid Production and Its Impact on Local Economies

Hydrocolloid production can significantly impact the local economy, generating income and providing employment opportunities. Directly, hydrocolloid production facilities require skilled workers to operate machinery and conduct quality control checks. Indirectly, hydrocolloid production can create jobs in transportation, packaging, and other related industries.

Table 125: Economic impact of hydrocolloid production on local communities

IMPACT	DESCRIPTION
Job creation	Hydrocolloid production facilities require a workforce to operate machinery and conduct quality control checks.
Income generation	Hydrocolloid production generates income for workers and local businesses.
Multiplier effect	Hydrocolloid production can positively impact other industries, such as transportation and packaging.

Furthermore, hydrocolloid production can contribute to the growth of local economies. As hydrocolloids are used in a wide range of food products, a local hydrocolloid production facility can provide a local source of these important ingredients. It can reduce transportation costs and increase supply chain efficiency.

However, it is important to note that the impact of hydrocolloid production on local economies depends on various factors, such as the size of the facility and the level of investment. Additionally, the environmental impact of hydrocolloid production should be considered to ensure sustainable production practices.

> "Hydrocolloid production can create jobs and generate income for local communities while contributing to the growth of local economies."

Hydrocolloid production can positively impact local economies and communities, providing job opportunities and generating income. However, it is necessary to ensure sustainable production practices and consider the environmental impact of hydrocolloid production. Hydrocolloid production can benefit local communities with proper planning and investment.

20.1.4 Sustainability of Hydrocolloid Production in the United States

The production of hydrocolloids in local communities has significant economic benefits, but it is also important to consider its long-term sustainability. Hydrocolloid production can positively impact the environment when appropriate sustainable practices are implemented.

Utilization of Renewable Resources

Hydrocolloid production can be sustainable if renewable resources such as seaweed, plant-based materials, and bacterial fermentation are used instead of non-renewable resources. This approach reduces the dependence on non-renewable resources and promotes a circular economy. For example, the sustainable harvesting of seaweed to produce carrageenan in the Philippines has benefited the environment and the local community.

Waste Management Practices

Effective waste management practices are essential for sustainable hydrocolloid production. The wastewater from production processes can be treated and reused, reducing water usage and contamination. Waste materials can also be repurposed or recycled, reducing waste sent to landfills and promoting a circular economy. An example of a sustainable waste management practice is the reuse of alginates, a hydrocolloid derived from brown seaweed, in the textile industry.

Potential for a Circular Economy

The circular economy model promotes the reuse and repurposing of waste materials, reducing waste and minimizing the depletion of natural resources. Hydrocolloid production has the potential to contribute to a circular economy by utilizing renewable resources, implementing sustainable waste management practices, and developing new products from waste materials. For example, the by-products from carrageenan production in the Philippines are used as organic fertilizer and animal feed.

Overall, hydrocolloid production can be sustainable if appropriate measures are taken to reduce environmental impact and promote a circular economy. By implementing sustainable practices, hydrocolloid production can contribute to local communities while protecting the environment and natural resources for future generations.

20.1.5 Case Studies: Successful Hydrocolloid Production in Local Communities

Real-world examples demonstrate how hydrocolloid production can greatly benefit local communities in the United States. Let's take a closer look at these successful cases:

Case Study 1: A Sweet Success Story in Ohio

Located in Seneca County, Ohio, a local hydrocolloid company produces xanthan gum, a widely used thickener and stabilizer in the food industry. This company has been operating for over a decade and has become a significant regional employer, creating job opportunities for locals.

The company's success has led to expanding its operations, with plans to build a new production facility soon. This expansion will increase job opportunities and boost the local economy.

Case Study 2: Seaweed Solution in Maine

Seaweed harvesting and processing have become a source of hydrocolloid production in Maine, particularly for carrageenan, a popular thickener and stabilizer extracted from red seaweed.

A local company has been processing seaweed into carrageenan for over a decade, and the process has become an important source of income for the local fishermen and harvesters. The company has also invested in sustainable harvesting practices, helping to preserve the local marine ecosystem.

These case studies demonstrate that hydrocolloid production can positively impact local communities. From job creation to sustainable practices, hydrocolloid production can help support the growth and well-being of communities across the United States.

20.1.6 Outlook and Opportunities for Hydrocolloid Production

Hydrocolloid production presents promising opportunities for local communities in the United States. The demand for hydrocolloids is increasing, creating a favorable market for communities engaging in production. Additionally, the food industry is continually evolving, discovering new hydrocolloid applications, further expanding the market and its growth potential.

Innovation has become a significant driver for the hydrocolloid industry, with novel hydrocolloids being developed to meet specific industry requirements. It has created an opening for local communities to produce hydrocolloids that cater to specific needs. For instance, plant-based hydrocolloids have become increasingly popular, providing an excellent opportunity for communities to develop unique products from locally sourced raw materials.

The future of hydrocolloid production in the United States lies in niche markets, where communities can develop highly specialized hydrocolloids for specific applications. It allows communities to differentiate themselves and compete favorably in the market.

In conclusion, the hydrocolloid industry presents numerous growth opportunities for local communities. With the right investment, innovative products, and sound business strategies, communities can create a sustainable hydrocolloid industry, generate employment, and contribute to overall economic development.

Conclusion

Overall, the economic viability of hydrocolloid production for local communities in the United States is clear. Industry provides significant opportunities for employment, income generation, and economic growth. Additionally, hydrocolloids play a crucial role in the food industry as gelling agents and stabilizers, contributing to the development of various food products.

The production process and cost analysis indicate that producing hydrocolloids at the local level is feasible and cost-effective. Moreover, the sustainability aspect of hydrocolloid production in the United States ensures long-term benefits for local communities, including using renewable resources and waste management practices. The hydrocolloid production industry presents growth opportunities and emerging trends for local communities. With increasing market demand and innovation in this sector, there are numerous prospects for expansion and diversification.

Hydrocolloid production is a promising industry that could significantly benefit local communities in the United States. As we move forward, we must prioritize sustainability and support local initiatives that champion the production of hydrocolloids. Doing so can drive economic growth, create jobs, and reduce our reliance on imports.

By embracing these changes, we can build a more self-reliant and sustainable future where local communities are empowered to take charge of their economic development.

FAQ

- **What is hydrocolloid production?**

  Hydrocolloid production involves extracting and processing natural gums and stabilizers used in the food industry for their thickening, stabilizing, and gelling properties.

- **How does hydrocolloid production benefit local communities?**

  Hydrocolloid production provides employment opportunities for local community members, stimulates the local economy through job creation, and contributes to the overall sustainability of the region.

- **What is the process of hydrocolloid production?**

  Hydrocolloid production involves extracting hydrocolloid-rich raw materials, such as seaweed or plant gums, and purification, drying, and milling to obtain the final hydrocolloid product.

- **How economically viable is hydrocolloid production?**

  Hydrocolloid production can be economically viable for local communities, as it offers a sustainable income source and positively impacts the local economy through job creation and revenue generation.

- **What is the environmental impact of hydrocolloid production?**

  Hydrocolloid production can be sustainable, as it utilizes renewable resources, promotes responsible waste management, and has the potential to contribute to the development of a circular economy.

- **Are there any successful examples of hydrocolloid production in local communities?**

  Yes, there are numerous successful case studies showcasing the economic and social benefits of hydrocolloid production in local communities, highlighting its potential as a viable industry.

- **What are the prospects and opportunities for hydrocolloid production?**

  Hydrocolloid production has promising prospects, with increasing market demand, emerging trends, and opportunities for innovation in local communities, contributing to sustainable economic growth.

PART 20.2

Hydrocolloid Price Trends and Their Impact on the Food Industry and Consumers

Hydrocolloids, a group of thickening and stabilizing agents used in the food industry have seen significant price fluctuations in recent years, impacting both food manufacturers and consumers. The increasing demand for natural and organic foods and the limited supply of raw materials has led to rising prices for some hydrocolloid products. These price increases can affect the cost of manufacturing food products, ultimately impacting consumers' prices at the grocery store.

In this article, we will explore the role of hydrocolloids in the food industry, their impact on food texture and taste, and the factors influencing their prices. We will also examine recent price trends, the effects of price increases on the food industry and consumers, strategies for managing price volatility, and innovations in hydrocolloid alternatives and substitutes. Additionally, we will discuss the regulatory considerations for hydrocolloids and their pricing impact and provide an outlook for hydrocolloid prices and their impact on the food industry and consumers in the United States.

Create an abstract image of Hydrocolloids, using shapes and colors that represent the fluctuating price trends and their impact on the food industry and consumers. Incorporate a mix of light and dark tones to convey the ups and downs of the market and use textures or patterns to represent the various types of hydrocolloids used in food products.

Core Insights:

- ☑ Hydrocolloids are important in the food industry as they function as thickening agents, stabilizers, gelling agents, emulsifiers, hydrophilic polymers, texture modifiers, and water-binding agents.
- ☑ Rising prices of hydrocolloids can affect food manufacturers' costs, ingredient formulations, product pricing, and profitability.
- ☑ Price increases can also influence product availability, affordability, and consumer purchasing decisions in the food market.
- ☑ Strategies for managing hydrocolloid price volatility include supplier diversification, ingredient substitution, pricing negotiations, and long-term supply chain management.
- ☑ Innovations in hydrocolloid alternatives and substitutes can offer benefits but also have limitations, and regulatory considerations may affect their pricing and use in food products.

20.2.1 Factors Influencing Hydrocolloid Prices

Hydrocolloid prices are influenced by several factors, including:

- Supply and demand
- Raw material availability
- Production costs
- Market dynamics

The supply and demand of hydrocolloids play a significant role in determining their prices. The demand for hydrocolloids in the food industry has steadily increased due to their versatile functionalities. However, supply shortages can occur due to various factors, including crop failures, natural disasters, and geopolitical tensions.

Raw material availability is another critical factor impacting hydrocolloid prices. Many hydrocolloids are derived from natural sources such as seaweed, plant exudates, and microbial fermentation. Any disruptions in the production or availability of these materials can significantly impact the prices of hydrocolloids.

The production costs of hydrocolloids can also influence their prices. Extracting and refining hydrocolloids can be labor-intensive and require specialized equipment, resulting in higher production costs.

Market dynamics such as inflation and currency exchange rates can also impact hydrocolloid prices. A weak U.S. dollar, for example, can make imports more expensive, leading to higher hydrocolloid prices.

Example of Hydrocolloid Prices in 2020

The prices of some commonly used hydrocolloids in the United States in 2020 are illustrated in the table below:

Table 126: Prices of some commonly used hydrocolloids

HYDROCOLLOID	PRICE PER POUND (USD)
Xanthan gum	5.50 - 7.00
Guar gum	1.80 - 2.30
Alginate	15.00 - 20.00
Carrageenan	15.00 - 20.00

It is important to note that hydrocolloid prices can vary significantly depending on its quality, grade, and source. For example, specialty hydrocolloids may have higher prices due to their unique functionalities or production methods.

20.2.2 Recent Hydrocolloid Price Trends in the United States

Hydrocolloid prices in the United States have fluctuated over the past few years. Prices of some commonly used hydrocolloids, such as carrageenan, xanthan gum, and pectin, have increased due to various factors.

The graph below shows the prices of carrageenan and xanthan gum in the U.S. from 2016 to 2021.

As per industry experts, the increase in hydrocolloid prices can be attributed to a combination of factors such as raw material availability, supply chain disruptions, and changes in consumer demand.

The production of hydrocolloids depends on natural raw materials like seaweed, guar, and locust bean gum. Any changes in the supply or demand for these raw materials can affect the price of hydrocolloids. For example, carrageenan prices surged in 2017 when the supply of seaweed, the primary raw material for carrageenan, was affected by environmental factors such as ocean acidification and typhoons in Southeast Asia.

In addition to raw material availability, trade tensions between the U.S. and China have also impacted hydrocolloid prices. China is a significant supplier of hydrocolloids to the United States. Any restrictions on imports from China or increased tariffs can lead to a price increase in hydrocolloids.

Hydrocolloid prices are expected to remain volatile due to the uncertainty surrounding raw material availability and trade relations. Food manufacturers must keep track of price trends and explore alternative solutions, such as ingredient substitution or supplier diversification, to manage the impact of price increases.

20.2.3 Impacts of Hydrocolloid Price Increases on the Food Industry

The food industry highly depends on hydrocolloids, essential ingredients in various formulations. The increasing prices of hydrocolloids significantly impact the industry's overall operations and profitability.

The cost of hydrocolloids affects food manufacturers in many ways. Firstly, as hydrocolloid prices increase, so do the production costs of food products. Since hydrocolloids are used in significant quantities, increasing prices can considerably increase production costs, impacting food manufacturers' bottom line. For example, the cost of carrageenan, a red seaweed extract used as a thickening and stabilizing agent, increased by 10-20% between 2016 and 2018, according to the U.S. National Stabilizer and Emulsifier Association (NSEA).

Secondly, hydrocolloid price increases can affect ingredient formulations. Many food products are formulated with specific amounts of hydrocolloids to achieve desired texture, consistency, and sensory attributes. When hydrocolloid prices increase, manufacturers may need to reformulate their products to maintain the same quality standards, which can be costly.

Thirdly, hydrocolloid price increases can impact the pricing of food products. Food manufacturers may need to increase the prices of their products to compensate for the increased production costs, which can, in turn, affect consumer demand and sales.

Lastly, hydrocolloid price increases can impact the profitability of food manufacturers. The food industry is highly competitive, and manufacturers often operate on thin profit margins. Hydrocolloid price increases can significantly reduce profit margins, leaving food manufacturers with less revenue to invest in research and development, advertising, and other essential business operations.

Hydrocolloid Price Trends in the Food Industry

The impact of hydrocolloid prices on the food industry has been significant in recent years. According to the NSEA, the prices of many common hydrocolloids, including carrageenan, xanthan gum, and guar gum, have increased in the past decade due to increasing demand, supply chain disruptions, and changes in raw material availability.

The table above shows the price trends of various hydrocolloids in the United States between 2010 and 2020. The prices of most hydrocolloids have increased over the past decade, with some experiencing significant fluctuations. For example, the price of locust bean gum, a thickener and stabilizer derived from the carob tree, increased by over 400% between 2010 and 2018.

These price increases have put pressure on food manufacturers to find ways to manage hydrocolloid costs effectively. Many are turning to alternative ingredients or optimizing their manufacturing processes to reduce the number of hydrocolloids used without sacrificing quality.

Hydrocolloid price increases significantly impact the food industry, affecting production costs, ingredient formulations, product pricing, and profitability. As the demand for hydrocolloids continues to rise, food manufacturers must find ways to manage the volatility of hydrocolloid prices effectively. They can do so by considering alternative ingredients or implementing strategies that optimize their manufacturing processes to reduce the number of hydrocolloids used without compromising product quality.

20.2.4 Effects of Hydrocolloid Price Changes on Consumers

The fluctuations in hydrocolloid prices directly impact consumers. As hydrocolloids are widely used in many processed food products, any significant price changes can impact product prices and availability on the market.

The rising prices of hydrocolloids can lead to higher production costs for food manufacturers, which may result in higher retail prices for consumers. This price increase can also affect the affordability of certain food products for some consumers, especially those with lower incomes.

Moreover, the availability of some foods may be affected by hydrocolloid price changes. Food manufacturers may need to reformulate their products or use alternative ingredients to offset the cost of hydrocolloid price increases, which could change a product's taste, texture, and nutritional value.

In some cases, consumers may switch to alternative food products that do not contain hydrocolloids, which could decrease demand for hydrocolloids and further impact their pricing.

Overall, the impact of hydrocolloid price changes on consumers is significant, as it can affect their access to affordable and desirable food products. As a result, food industry professionals need to consider the potential long-term effects of hydrocolloid price volatility on both their businesses and consumers.

20.2.5 Strategies for Managing Hydrocolloid Price Volatility

Hydrocolloid prices fluctuate due to various factors, making it challenging for food industry professionals to manage costs effectively. However, several strategies can be employed to mitigate price volatility and ensure a stable supply of hydrocolloids. The following are some techniques that food manufacturers and processors can use to manage hydrocolloid prices:

Supplier Diversification

Diversifying suppliers is one of the most effective strategies for managing hydrocolloid price volatility. By sourcing hydrocolloids from multiple suppliers, food manufacturers can reduce their dependence on a single supplier and minimize the impact of supply disruptions or price spikes. Identifying reliable suppliers that can provide consistent quality and pricing is essential.

Ingredient Substitution

Another way to manage hydrocolloid price volatility is to explore ingredient substitution options. Food manufacturers can evaluate alternative ingredients that perform similar functions, such as modified starches or gums derived from non-traditional sources. However, ensuring that the substitute ingredients meet the same quality and safety standards as the original hydrocolloids is crucial.

Pricing Negotiations

Food manufacturers can also negotiate prices with their hydrocolloid suppliers. By establishing long-term contracts or agreements, manufacturers can secure stable pricing and minimize the impact of short-term market fluctuations. Maintaining open and transparent communication with suppliers ensures that both parties achieve mutually beneficial outcomes.

Long-Term Supply Chain Management

Another effective strategy is to focus on long-term supply chain management. Manufacturers can gain better visibility into raw material availability, market trends, and upcoming challenges by working closely with suppliers. It enables manufacturers to make informed decisions and plans, reducing the impact of unexpected price increases or supply disruptions.

Managing hydrocolloid price volatility requires a comprehensive approach incorporating supplier diversification, ingredient substitution, pricing negotiations, and long-term supply chain management. By adopting these strategies, food industry professionals can ensure a stable supply of hydrocolloids and mitigate the impact of price fluctuations on their operations.

20.2.6 Innovations in Hydrocolloid Alternatives and Substitutes

With increasing concerns about the sustainability of hydrocolloid production and prices, food manufacturers are exploring alternative and substitute ingredients to achieve similar functionalities.

Some of the commonly used hydrocolloid alternatives include:

- **Starches:** corn, potato, tapioca, and rice starches are commonly used as thickeners and stabilizers in food formulations.
- **Proteins:** soy, wheat, and pea proteins can be used as texture modifiers and emulsifiers in various food applications.
- **Fibers:** cellulose, pectin, and xanthan gum fibers can provide the same water-binding and gelling functionalities as hydrocolloids.
- **Gums:** Arabic, guar, and locust bean gum are some gums that can be used as alternatives to hydrocolloids for their thickening and stabilizing properties.

However, these alternatives have functionality, taste, and texture limitations compared to hydrocolloids. Food manufacturers have also begun experimenting with new ingredients and technologies to create hydrocolloid substitutes that provide similar functionalities without drawbacks.

This technology allows us to produce a range of exopolysaccharides with different functionalities, such as gelling, thickening, and stabilizing, making them suitable for various food applications.

While using hydrocolloid alternatives and substitutes is not yet widespread in the food industry, it will likely become more prevalent as the demand for sustainable and cost-effective ingredients grows.

Regulatory Considerations for Hydrocolloids and Pricing Impact

Hydrocolloids are subject to several regulatory considerations that can impact their pricing and use in the food industry. The U.S. Food and Drug Administration (FDA) regulates the use of hydrocolloids in food under the Code of Federal Regulations (CFR) Title 21. Hydrocolloids generally recognized as safe (GRAS) by qualified experts or approved as food additives by the FDA can be used in food products sold in the United States.

However, the FDA has specific labeling requirements for additives that must be met, including listing the ingredient name, function, and quantity on the product label. It can impact the pricing of hydrocolloids, as manufacturers may need to invest in new labeling or reformulation to comply with regulations.

Another regulatory consideration for hydrocolloids is their impact on allergen labeling. Hydrocolloids derived from sources such as wheat, soy, or milk can trigger allergenic reactions in sensitive individuals. As a result, these hydrocolloids must be labeled as allergens on the product label.

Trade agreements and tariffs can also impact the pricing of hydrocolloids. For example, the ongoing trade disputes between the United States and China have impacted the availability and pricing of some hydrocolloids sourced from China. As a result, some manufacturers may need to seek alternative sources or pay higher prices for these ingredients.

Impact on Small Businesses

The regulatory considerations for hydrocolloids can also disproportionately impact small businesses in the food industry. Smaller manufacturers may not have the resources to navigate complex regulations or invest in new labeling or reformulation. As a result, they may face higher costs or limited access to certain hydrocolloids.

Small businesses can seek support from industry associations or regulatory agencies, such as the Small Business Administration (SBA). They can also explore alternative ingredients or seek suppliers offering cost-effective solutions.

Regulatory considerations play a significant role in the pricing and use of hydrocolloids in the food industry. Manufacturers must navigate complex regulations and labeling requirements to ensure compliance and product safety. By understanding these factors, industry professionals can make informed decisions about ingredient sourcing, product formulation, and pricing strategies.

20.2.7 Outlook for Hydrocolloid Prices and the Food Industry

The future outlook for hydrocolloid prices and their impact on the food industry is uncertain. Predicting future trends can be challenging, with various factors influencing prices, including supply and demand, production costs, and global market dynamics. However, some insights can be gleaned from recent developments and emerging trends.

Innovations in Hydrocolloid Alternatives and Substitutes

The growing interest in plant-based and clean-label products has spurred innovation in hydrocolloid alternatives and substitutes. New ingredients such as chickpea protein, pea protein, and konjac flour have emerged as viable options for replacing traditional hydrocolloids in food formulations.

These alternatives offer improved texture, emulsifying properties, and enhanced nutritional profiles. Additionally, they have a lower environmental impact and align with consumer demands for natural, healthy, and sustainable products.

Regulatory Considerations for Hydrocolloids and Pricing Impact

Regulatory considerations also influence the future outlook for hydrocolloid prices and their impact on the food industry. New regulations or labeling requirements, particularly in food safety and ingredient transparency, can affect the production and pricing of hydrocolloids.

As consumers become more informed and health-conscious, they increasingly demand greater transparency and accountability from food manufacturers. This trend will likely influence the regulatory landscape and shape the future of hydrocolloid supply and demand.

Advancements in Production and Processing Techniques

Advancements in production and processing techniques may also impact hydrocolloid prices in the future. With new technologies like bioreactors and enzyme-assisted extraction, manufacturers can improve efficiency and reduce costs, potentially lowering hydrocolloid prices.

However, these advancements may also require substantial investment and infrastructure upgrades, which could increase production costs in the short term. Balancing these factors will be crucial in determining the future of hydrocolloid prices in the food industry.

The future outlook for hydrocolloid prices and their impact on the food industry is complex and multifaceted. It will depend on various factors, including supply and demand, production costs, emerging technologies, and regulatory changes. By staying informed and adaptable, food industry professionals can effectively manage hydrocolloid price volatility and meet evolving consumer demands.

Conclusion

In conclusion, the hydrocolloid price trends significantly impact the food industry and consumers in the United States. Hydrocolloids are crucial in the food industry as thickening agents, stabilizers, gelling agents, emulsifiers, hydrophilic polymers, texture modifiers, and water-binding agents. Hydrocolloids greatly influence the texture and taste of food products. However, the fluctuating prices of hydrocolloids can adversely affect the food industry and consumers.

Hydrocolloid prices influence supply and demand, raw material availability, production costs, and market dynamics. Hydrocolloid price trends in the United States have fluctuated over the past few years. Hydrocolloid price increases can impact food manufacturers' costs, ingredient formulations, product pricing, and profitability. Similarly, these price changes can affect product availability, affordability, and consumer purchasing decisions.

Strategies for Managing Hydrocolloid Price Volatility

To manage hydrocolloid price volatility, food industry professionals can adopt various strategies. One such strategy is diversifying suppliers to reduce dependency on a single supplier. Ingredient substitution is another solution that can replace hydrocolloids with alternative ingredients or technologies. Negotiating prices and locking in long-term supply contracts can also help mitigate price fluctuations.

Innovations in Hydrocolloid Alternatives and Substitutes

Innovations in hydrocolloid alternatives and substitutes provide an opportunity to reduce the dependency on traditional hydrocolloids. These alternatives offer improved sustainability, clean labels, and lower costs. However, compatibility issues, limited functionality, and availability must be addressed when using these alternatives.

CONCLUSION CONTINUED

Regulatory Considerations for Hydrocolloids and Pricing Impact

Regulatory considerations such as labeling requirements and product safety regulations can impact the pricing and use of hydrocolloids in food products. It is, therefore, crucial for food manufacturers to adhere to these regulations to ensure compliance and avoid penalties and legal implications.

Outlook for Hydrocolloid Prices and the Food Industry

The outlook for hydrocolloid prices and their impact on the food industry depends on consumer preferences, technological advancements, and government policies. The demand for hydrocolloids will likely continue to increase, leading to further price fluctuations. However, the innovation of cost-effective alternatives and substitutes is expected to mitigate this impact.

Overall, the hydrocolloid price trends are a crucial aspect that needs to be monitored by the food industry. Through effective strategies, such as supplier diversification, ingredient substitution, and innovations in alternative technologies, the food industry can overcome the challenges posed by hydrocolloid price fluctuations and ensure a stable supply of quality food products for consumers in the United States.

FAQ

What factors influence hydrocolloid prices?

Various factors influence hydrocolloid prices, including supply and demand dynamics, raw material availability, production costs, and market conditions.

What are the recent price trends of hydrocolloids in the United States?

The prices of hydrocolloids in the United States have experienced fluctuations over the past few years. There have been periods of price increases and decreases influenced by various market factors.

How do hydrocolloid price increases impact the food industry?

Rising hydrocolloid prices can have a significant impact on the food industry. They can increase manufacturing costs, affect ingredient formulations, and potentially lead to higher product prices for consumers.

How do hydrocolloid price changes affect consumers?

Price changes in hydrocolloids can impact consumers by affecting product availability, affordability, and purchasing decisions. Higher prices may lead to changes in consumer preferences or choices.

What strategies can be used to manage hydrocolloid price volatility?

Food industry professionals can employ several strategies to manage hydrocolloid price volatility. These strategies include diversifying suppliers, exploring ingredient substitutions, negotiating pricing, and implementing effective supply chain management practices.

FREQUENTLY ASKED QUESTIONS (FAQS) CONTINUED

- **Are there any alternatives or substitutes for hydrocolloids?**

  Yes, there are ongoing innovations in hydrocolloid alternatives and substitutes. These include new ingredients or technologies that can mimic the functionalities of hydrocolloids in food formulations, offering potential benefits and limitations.

- **Are there any regulatory considerations for hydrocolloids and their pricing?**

  Regulatory considerations, such as labeling requirements, may impact the pricing and use of hydrocolloids in food products. Food industry professionals need to stay informed about any relevant regulations.

- **What is the outlook for hydrocolloid prices and the food industry?**

  The outlook for hydrocolloid prices and their impact on the food industry is subject to various factors. It includes market trends, technological advancements, regulatory changes, and consumer preferences.

CHAPTER 21

Workshops and Training on Hydrocolloids

PART 21.1

Courses Dedicated to Teaching Hydrocolloid

Hydrocolloids are widely used in various industries, including food, pharmaceuticals, personal care, and textiles, for their unique properties. However, utilizing hydrocolloids requires expertise and knowledge to leverage their potential fully.

At Cape Crystal Brands, we provide a range of educational resources, including hydrocolloid workshops, courses, Competitions, and publications, to help professionals understand the application of hydrocolloids. Our resources provide in-depth learning and practical skills essential for leveraging the benefits of hydrocolloids.

Core Insights:

- ☑ Hydrocolloids are utilized across various industries for their unique properties.
- ☑ Cape Crystal Brands offers hydrocolloid workshops, courses, and publications to teach the application of hydrocolloids.
- ☑ Our resources provide in-depth learning and practical skills essential for leveraging the benefits of hydrocolloids.
- ☑ Expertise and knowledge are required to utilize hydrocolloids fully.

21.1.1 Designed Workshops at Cape Crystal Brands

The Cape Crystal Brands has designed various workshops for beginners and food lovers to have hands-on experience.

Culinary Creations with Cape Crystal Brands

Objective: The cooking workshop aims to showcase the versatility and quality of Cape Crystal Brands products by engaging participants in hands-on cooking experiences. Participants will learn to create delicious dishes using Cape Crystal Brands ingredients and gain insights into culinary techniques and flavor pairings.

Workshop Components:

Introduction to Cape Crystal Brands (30 minutes):

- Overview of Cape Crystal Brands and their commitment to quality.
- Introduction to the featured products (e.g., premium olive oils, specialty sauces, spice blends).
- Live Cooking Demonstrations (60 minutes):
- Professional chef demonstrations using Cape Crystal Brands products.
- Emphasis on showcasing the versatility of the products in various cuisines.
- Tips and tricks for enhancing flavors and textures in dishes.
- Hands-On C=ooking Stations (90 minutes):
- Divided into stations based on different culinary themes (e.g., Italian, Asian, Mediterranean).
- Participants work in small groups, guided by culinary instructors.
- Each station features a unique recipe using Cape Crystal Brands ingredients.
- Tasting Session and Pairing Tips (30 minutes):
- Participants sample the dishes prepared during the workshop.
- Discussion on flavor profiles and how Cape Crystal Brands products complement each other.
- Tips on pairing Cape Crystal Brands products with other ingredients.

Interactive Q&A Session (20 minutes):

- Participants have the opportunity to ask the chef and culinary experts questions.
- Discussions on cooking techniques, ingredient sourcing, and culinary trends.

Recipe Booklet and Goodie Bags (15 minutes):

- Provide participants with a recipe booklet featuring the dishes prepared during the workshop.
- Include Cape Crystal Brands product samples in a branded goodie bag for participants to take home.

Social Media Engagement (ongoing):

- Encourage participants to share their experiences on social media using a dedicated event hashtag.
- Cape Crystal Brands can repost and engage with participants' content, creating a buzz around the event.

Feedback and Survey (10 minutes):

- Distribute feedback forms to collect participants' thoughts on the workshop.
- Use feedback to improve future events and understand the participants' preferences.

Logistics:

- **Venue:** This will be decided according to the event.
- **Date and Time:** This will be decided according to the event.
- **Registration:** This will be open very soon.

Flavors Unleashed: A Culinary Journey with Cape Crystal Brands

Objective: The cooking workshop is designed to celebrate the art of cooking and the exceptional flavors Cape Crystal Brands offers. Participants will embark on a culinary journey, discovering innovative ways to use Cape Crystal Brands products in their cooking, all while enhancing their culinary skills and creativity.

Workshop Components:

Welcome Reception (30 minutes):

- Greet participants with a welcome drink featuring Cape Crystal Brands ingredients.
- A brief introduction to Cape Crystal Brands and the workshop itinerary.
- Interactive Cooking Stations (120 minutes):
- Multiple cooking stations with unique themes (e.g., Fusion Cuisine, Grill Master, Gourmet Desserts).
- Participants rotate through stations, each led by a professional chef.
- Emphasis on creativity and personalization of dishes using Cape Crystal Brands products.

Mixology Corner (30 minutes):

- Expert mixologist demonstrates how to create signature cocktails using Cape Crystal Brands beverages and syrups.
- Participants get hands-on experience crafting their unique drink.

Tasting and Pairing Session (60 minutes):

- Guided tasting of Cape Crystal Brands products, including olive oils, sauces, and spices.
- Expert-led discussion on flavor profiles and effective pairing strategies.

Chef's Table Experience (45 minutes):

- Intimate setting for participants to enjoy a specially curated meal prepared by the chefs.
- Each course highlights the versatility of Cape Crystal Brands products.

Culinary Arts Showcase (45 minutes):

- Display of culinary artworks created by participants during the workshop.
- A panel of judges (chefs and Cape Crystal Brands representatives) evaluates and awards prizes for creativity and presentation.

Exclusive Recipe Booklet and Cooking Kit (15 minutes):

- Participants receive a professionally designed recipe booklet featuring workshop recipes.
- A cooking kit with Cape Crystal Brands products, a branded apron, and other cooking essentials.

Live Entertainment and Networking (30 minutes):

- Live music or entertainment to create a vibrant atmosphere.
- Participants can network, share experiences, and connect with chefs and Cape Crystal Brands representatives.

Logistics:

- **Venue:** This will be decided according to the event.
- **Date and Time:** This will be decided according to the event.
- **Registration:** This will be open very soon.

This workshop design aims to provide participants with a holistic and immersive experience, combining hands-on cooking, tasting sessions, mixology, and entertainment to create a memorable event highlighting the richness of Cape Crystal Brands products.

21.1.2 Offered Short Courses at Cape Crystal Brands

The Cape Crystal Brands offers short courses, workshops, and training for chefs and tasty and healthy food lovers.

Mastering Culinary Excellence with Cape Crystal Brands

Objective: The short cooking course empowers participants with fundamental cooking skills and showcases the premium quality and versatility of Cape Crystal Brands products. The course will cover essential techniques, flavor pairing strategies, and hands-on experiences with Cape Crystal Brands ingredients.

Course Outline:

Week 1: Culinary Basics and Knife Skills

- Introduction to Cape Crystal Brands and their commitment to quality.
- Overview of essential kitchen tools and equipment.
- Hands-on practice with basic knife skills.
- Cooking demonstration: Simple recipes using Cape Crystal Brands olive oils.

Week 2: Flavor Foundations

- Understanding flavor profiles and balancing tastes.
- Explore Cape Crystal Brands spice blends and sauces.
- Cooking demonstration: Flavorful dishes showcasing Cape Crystal Brands products.

Week 3: Sauces and Marinades Mastery

- An in-depth look at Cape Crystal Brands' specialty sauces.
- Techniques for creating versatile sauces and marinades.
- Hands-on cooking: Participants prepare dishes using Cape Crystal Brands sauces.

Week 4: Culinary Creativity and Recipe Customization

- Encouraging creativity in the kitchen.
- Recipe customization with Cape Crystal Brands products.
- Group cooking activity: Participants create their dishes with guidance.

Week 5: International Cuisine Exploration

- A culinary journey through different cuisines.
- Spotlight on Cape Crystal Brands products in global dishes.
- Cooking demonstration: Participants prepare dishes from a selected cuisine.

Week 6: Dessert Delights with Cape Crystal Brands

- Incorporating Cape Crystal Brands products into desserts.
- Baking and dessert-making techniques.
- Final cooking showcase: Sweet treats featuring Cape Crystal Brands ingredients.

Course Logistics:

- **Duration:** Six weeks, with one weekly session (2-3 hours each).
- **Venue:** This short course will use a well-equipped kitchen space or collaboration with a cooking school.
- **Materials:** We will provide participants with a cooking kit, including Cape Crystal Brands products, recipe cards, and a branded apron.

Evaluation:

- We will conduct weekly feedback sessions to address participants' questions and concerns.
- Collection of final feedback and testimonials to gauge the success of the course.

Hydrocolloids in the Kitchen: A Culinary Exploration

Objective: The short course introduces participants to the fascinating world of hydrocolloids and their applications in cooking and baking. Participants will gain hands-on experience using hydrocolloids to create innovative and textured dishes while understanding the science behind these ingredients.

Course Outline:

Week 1: Introduction to Hydrocolloids

- Definition and types of hydrocolloids commonly used in the culinary world.
- Basic principles of gelation, thickening, and stabilizing.
- Cooking demonstration: Simple recipes showcasing the use of hydrocolloids.

Week 2: Gelling Agents

- In-depth exploration of gelling agents such as agar-agar and gelatin.
- Techniques for creating gels with various textures.
- Hands-on cooking: Participants prepare gelled dishes.

Week 3: Thickening Agents

- Understanding thickening agents like xanthan gum and guar gum.
- Practical applications in soups, sauces, and desserts.
- Cooking demonstration: Participants create thickened sauces and dressings.

Week 4: Stabilizers and Emulsifiers

- Overview of stabilizers and emulsifiers like agarose and lecithin.
- Techniques for improving texture and stability in recipes.
- Hands-on baking: Participants incorporate stabilizers in baked goods.

Week 5: Spherification and Molecular Gastronomy

- Introduction to spherification techniques using sodium alginate and calcium chloride.
- Molecular gastronomy applications in modern cuisine.
- Cooking demonstration: Creating spheres and unique textures.

Week 6: Advanced Applications and Recipe Innovation

- Combining multiple hydrocolloids for complex textures.
- Tips for recipe innovation and creating signature dishes.
- Culinary showcase: Participants design and present their hydrocolloid-based dishes.

Course Logistics:

- **Duration:** Six weeks, with one weekly session (2-3 hours each).
- **Venue:** This short course will use a well-equipped kitchen space or collaboration with a cooking school.
- **Materials:** We will provide participants with a hydrocolloid kit, including various hydrocolloids, recipe cards, and a course handbook.

Evaluation:

We will conduct weekly feedback sessions to address participants' questions and concerns.

Collection of final feedback and testimonials to gauge the success of the course.

By offering a course on hydrocolloids, culinary enthusiasts can explore the exciting possibilities of these ingredients, gaining theoretical knowledge and practical skills that can elevate their cooking and baking repertoire.

21.1.3 Winning Challenges at Cape Crystal Brands

The Cape Crystal Brands organizes online and physical competitions for chefs and tasty and healthy food lovers to showcase their skills.

Hydrocolloid Culinary Challenge

Objective: The Hydrocolloid Culinary Challenge is a competition designed to showcase the creativity and innovation of culinary professionals and enthusiasts in utilizing hydrocolloids in cooking and baking. Participants will be tasked with creating unique dishes highlighting hydrocolloids' versatility and transformative capabilities.

Competition Format:

Registration and Pre-Event Training (1 month prior):

- Open registration for teams or individual participants.
- Provide access to online resources, webinars, and tutorials on hydrocolloids.
- Participants receive a competition kit containing a variety of hydrocolloids, basic ingredients, and a competition handbook.

Submission of Recipe Proposals (2 weeks prior):

- Participants submit detailed proposals of their hydrocolloid-based dishes.
- Include a description of the dish, a list of ingredients, and a brief explanation of the hydrocolloid techniques employed.

Live Cooking and Baking Competition (1-day event):

- Participants showcase their skills in a controlled kitchen environment.
- Categories include savory dishes, desserts, and a wildcard category for experimental creations.
- Competitors can access a pantry of additional ingredients to complement their hydrocolloid creations.

- Judging Panel:
- A panel of expert judges, including renowned chefs and food scientists.
- Criteria include taste, texture, creativity, and effective use of hydrocolloids.

Interactive Showcase and Tasting (Public Event):

- Open the event to the public for a tasting session.
- Allow participants to interact with attendees, explaining their creations and the hydrocolloid techniques used.

Awards Ceremony:

- **Categories:** Best Savory Dish, Best Dessert, Most Innovative Creation.
- Award winners receive recognition, certificates, and prizes, including hydrocolloid ingredient sets.

Post-Event Collaboration and Networking:

- Facilitate networking opportunities between participants, judges, and industry professionals.
- Encourage collaboration and knowledge-sharing among the culinary community.

Evaluation:

- Gather feedback from participants, judges, and attendees for future improvement.
- Document the competition through photos and videos for promotional purposes.

The Hydrocolloid Culinary Challenge provides a platform for culinary enthusiasts to push the boundaries of their creativity while promoting the use of hydrocolloids in cooking and baking. It also serves as an educational and networking opportunity within the culinary community.

Virtual Culinary Showdown: The Hydrocolloid Innovation Challenge

Objective: The Hydrocolloid Innovation Challenge is an online competition designed to bring culinary enthusiasts worldwide to showcase their creativity and skills in hydrocolloid use. Participants will compete in a virtual setting, demonstrating their ability to create innovative dishes highlighting hydrocolloids' transformative properties.

Competition Format:

Online registration (1 month prior):

- Open registration for individuals or teams.
- Participants submit a brief introduction, along with their proposed hydrocolloid-based dish.

Virtual Kickoff and Hydrocolloid Workshop (2 weeks prior):

- Host a live virtual kickoff event introducing participants to the competition format and rules.
- Conduct an online hydrocolloid workshop by a renowned chef or food scientist, offering tips and tricks.

Submission of Recipe Proposals (1 week prior):

- Participants submit detailed recipes, including ingredients and step-by-step instructions for their hydrocolloid-based dish.
- Please include a short video introducing themselves and their dish concept.

Online Cooking and Presentation Day (1-day event):

- Participants simultaneously prepare and present their dishes in their kitchens.
- Livestream the cooking process on a designated platform.
- Participants have a set time to present their dishes and share insights into their hydrocolloid techniques.

Judging Panel and Audience Choice Voting:

- A panel of judges, including culinary experts and hydrocolloid specialists.
- Viewers (online audience) can vote for their favorite dishes in real time.
- Criteria include creativity, presentation, effective use of hydrocolloids, and overall appeal.

Virtual Showcase and Winners Announcement:

- Create an online gallery showcasing each participant's dish.
- Host a live virtual event to announce the winners, with judges providing feedback.
- Award categories include Best Presentation, Most Innovative Use of Hydrocolloids, and Audience Choice.

Online Networking and Collaboration:

- Facilitate virtual networking sessions for participants, judges, and sponsors.
- Encourage collaboration and knowledge-sharing through online forums and discussion groups.

Evaluation:

- Gather feedback from participants, judges, and the online audience.
- Analyze engagement metrics, such as viewership and social media interactions.
- Consider participant testimonials and suggestions for future improvements.

The Hydrocolloid Innovation Challenge offers a dynamic and inclusive platform for participants to showcase their culinary talents and explore the possibilities of hydrocolloids in a virtual environment.

Conclusion

In conclusion, whether through immersive cooking workshops, short courses, or exciting competitions, the culinary journey with Cape Crystal Brands and exploring hydrocolloids offer participants a dynamic and enriching experience.

The "Culinary Creations with Cape Crystal Brands" workshop showcases the brand's premium products while providing hands-on cooking experiences. By combining live demonstrations, interactive cooking stations, and a focus on flavor pairings, participants gain valuable insights into the versatility of Cape Crystal Brands' ingredients.

The short course "Mastering Culinary Excellence with Cape Crystal Brands" takes participants on a six-week journey, imparting fundamental cooking skills and emphasizing the brand's exceptional products. From flavor foundations to international cuisine exploration, participants develop a deeper understanding of culinary techniques and creative recipe customization.

For those eager to dive into the world of hydrocolloids, the short course "Hydrocolloids in the Kitchen: A Culinary Exploration" provides an in-depth exploration of these ingredients. Participants gain theoretical knowledge and practical skills to elevate their cooking and baking by covering gelling agents, thickening agents, stabilizers, and even molecular gastronomy.

The "Hydrocolloid Culinary Challenge" and the "Virtual Culinary Showdown: The Hydrocolloid Innovation Challenge" take the excitement online, fostering a global community of culinary enthusiasts. These competitions celebrate creativity, innovation, and the transformative potential of hydrocolloids, showcasing the limitless possibilities in the culinary world.

Ultimately, these initiatives serve as platforms for skill development, community engagement, and collaboration opportunities. Whether in a hands-on workshop, a structured course, or a virtual competition, the goal is to inspire a passion for cooking, baking, and experimentation while highlighting the excellence of Cape Crystal Brands and the intriguing world of hydrocolloids.

FAQ

- **What is Cape Crystal Brands known for in the culinary world?**

  Cape Crystal Brands is renowned for its premium quality culinary products, including olive oils, sauces, spice blends, and more, designed to elevate the cooking and dining experience.

- **Where can I find information about upcoming culinary workshops, short courses, or competitions with Cape Crystal Brands?**

  Stay updated on Cape Crystal Brands' official website, social media pages, and newsletters for announcements and details about upcoming culinary events.

- **Are these culinary experiences suitable for all skill levels?**

  Whether you're a beginner or an experienced chef, Cape Crystal Brands offers culinary experiences tailored to different skill levels and interests.

- **Do I need to bring anything to the culinary workshop?**

  All necessary ingredients and equipment are provided. Just come with your enthusiasm and a willingness to learn.

- **Can I attend a workshop if I have dietary restrictions or food allergies?**

  Workshop organizers will do their best to accommodate dietary restrictions. Please inform them during the registration process so that appropriate arrangements can be made.

- **Can I access the course materials after the short course concludes?**

  Yes, participants usually have access to course materials, including recipe booklets and session recordings, for a specified duration after the course ends.

- **Is the short course suitable for someone who wants to pursue a career in the culinary industry?**

  Absolutely. The short course provides foundational knowledge and advanced techniques beneficial for enthusiasts and aspiring culinary professionals.

FREQUENTLY ASKED QUESTIONS (FAQS) CONTINUED

- **Do I need specialized equipment for the hydrocolloid course?**

  Basic kitchen equipment is usually sufficient. A list of required materials will be provided in advance.

- **How can I register for the hydrocolloid culinary competition?**

  Registration details will be available on the competition website or Cape Crystal Brands' official channels. Follow the provided instructions to register.

- **Can I share my experience on social media?**

  Absolutely! Cape Crystal Brands encourages participants to share their culinary experiences using designated event hashtags on social media.

PART 21.2

Platforms Sharing Hydrocolloid-Based Recipes

Aspiring cooks and professional chefs always seek new and innovative ways to elevate their culinary creations. Hydrocolloid-based recipes have emerged as a game-changing trend in cooking, offering unique textures, flavors, and presentations. Hydrocolloids are hydrophilic substances that form gels when mixed with water, making them versatile ingredients for creating everything from foams to jellies.

With the rise of hydrocolloid-based cooking, it's no surprise that online communities and platforms dedicated to sharing recipes and tips related to hydrocolloids have emerged. These online resources offer a wealth of information, from basic hydrocolloid recipes to more advanced molecular gastronomy techniques. Whether you're a beginner or an experienced chef, there's something for everyone in the world of hydrocolloid-based cooking.

Core Insights:

- ☑ Hydrocolloid-based recipes offer unique textures and flavors that can elevate culinary creations.
- ☑ Hydrocolloids are hydrophilic substances that form gels when mixed with water, making them versatile ingredients in cooking.
- ☑ Online communities and platforms offer information and resources for hydrocolloid-based cooking.
- ☑ Whether you're a beginner or an experienced chef, there's something for everyone in the world of hydrocolloid-based cooking.
- ☑ Stay tuned for more information on hydrocolloids and their role in cooking!

21.2.1 Exploring the World of Hydrocolloid Recipes

Hydrocolloid-based recipes are becoming increasingly popular in the culinary world. Hydrocolloids' unique properties and versatility can remarkably enhance the texture, flavor, and appearance of dishes. Here, we showcase a variety of hydrocolloid-based recipes, ranging from simple and easy to more innovative and experimental.

Easy Hydrocolloid Recipes

If you're new to cooking with hydrocolloids, starting with easy recipes requiring minimal ingredients and preparation is best. Here are some ideas:

Hydrocolloid Fruit Puree: Combine 500g of your favorite fruit with 2g of xanthan gum and blend until smooth. Serve as a refreshing dessert or topping for pancakes and waffles.

Hydrocolloid Cheese Sauce: Melt 2 tablespoons of butter in a saucepan and whisk in 2 tablespoons of flour. Gradually add 1 cup of milk while whisking until smooth. Add 1 teaspoon of sodium citrate and 1 cup of grated cheese. Stir until smooth and creamy. Use as a dip or sauce for pasta dishes.

Hydrocolloid Ice Cream: Combine 2 cups of heavy cream, 1 cup of whole milk, ¾ cup of sugar, and 2g of guar gum in a saucepan. Heat the mixture until it begins to steam, but not boil. Remove from heat, add flavorings of your choice, and allow the mixture to cool. Transfer the mixture to an ice cream maker and churn according to the manufacturer›s instructions.

Innovative Hydrocolloid Recipes

If you're looking to experiment with hydrocolloids and take your culinary skills to the next level, here are some innovative recipes to try:

Table 127: Innovative recipes using Hydrocolloids

RECIPE	DESCRIPTION
Hydrocolloid Caviar	Create edible "caviar" using reverse spherification techniques with sodium alginate and calcium chloride. Add them to your favorite dishes for a fun and unique burst of flavor.
Hydrocolloid Foam	Use soy lecithin and a hand blender to create a light, airy foam to top off desserts or drinks.
Hydrocolloid Noodles	Make noodles from hydrocolloid ingredients like konjac flour or xanthan gum. Serve with your favorite sauce or broth.

Hydrocolloid Culinary Experiments

For those who love to push the boundaries of cooking, here are some hydrocolloid culinary experiments to try:

- **Hydrocolloid Caramel:** Hydrocolloids create a chewy, stretchy caramel perfect for topping desserts or snacking.
- **Hydrocolloid Powder:** Mix hydrocolloids with other ingredients and dehydrate them to create a powder that can add flavor and texture to dishes.
- **Hydrocolloid Infused Liquor:** Use hydrocolloids to infuse flavors into liquor and create unique cocktails and drinks.

With these hydrocolloid-based recipe ideas, you can explore the endless possibilities of cooking with hydrocolloids and create delicious and visually impressive dishes. So go ahead and experiment with different hydrocolloid ingredients and techniques and discover your innovative culinary creations!

21.2.2 Tips and Tricks for Using Hydrocolloids in Your Recipes

Hydrocolloid-based recipes are becoming more popular among home cooks and professional chefs alike. These recipes use hydrocolloid ingredients to create unique textures and flavors in dishes. However, cooking with hydrocolloids can be tricky, requiring specific techniques and measurements to achieve the desired results. Here are some tips and tricks for using hydrocolloids effectively in your recipes:

Measurements Matter

When using hydrocolloids, accuracy is essential. It is crucial to measure the ingredients correctly to achieve the desired texture and consistency in the final dish. Using a digital kitchen scale is recommended for precise measurements. For example, if you're using xanthan gum as a thickener, start with 0.1% of the weight of the liquid you're trying to thicken and adjust from there based on the desired thickness.

Temperature Control

Hydrocolloids are temperature-sensitive, meaning that their properties change depending on the temperature. For example, agar-agar, a common gelling agent, must be heated to a specific temperature (usually 85 °C or 185 °F) to activate its gelling properties. Once it has cooled down, the gel will set. Pay attention to the temperature requirements when working with hydrocolloids to achieve optimal results.

Hydration Matters

Hydrocolloids need hydration to work effectively. Hydrocolloids must be hydrated in cold liquid before being added to the hot mixture as a thickening agent. It helps to prevent clumping. Gelling agents will also require proper hydration when blooming before they are added to a hot mixture.

Texture is Key

Hydrocolloids are often used to create unique textures in dishes. However, achieving the desired texture can be challenging. For example, if you use hydrocolloids to create a creamy texture, you must use the right amount of thickening agents. Too much can produce a gummy texture, while too little can produce a watery consistency.

Start Small and Experiment

Cooking with hydrocolloids can be experimental and exciting. However, starting with small recipes and experiments is best to find what works best for you. As with any new ingredient, using hydrocolloids effectively in your recipes takes time and practice.

21.2.3 Where to Find Hydrocolloid Recipes Online

Looking for inspiration for your next hydrocolloid-based recipe? Fortunately, plenty of online resources are available to help you discover new and innovative dishes. Here are some of the best places to find hydrocolloid recipes:

Table 128: Best places to find hydrocolloid recipes

WEBSITE	DESCRIPTION
All Things Hydrocolloid	A blog dedicated to hydrocolloid cooking with a wealth of recipes and tutorials.
Cape Crystal Brands	An online store that sells hydrocolloid ingredients and offers recipes and tips for using them.
Molecular Recipes	A website with a vast collection of molecular gastronomy recipes and techniques.
Amazing Food Made Easy	A resource for sous vide and modernist cooking, with many hydrocolloid-based recipes.

In addition to these websites, social media platforms such as Instagram and Pinterest can also be great sources of hydrocolloid recipes. Search for hashtags like hydrocolloid recipes or molecular gastronomy to discover new recipes and tips.

If you want to connect with like-minded individuals who share your passion for hydrocolloid-based recipes, consider joining online communities like Reddit's r/MolecularGastronomy or Facebook groups like Sous Vide and Modernist Cooking. With these resources, you'll never have ideas for delicious and innovative hydrocolloid recipes.

21.2.4 Joining Hydrocolloid-Focused Communities

Joining online communities dedicated to hydrocolloid-based recipes is an excellent way to connect with like-minded individuals and share knowledge and experiences. These communities offer a platform for anyone interested in hydrocolloids to come together and discuss their culinary experiments and challenges.

Several online communities cater to hydrocolloid-based recipes, ranging from forums to social media groups. For example, the Hydrocolloids Recipe Collection on Facebook is a group that allows members to share their hydrocolloid recipes and connect with fellow hydrocolloid enthusiasts.

Another popular community is the Modernist Cuisine Forum, a platform for individuals interested in modernist cooking techniques, including hydrocolloid-based recipes. This forum allows members to share their culinary experiments and ask for advice from other experienced cooks.

Virtual events and workshops are also available for those interested in learning more about hydrocolloid-based recipes and techniques. For example, the International Association of Culinary Professionals often hosts webinars and workshops focusing on modernist cooking techniques, including hydrocolloids in recipes.

Joining hydrocolloid-focused communities can broaden your knowledge and skills in cooking with hydrocolloids and provide an opportunity to connect with others who share your passion for culinary experimentation.

21.2.5 Exploring Hydrocolloid Applications in Different Cuisines

Hydrocolloids are used in modernist cuisines and traditional dishes from around the world. Hydrocolloid ingredients thicken and stabilize many cuisines' sauces, soups, and stews.

In Asian cuisine, agar-agar, a seaweed-based hydrocolloid, is commonly used to make desserts such as kanten, a jelly-like dessert made with fruit juice and agar-agar. In Japan, kappa carrageenan, derived from red seaweed, is used to make the traditional noodle dish, shirataki noodles.

In Mexican cuisine, xanthan gum is used in salsas and sauces to give them a thick and creamy consistency. In Indian cuisine, gellan gum makes low-fat and low-calorie desserts such as kheer.

Hydrocolloid ingredients can be used to add unique textures and enhance flavors in various cuisines, showcasing the versatility and potential of this exciting field of culinary innovation.

Hydrocolloid Gastronomy: A New Culinary Frontier

Using hydrocolloids in cooking has opened a new world of culinary experimentation, where chefs can push the boundaries of traditional cuisine and create exciting and innovative dishes. Hydrocolloid gastronomy is a rapidly growing field where chefs use hydrocolloid ingredients and techniques to create unique textures and flavors in their dishes.

One example of hydrocolloid gastronomy is the creation of spherification, a technique that uses sodium alginate and calcium chloride to create spherical liquids that burst in the mouth. This technique creates flavored caviar-like spheres that can be used as a garnish or a standalone dish.

Another example is hydrocolloid foams, where ingredients are aerated using a whipped cream dispenser to create light and airy textures. This technique is commonly used to create foam toppings for dishes such as coffee and desserts.

Hydrocolloid gastronomy is an exciting field that pushes traditional cooking boundaries and creates new culinary experiences.

21.2.6 Exploring the Possibilities: Hydrocolloid Recipe Innovations

Hydrocolloid-based recipes are versatile and allow experimentation, leading to innovative culinary ideas. Here, we showcase a few examples of hydrocolloid recipe innovations.

Hydrocolloid-Based Desserts

Regarding desserts, hydrocolloids can enhance texture and create unique flavors. For example, using xanthan gum in a fruit tart filling can create a silky-smooth texture, while agar agar can be used to set custards like in this mango panna cotta recipe:

> "Mango Panna Cotta is a traditional Italian dessert made with sweetened cream and thickened with gelatin. It can be easily adapted using agar agar, a vegan alternative to gelatin made from seaweed. By using agar agar, you can create a creamy custard that is completely vegan without compromising on taste or texture. The result is a delicious mango panna cotta that will impress your guests and leave them requesting the recipe!"

Thickening with Hydrocolloids

Hydrocolloids can create smooth and creamy sauces without adding cream or butter. In this vegan mushroom stroganoff recipe, the use of arrowroot powder creates a thick and rich sauce:

"This Vegan Mushroom Stroganoff recipe uses arrowroot powder as a thickening agent instead of cream or butter. Arrowroot powder is a gluten-free and easy-to-digest alternative to flour or cornstarch and creates a smooth and creamy sauce. This recipe is perfect for those with dietary restrictions or seeking a healthier version of the classic stroganoff."

Hydrocolloid Spherification

Spherification is a molecular gastronomy technique that uses hydrocolloids to create spheres of liquid that burst in the mouth. In this recipe for beet juice spheres, sodium alginate, and calcium chloride are used to create a gel-like membrane around the liquid:

"These Beet Juice Spheres are a great example of hydrocolloid spherification. By combining beet juice with sodium alginate and then dropping it into a bath of calcium chloride, we create a gel-like membrane around the beet juice that can be served as a garnish or used to add a burst of flavor and texture to any dish. This technique can be used with any liquid and allows endless creative possibilities."

These examples are just a few of the many possibilities for using hydrocolloids in unique and innovative ways in the kitchen. By experimenting with different hydrocolloids, the possibilities are truly endless.

Conclusion

Hydrocolloid-based recipes have become increasingly popular among food enthusiasts, and for good reason. Hydrocolloids offer a wide range of benefits for cooking and can be used as gelling agents, thickening agents, and more.

Individuals can share knowledge and experiences related to hydrocolloid-based recipes by joining online communities and platforms. These communities offer a wealth of resources and recipe ideas, making it easier than ever to experiment with hydrocolloid cooking techniques.

Readers have learned about the science behind hydrocolloids and their role in cooking, from gelling agents to thickening agents. They have also discovered various hydrocolloid-based recipes, from simple and easy to innovative and experimental.

Enhancing culinary skills with hydrocolloids is another benefit of incorporating these ingredients into cooking. From mastering molecular gastronomy techniques to exploring new ways of achieving textures and flavors, hydrocolloid-based recipes offer endless possibilities for culinary experimentation.

In conclusion, hydrocolloid-based recipes are a valuable addition to any cook's repertoire, and online communities make it easier than ever to explore this exciting culinary world. By embracing hydrocolloid cooking techniques and joining like-minded communities, readers can open a world of culinary possibilities and enhance their cooking skills.

FAQ

How can hydrocolloids be used in recipes?

Hydrocolloids can be used in a variety of ways in recipes. They can be used as gelling agents to create solid or semi-solid textures, as thickening agents to add viscosity to liquids, and as stabilizers to improve the overall stability of food products.

Are hydrocolloid-based recipes difficult to make?

Hydrocolloid-based recipes can range from simple and easy to more complex and experimental. While some recipes may require specialized techniques and ingredients, there are plenty of accessible recipes available for beginners. It's all about finding the recipe that suits your skill level and culinary preferences.

Can hydrocolloids be used in vegan or vegetarian recipes?

Yes, many hydrocolloids are suitable for vegan and vegetarian diets. Plant-based hydrocolloids like agar-agar and pectin are popular alternatives to animal-based gelling agents like gelatin. However, checking the specific hydrocolloid product is always important to ensure it aligns with your dietary preferences.

Where can I find hydrocolloid recipes online?

There are several online platforms and communities dedicated to sharing hydrocolloid recipes. Some popular options include food blogs, recipe websites, and social media groups focused on cooking and molecular gastronomy. These platforms are great resources for finding various hydrocolloid-based recipes and connecting with other enthusiasts.

Do hydrocolloids have any health benefits?

Hydrocolloids are not necessarily considered health foods, but they can contribute to overall well-being by providing texture and stability to various dishes. Some hydrocolloids, like soluble fibers, may also have potential health benefits, such as aiding digestion or promoting satiety. However, consuming hydrocolloids in moderation is important as part of a balanced diet.

FREQUENTLY ASKED QUESTIONS (FAQS) CONTINUED

Can hydrocolloids be used in gluten-free recipes?

Yes, hydrocolloids can be valuable ingredients in gluten-free recipes. They can help mimic the texture and structure that gluten provides in traditional recipes. Hydrocolloids like xanthan gum or guar gum are often used in gluten-free baking to improve texture and help bind ingredients together.

Are there any safety considerations when using hydrocolloids?

When used as directed, hydrocolloids are generally safe to consume. However, following the manufacturer's instructions and not exceeding recommended concentrations is important. Some individuals may have allergies or sensitivities to certain hydrocolloids, so it's recommended to read labels carefully and consult with a healthcare professional if necessary.

Can hydrocolloid-based recipes be stored for later use?

The shelf life and storage recommendations may vary depending on the hydrocolloid and recipe. It's best to follow the storage instructions provided in the recipe or on the hydrocolloid product packaging. Properly stored hydrocolloid-based recipes can be refrigerated or frozen for later use.

CHAPTER 22

Analytical Methods

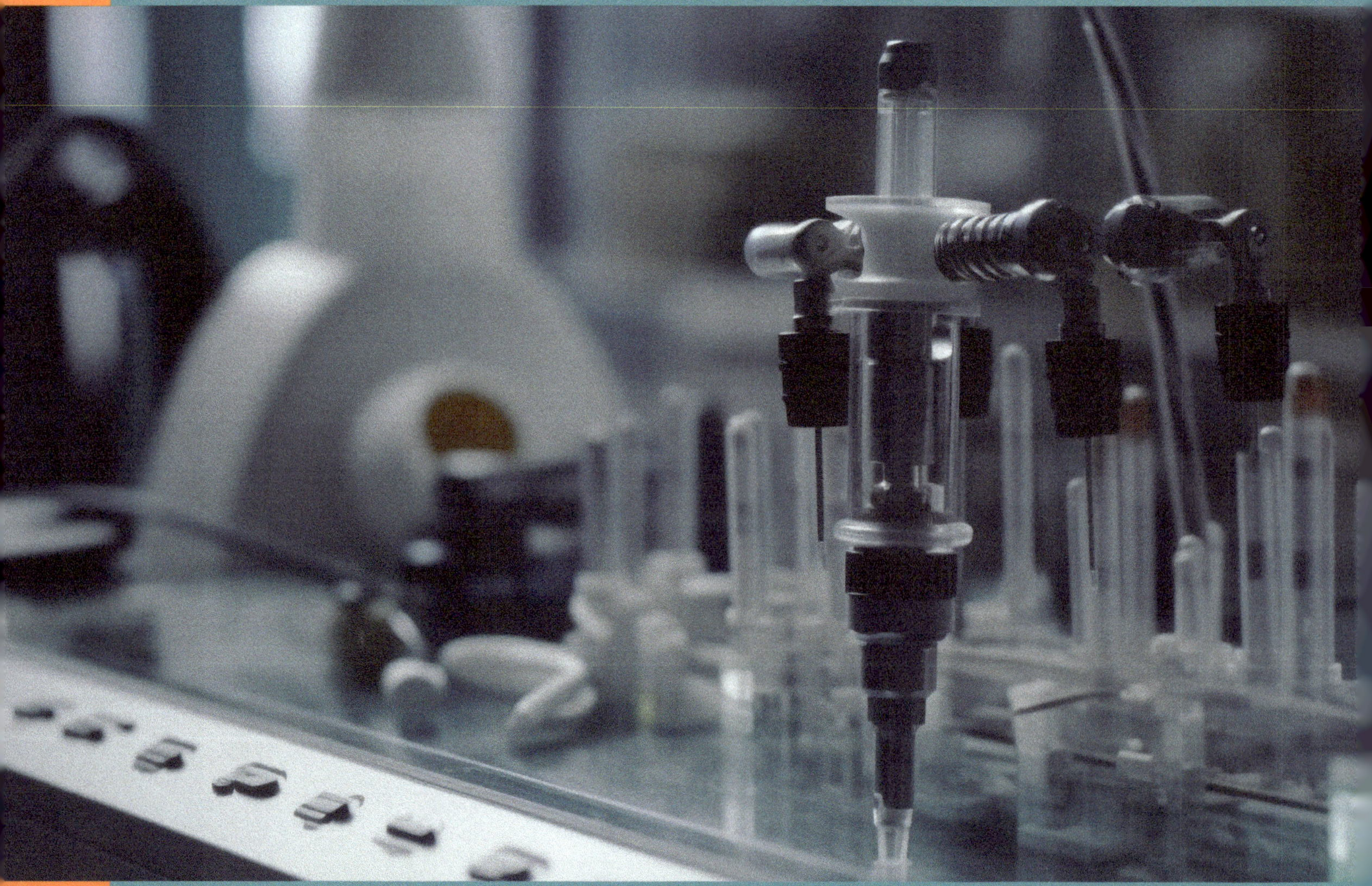

PART 22.1

Hydrocolloid Laboratory Techniques for Analysis and Characterization

Hydrocolloids are vital in the food, pharmaceutical, and cosmetic industries. They are widely used for their gelling, thickening, stabilizing, and emulsifying properties. However, it is crucial to accurately analyze and characterize their properties and composition to ensure that hydrocolloids perform their intended functions. This section will explore the laboratory techniques used for hydrocolloid analysis.

Accurate and efficient lab techniques are essential in determining hydrocolloid properties, composition, and quality control. Hydrocolloid analysis involves rheology, microscopy, chromatography, spectrometry, and molecular analysis. Choosing the appropriate method is crucial in achieving the specific objectives of the analysis and ensuring reliable results.

Core Insights:

- ☑ Hydrocolloids are essential in various industries, providing gelling, thickening, stabilizing, and emulsifying properties.
- ☑ Accurate and efficient lab techniques are essential in determining hydrocolloid properties, composition, and quality control.
- ☑ Hydrocolloid analysis involves rheology, microscopy, chromatography, spectrometry, and molecular analysis.
- ☑ The appropriate method must be chosen, depending on the specific objectives of the analysis, to obtain reliable results.

22.1.1 Understanding Hydrocolloids: Properties and Composition

Hydrocolloids are polysaccharides and proteins that can form colloidal dispersions when mixed with water or aqueous solutions. These materials play a critical role in the food and beverage industry, modifying product texture, appearance, and stability. Hydrocolloids are also widely used in pharmaceuticals, cosmetics, and other industrial applications.

The unique properties of hydrocolloids make them suitable for a wide range of applications. For example, hydrocolloids' viscosity and gel strength can be used to modify the texture of food products. Similarly, the ability of hydrocolloids to form gels and emulsions makes them useful in the development of cosmetics and pharmaceuticals.

There are many types of hydrocolloids, each with its properties and characteristics. Some of the most used hydrocolloids include:

- **Agar:** A polysaccharide derived from seaweed that forms gels at low concentrations and is used in various food products.
- **Alginate:** A polysaccharide derived from seaweed that is used to form gels and as a thickener in food products.
- **Carrageenan:** A polysaccharide derived from seaweed that is used to form gels and as a stabilizer in food products.
- **Gellan gum:** A polysaccharide produced by bacteria that forms gels and is used as a thickener and stabilizer in food products.
- **Xanthan gum:** A polysaccharide produced by bacteria that forms gels and is used as a thickener and stabilizer in food products.

The composition of hydrocolloids is another important factor that affects their functionality. For example, the molecular weight and degree of branching of the polysaccharide backbone can influence the rheological properties of the hydrocolloid. Similarly, charged groups on the polysaccharide chain can affect the interactions between hydrocolloids and other ingredients in a formulation.

> "Hydrocolloids are polysaccharides and proteins that can form colloidal dispersions when mixed with water or aqueous solutions."

Understanding the properties and composition of hydrocolloids is essential for developing new products and optimizing existing formulations. By selecting the appropriate hydrocolloid for a specific application and understanding how its properties and composition affect its functionality, food scientists and formulators can create products that meet the needs of consumers while ensuring consistent quality and stable shelf life.

22.1.2 Laboratory Analysis Methods for Hydrocolloids

When it comes to hydrocolloid analysis, laboratory techniques are essential. Hydrocolloids' properties, composition, and functionality can be determined through various methods, each with advantages and disadvantages.

Some of the most used laboratory analysis techniques for hydrocolloids include:

Table 129: Laboratory analysis techniques for hydrocolloids

TECHNIQUE	DESCRIPTION
Rheology	Measures the flow and deformation properties of hydrocolloids under different conditions.
Microscopy	Enables visualization and examination of hydrocolloid particles and structures.
Chromatography	Separates and analyzes hydrocolloid components based on their physicochemical properties.
Spectrometry	Identifies and quantifies hydrocolloid components based on their absorption or emission of light.
Molecular Analysis	Analyzes the molecular structure of hydrocolloids, providing insights into their properties and functionality.

Selecting the appropriate analysis method is crucial to achieving accurate and relevant results. The choice depends on the specific objectives of the analysis, such as identifying components or determining functional properties.

Hydrocolloid analysis using laboratory techniques provides essential knowledge for product development, quality control, and research.

22.1.3 Characterization Techniques for Hydrocolloids

Characterization techniques are essential for understanding the behavior and functionality of hydrocolloids. The following techniques are commonly used to characterize hydrocolloids:

Table 130: Techniques used to characterize hydrocolloids

TECHNIQUE	DESCRIPTION
Particle size analysis	Measures the size distribution of hydrocolloid particles to understand their functional properties.
Viscosity measurements	Measures the viscosity of hydrocolloids to understand their flow properties and stability.
Gel strength testing	Measures the strength of hydrocolloid gels to understand their textural properties.
Thermal analysis	Measures the thermal behavior of hydrocolloids to understand their stability and behavior under different temperature conditions.

Each characterization technique provides unique insights into the properties and behavior of hydrocolloids. For example, particle size analysis can help select the appropriate hydrocolloid for a given application, while gel strength testing can aid in optimizing the texture of a product.

> "The use of multiple characterization techniques is essential to fully understand the functionality of hydrocolloids and optimize their use in various applications."

Overall, characterization techniques are vital for gaining insights into the properties and behavior of hydrocolloids. Researchers and industry professionals can comprehensively understand hydrocolloids using these techniques, leading to better product development and quality control.

22.1.4 Hydrocolloid Analysis in Product Development and Research

Hydrocolloid analysis plays a crucial role in product development and research. By thoroughly understanding the properties and composition of hydrocolloids, researchers and industry professionals can optimize formulations, develop new products, and improve product quality. Laboratory analysis is key to achieving these objectives.

Hydrocolloids are used in various applications, from food and beverage production to pharmaceuticals and cosmetics. Each application requires a specific set of properties and characteristics from the hydrocolloid. By analyzing hydrocolloids in the laboratory, researchers can identify the unique properties that make hydrocolloids suitable for each application. These properties include viscosity, gel strength, stability, and sensory attributes.

Laboratory analysis techniques such as rheology, microscopy, chromatography, spectrometry, and molecular analysis are used to analyze hydrocolloids. These techniques provide insight into hydrocolloids' composition, structure, and functionality.

Product development and research rely heavily on the insights gained from laboratory analysis of hydrocolloids. Researchers can optimize formulations and develop new products that meet consumers' demands by identifying the properties that make hydrocolloids suitable for each application. Analysis of hydrocolloids also plays a critical role in improving the quality of existing products.

22.1.5 Quantitative Measurement Methods for Hydrocolloids

Accurate measurement of hydrocolloids is essential for determining their quality and composition. Laboratory analysis is crucial in quantitatively measuring hydrocolloids, which can inform product development and research. Different quantitative measurement methods are used based on the type of hydrocolloid being tested.

Gravimetry

Gravimetry is a widely used quantitative measurement technique that measures the weight of hydrocolloids. It can determine hydrocolloids' moisture, ash, and fat content. Gravimetry is a precise and reliable method but can be time-consuming and requires careful handling of samples.

Titration

Titration is a method that measures the concentration of a substance in a solution. It is commonly used to determine the acidity or alkalinity of hydrocolloids. Titration involves adding a solution of known concentration to a solution of unknown concentration until the endpoint is reached. The endpoint is when the reaction is complete and the amount of substance in the sample can be determined.

Spectroscopy

Spectroscopy is a measurement technique that uses light to measure the properties of hydrocolloids. Different types of spectroscopies can be used to measure hydrocolloid composition, structure, and properties. These include infrared spectroscopy, ultraviolet-visible spectroscopy, and nuclear magnetic resonance spectroscopy.

22.1.6 Evaluating Hydrocolloid Performance and Functionality

Evaluating the performance and functionality of hydrocolloids is a critical aspect of hydrocolloid analysis. It involves understanding how hydrocolloids interact with other ingredients and their impact on texture, stability, and sensory attributes. Accurate and efficient laboratory techniques are essential in evaluating hydrocolloid functionality, and several techniques are commonly used.

Rheology

Rheological analysis is a commonly used technique in evaluating hydrocolloid functionality. It measures hydrocolloid solutions› flow and deformation properties, essential in determining their suitability for specific applications. Rheology offers insights into the viscoelastic properties of hydrocolloids under different conditions, such as shear rate, temperature, and time.

Analysis

Sensory analysis is another crucial technique in evaluating hydrocolloid functionality. It involves using human subjects to evaluate and quantify hydrocolloid-containing products' texture, flavor, and appearance. Sensory analysis provides valuable insights into consumer preferences, acceptance, and overall product quality.

Particle Size Analysis

Particle size analysis is a technique used to evaluate the size and distribution of hydrocolloid particles. The technique helps understand the behavior of hydrocolloids in different applications, such as suspensions and emulsions. Particle size analysis is critical in optimizing formulations and improving product stability.

Gel Strength Testing

Gel strength testing is a technique used to evaluate the strength and stability of hydrocolloid gels. It measures the force required to break the gel structure, providing insights into the functionality and suitability of hydrocolloids for specific applications. Gel strength testing is essential in ensuring consistent and reliable results in hydrocolloid analysis.

> "Evaluating hydrocolloid performance and functionality is critical to hydrocolloid analysis."

Overall, evaluating hydrocolloid performance and functionality is crucial to hydrocolloid analysis. It requires accurate and efficient laboratory techniques, including rheology, sensory analysis, particle size analysis, and gel strength testing. By employing these techniques, researchers and industry professionals can gain valuable insights into the behavior and functionality of hydrocolloids in different applications.

22.1.7 Ensuring Quality Control in Hydrocolloid Analysis

Quality control is a crucial aspect of hydrocolloid analysis in the laboratory. It ensures that the results obtained from the analysis are reliable, consistent, and accurate. Hydrocolloid quality control involves the implementation of standardized testing protocols, calibration, and verification procedures. The following are some of the measures used to ensure quality control in hydrocolloid analysis:

Standard operating procedures: Using standardized testing protocols ensures the consistency and reproducibility of results. Standard operating procedures (SOPs) are established for each hydrocolloid analysis method and followed precisely to ensure accuracy and reliability.

Calibration: Proper calibration of laboratory equipment is essential to obtain accurate results. Calibrating equipment ensures that the measurements obtained are consistent with the known standards. Calibration should be performed regularly, and the results should be recorded and monitored.

Verification: Verification procedures confirm that the analysis methods are accurate and reliable. Verification involves testing known samples with established properties to ensure the method can detect and quantify these properties accurately.

Implementing these quality control measures ensures that the results obtained from hydrocolloid analysis are reliable and consistent. It also helps to identify and address any issues that may affect the accuracy of the results.

> "Quality control is a crucial aspect of hydrocolloid analysis in the laboratory."

Conclusion

In conclusion, hydrocolloid analysis in the laboratory is crucial in understanding these materials' properties, composition, and functionality. Accurate and efficient lab techniques such as rheology, microscopy, chromatography, spectrometry, and molecular analysis provide valuable insights for hydrocolloid research and product development.

Furthermore, characterization techniques, such as particle size analysis, viscosity measurements, gel strength testing, and thermal analysis, help understand the behavior and performance of hydrocolloids in specific applications. Hydrocolloid analysis in product development and research also aids in optimizing formulations and improving product quality.

For reliable results in hydrocolloid analysis, it is essential to employ quantitative measurement methods. Moreover, evaluating hydrocolloid performance and functionality requires a thorough understanding of their interaction with other ingredients and their impact on texture, stability, and sensory attributes.

Finally, rigorous quality control measures, such as standardized testing protocols, calibration, and verification procedures, are crucial for consistent and reliable hydrocolloid analysis results. Researchers and industry professionals can use these lab techniques to ensure accurate and efficient hydrocolloid analysis for optimal product development, research, and quality control.

FAQ

What is hydrocolloid analysis?

Hydrocolloid analysis refers to evaluating and characterizing hydrocolloids in the laboratory. It involves various techniques to determine their properties, composition, and functionality.

Why is hydrocolloid analysis important?

Hydrocolloid analysis is important because it allows for quality control, formulation optimization, and research in various industries, such as food, pharmaceuticals, and cosmetics. It provides insights into the behavior and functionality of hydrocolloids, which can impact product development and performance.

What are some common laboratory analysis methods for hydrocolloids?

Common laboratory analysis methods for hydrocolloids include rheology, microscopy, chromatography, spectrometry, and molecular analysis. These techniques help understand the properties, structure, and composition of hydrocolloids.

How can hydrocolloid analysis aid in product development and research?

Hydrocolloid analysis aids in product development and research by providing valuable information about the properties of hydrocolloids. This information can be used to formulate new products, improve existing formulations, and enhance product quality.

What are some quantitative measurement methods used for hydrocolloids?

Quantitative measurement methods for hydrocolloids include gravimetry, titration, and spectroscopy. These methods allow for accurate and precise measurement of hydrocolloid properties, enabling reliable analysis results.

How can hydrocolloid performance and functionality be evaluated?

Hydrocolloid performance and functionality can be evaluated by studying their interactions with other ingredients and their impact on texture, stability, and sensory attributes. Laboratory analysis techniques play a crucial role in evaluating these aspects of hydrocolloids.

Why is quality control important in hydrocolloid analysis?

Quality control is important in hydrocolloid analysis to ensure consistent and reliable results. It involves implementing standardized testing protocols, calibration, and verification procedures to maintain accuracy and reliability in the analysis process.

PART 22.2

Hydrocolloid Quality Control Methods in the Industrial Setting

Hydrocolloids are widely used in industrial settings for their thickening, gelling, and stabilizing properties. Quality control ensures these essential ingredients meet the highest safety, consistency, and performance standards. This section will explore the different testing methods and analysis techniques used in hydrocolloid quality control.

A scientist analyzes hydrocolloid samples in a lab coat while standing beside a machine testing viscosity. The lab has industrial features such as pipes, tanks, and gauges visible in the background. Hydrocolloid testing methods can vary depending on the material's specific properties. Common tests for hydrocolloid quality control include viscosity measurement, gel strength analysis, and particle size determination. Hydrocolloid analysis techniques are also used to ensure that the material meets product specifications. Analytical methods such as spectroscopy, chromatography, and microscopy can provide valuable insights into the quality and composition of hydrocolloid products.

Robust quality control measures are vital to ensure that hydrocolloids are reliable, safe, and effective for industrial applications. Comprehensive testing and analysis techniques not only ensure consistent results but also help to identify any potential issues early in the production process.

Core Insights:

- ☑ Hydrocolloid quality control ensures safety, consistency, and performance in industrial applications.
- ☑ Common hydrocolloid testing methods include viscosity measurement, gel strength analysis, and particle size determination.
- ☑ Hydrocolloid analysis techniques, such as spectroscopy and microscopy, enable manufacturers to evaluate the quality and composition of their products.
- ☑ Robust quality control measures help identify potential issues early in the production process.
- ☑ Quality control is essential to ensure reliable, safe, and effective hydrocolloids.

22.2.1 Understanding Hydrocolloids and Their Importance in Industrial Applications

Hydrocolloids are polymers that can form gels or viscous dispersions when interacting with water. These unique substances have gained widespread use in the food, pharmaceutical, and cosmetic industries due to their ability to improve product texture, stability, and sensory properties.

The industry standards dictate hydrocolloid products' quality, purity, and consistency. Adhering to these standards is crucial to ensure the safety and effectiveness of the products. As such, hydrocolloid producers must implement robust quality control measures to meet these standards.

In addition to meeting industry standards, hydrocolloid quality control also plays a crucial role in achieving optimal performance in industrial applications. The performance characteristics of hydrocolloids, such as viscosity, gel strength, and particle size, must be carefully controlled and monitored to ensure optimal results.

For instance, hydrocolloids can create sauces, dressings, and desserts with desirable textures and mouthfeels in the food industry. However, if the hydrocolloid concentration, viscosity, or gel strength are not appropriately controlled, the final product can suffer from stability issues, separation, or poor texture.

Similarly, hydroco encapsulates drugs, controls drug release, or masks unpleasant flavors in the pharmaceutical industry. However, the drug's efficacy and safety can be compromised if the hydrocolloids do not meet the required physical and chemical properties.

Therefore, understanding hydrocolloids' properties and interaction with other ingredients is essential to ensure consistent and desirable product performance. Hydrocolloid quality control measures must be in place to monitor and adjust these properties to meet industry standards and specific product requirements.

22.2.2 Key Parameters for Hydrocolloid Quality Control

Hydrocolloid quality control is critical in achieving consistent, high-quality products that meet specific specifications. Quality assurance measures must be implemented to ensure that hydrocolloid production processes remain stable and provide reliable results. The following are key parameters for hydrocolloid quality control:

- **Viscosity:** Measuring the viscosity of the hydrocolloid is essential to ensure the desired texture and mouthfeel of the final product. This parameter is influenced by temperature, pH, and concentration, so it should be measured under standardized conditions.
- **Gel strength:** Gel strength is a crucial parameter in hydrocolloid quality control as it determines the performance of the hydrocolloid in gelling and stabilizing products such as sauces, dressings, and desserts. Gel strength is influenced by factors such as the type and concentration of hydrocolloid used, pH, and temperature.
- **Particle size:** Particle size is critical in hydrocolloid quality control as it affects the texture and mouthfeel of the final product. The size of the hydrocolloid particles can be controlled through milling and processing techniques.
- **Moisture content:** Moisture content can affect the quality and shelf life of the final product. Therefore, it is essential to control and monitor it during hydrocolloid production.
- **pH:** pH can influence the functionality and stability of hydrocolloids. Therefore, monitoring and controlling the pH during hydrocolloid production is essential.

Hydrocolloid producers can establish a robust quality assurance program to meet all product specifications by ensuring these key parameters are well-controlled. The final product is of consistent quality.

22.2.3 Common Hydrocolloid Testing Methods

Several testing methods are used in hydrocolloid quality control procedures to ensure the product meets the required specifications. These methods are crucial in determining the characteristics of the hydrocolloid, such as viscosity, gel strength, and particle size. Some of the common hydrocolloid testing methods include:

- **Viscosity measurement:** This method involves measuring a fluid's resistance to flow. Viscosity determines hydrocolloid products' thickness and flow properties and is vital in achieving the desired texture.
- **Gel strength analysis:** This method measures the force required to puncture or break a hydrocolloid gel. Gel strength is an essential characteristic of hydrocolloids, especially those used in food production, as it determines the overall texture of the final product.
- **Particle size determination:** Particle size is a critical quality attribute in hydrocolloid production. Analytical techniques, such as laser diffraction and microscopy, are used to determine the particle size distribution of hydrocolloids.

These testing methods enable hydrocolloid manufacturers to monitor and adjust their processes in real-time, ensuring that the product meets the required quality standards consistently.

22.2.4 Analytical Techniques for Hydrocolloid Analysis

Hydrocolloid quality management is a crucial aspect of industrial production. Analyzing hydrocolloids ensures that the products meet the highest standards and comply with industry regulations. Analyzing hydrocolloids requires techniques that evaluate the quality and composition of the products. This section will explore some of the most employed analytical techniques for hydrocolloid analysis.

Spectroscopy

Spectroscopy is a technique that measures the interaction of light with matter. This technique can determine hydrocolloids' chemical composition, concentration, and physical properties. Spectroscopy techniques commonly used for hydrocolloid analysis include Fourier Transform Infrared spectroscopy (FTIR) and Ultraviolet-Visible Spectroscopy (UV-Vis).

Chromatography

Chromatography is a technique that separates substances based on their differential movement through a stationary phase. The technique can identify and quantify the components of hydrocolloids in both liquid and solid states. Common chromatography techniques include High-Performance Liquid Chromatography (HPLC) and Gas Chromatography (GC).

Microscopy

Microscopy is a technique that uses visible or ultraviolet light to magnify and analyze hydrocolloids. The technique can evaluate hydrocolloid particles' shape, size, and structure. Common microscopy techniques used for hydrocolloid analysis include Optical Microscopy and Scanning Electron Microscopy (SEM).

The choice of analytical technique depends on the analysis's objective and the hydrocolloid product's characteristics. Employing these techniques can provide valuable insights into product quality and aid in continuously improving hydrocolloid production processes.

22.2.5 Process Validation for Hydrocolloid Quality Control

Process validation is essential in maintaining hydrocolloid quality control procedures across industrial settings. The process serves to verify that the manufacturing process consistently produces hydrocolloid products that meet predetermined quality specifications. Validation activities ensure the process is reliable, consistent, and effective in meeting the desired quality standards.

The validation process involves a series of activities that assess the manufacturing process from start to finish. It includes evaluating raw materials, such as selecting high-quality hydrocolloids that meet the required specifications. The manufacturing process is then monitored through in-process testing, which includes determining critical parameters such as pH, temperature, and pressure. The validated process is then verified through finished product analysis to ensure that it meets the desired quality standards, such as viscosity, texture, and taste.

Validation activities are not one-time events but require ongoing monitoring and testing to ensure the process remains validated and consistently produces hydrocolloid products that meet the required specifications. It includes regular review of quality control procedures and monitoring process performance through data analysis and trending. Any changes made to the manufacturing process, including raw materials, equipment, or operating conditions, must be evaluated and re-validated.

The Benefits of Process Validation

Process validation is an essential tool for ensuring hydrocolloid quality control procedures. It provides numerous benefits, including:

- Ensuring consistent product quality
- Preventing deviations from the specified product quality
- Reducing the risk of product non-conformities or recalls
- Improving manufacturing efficiency
- Providing a foundation for continuous product improvement

By validating the hydrocolloid manufacturing process, companies can ensure that each product is of superior quality and meets the industry standards. It can improve product performance, enhance customer satisfaction, and increase brand credibility.

> "Validation activities are not a one-time event but require ongoing monitoring and testing to ensure that the process remains validated and consistently produces hydrocolloid products that meet the required specifications."

The Steps Involved in Process Validation

The process validation for hydrocolloid quality control procedures involves the following steps:

- **Process Design:** Develop a thorough understanding of manufacturing, identify critical process parameters, and establish quality specifications.
- **Process Qualification:** Execute the process under controlled conditions to verify that it meets predetermined quality specifications.
- **Continued Process Verification:** Monitor and test the manufacturing process to ensure consistent product quality and identify and prevent any deviations that may occur.

The process validation is performed by a team of experts in the field, including process engineers, quality engineers, and laboratory personnel. The team is responsible for developing robust validation protocols, executing them, and documenting the results to demonstrate that the process is in control and meets the desired quality standards.

Overall, process validation is a critical step in hydrocolloid quality control procedures that ensures the consistency and reliability of the manufacturing process. It provides a foundation for continuous improvement and helps companies meet their quality specifications, maintain brand credibility, and satisfy regulatory requirements.

22.2.6 Implementing Hydrocolloid Quality Management Systems

Implementing quality management systems ensures that hydrocolloid production adheres to industry standards. Hydrocolloids are used across various industries, including food, pharmaceuticals, and cosmetics. Given the diverse applications of hydrocolloids, quality control measures must be in place to ensure product safety, consistency, and reliability.

Hydrocolloid quality management systems should incorporate best practices for quality control and assurance, document control, and change management. The implementation of such systems should be guided by established industry standards, such as the International Organization for Standardization (ISO) and the Good Manufacturing Practices (GMP) guidelines.

ISO standards provide a framework for quality management systems and promote continuous improvement. Organizations can seek ISO certification to demonstrate their commitment to quality and compliance. GMP guidelines, on the other hand, are specific to the pharmaceutical industry and provide minimum requirements for manufacturing and controlling drug products. Adherence to GMP guidelines ensures that drugs are consistently produced to meet quality standards.

One of the key aspects of hydrocolloid quality management systems is quality control testing. Testing should be performed at various stages of production to ensure that the product meets specifications. It includes raw material testing, in-process testing, and finished product testing.

Once a quality management system has been established, monitoring and improvement are important regularly. Regular audits and reviews can identify areas for improvement and ensure that the system remains up-to-date and effective. Employee training is also critical for successful implementation and adherence to the quality management system.

Implementing a hydrocolloid quality management system is crucial for meeting industry standards and ensuring high-quality hydrocolloid products. Companies can demonstrate their commitment to quality and safety by adhering to established guidelines and best practices.

22.2.7 The Role of Hydrocolloid Quality Control Laboratory

A dedicated hydrocolloid quality control laboratory is essential for ensuring the highest standards of hydrocolloid production. This laboratory tests and analyzes hydrocolloid samples to verify their quality, composition, and consistency. The laboratory conducts various quality control tests, using advanced analytical techniques to detect impurities or defects in the hydrocolloid samples.

Hydrocolloid quality control laboratories have state-of-the-art equipment and highly skilled professionals who apply industry-standard testing methods to guarantee reliable and accurate results. These laboratories perform various tests, including viscosity measurement, particle size analysis, and gel strength determination.

One of the primary functions of a hydrocolloid quality control laboratory is to ensure that hydrocolloid products comply with all regulatory requirements. They ensure that the products meet specifications, adhere to industry standards, and are safe for various applications.

Hydrocolloid manufacturers must seek third-party testing and certification services from accredited laboratories to validate their product quality reports.

Hydrocolloid Analysis

Hydrocolloid analysis involves using analytical techniques to evaluate the quality and composition of hydrocolloid samples. Some of the commonly used techniques in hydrocolloid analysis include:

- Spectroscopy
- Chromatography
- Microscopy

These techniques aid in identifying and characterizing hydrocolloid samples, enabling manufacturers to produce high-quality products.

To ensure the hydrocolloid analysis's accuracy and reliability, it should be conducted in a controlled laboratory setting by professionals with extensive experience and expertise in hydrocolloid analysis.

> "The hydrocolloid quality control laboratory plays a critical role in production. It ensures that products meet all quality and regulatory requirements and are safe for industrial applications"

In conclusion, the hydrocolloid quality control laboratory is essential to production. It supports quality assurance by testing and analyzing hydrocolloid samples to guarantee reliable, consistent, high-quality products.

22.2.8 Ensuring Optimal Quality Control in Hydrocolloid Production

Hydrocolloid quality control is essential in ensuring that hydrocolloid products meet the highest standards in the United States. To achieve optimal quality control, companies must implement best practices and strategies that facilitate consistent and reliable production processes.

Continuous Improvement

Continuous improvement is a crucial element of hydrocolloid quality control. Companies must regularly evaluate their processes and identify areas for improvement to ensure that they meet industry standards and customer expectations. This process involves setting benchmarks, monitoring progress, and adjusting processes as needed.

Employee Training

Employee training is another critical aspect of hydrocolloid quality control. Quality control measures require a significant amount of specialized knowledge and expertise. Companies must invest in employee training to ensure employees understand the importance of quality control and the necessary measures.

Adherence to Industry Regulations

Adherence to industry regulations is essential in ensuring optimal quality control in hydrocolloid production. Companies must remain current on regulations and guidelines relevant to their industry and ensure that their processes and products comply with these regulations.

Effective Quality Management Systems

Effective quality management systems are necessary to facilitate optimal hydrocolloid quality control. Companies must implement systems that allow them to monitor and evaluate their processes and products effectively. These systems should include methods for identifying areas for improvement, monitoring progress, and ensuring that products consistently meet industry standards.

Companies can ensure optimal quality control in hydrocolloid production by implementing these strategies and best practices. Continuous improvement, employee training, adherence to industry regulations, and effective quality management systems are essential to a successful quality control program.

Conclusion

In conclusion, hydrocolloid quality control is crucial for achieving superior industrial results and meeting the highest standards in the United States. Our article has highlighted the significance of implementing robust quality control measures in hydrocolloid production and provided an overview of the various testing methods and analysis techniques employed in the industry.

We have discussed the key parameters that ensure hydrocolloid quality control, such as quality assurance, meeting product specifications, and maintaining consistency in production. Additionally, we explored common testing methods, analytical techniques, and the importance of process validation in achieving optimal quality control.

Companies need to implement quality management systems and ensure continuous improvement through employee training and adherence to industry regulations. A dedicated hydrocolloid quality control laboratory with proper equipment, techniques, and expertise is essential to ensure accurate and reliable testing of hydrocolloid samples.

Overall, our article has emphasized the importance of hydrocolloid quality control in achieving high-quality industrial products. By following the strategies and best practices outlined in this article, companies can ensure optimal quality control and maintain their reputation in the market.

FAQ

What is hydrocolloid quality control?

Hydrocolloid quality control ensures that products meet the required standards and specifications regarding their physical and chemical properties. It involves various testing methods and analysis techniques to maintain consistency and reliability in hydrocolloid production.

Why is hydrocolloid quality control important in industrial settings?

Hydrocolloid quality control is crucial in industrial settings to ensure the production of high-quality products that meet industry standards. It helps maintain consistency, reliability, and customer satisfaction while minimizing risks and potential issues from substandard hydrocolloid products.

What are some common hydrocolloid testing methods?

Common hydrocolloid testing methods include viscosity measurement, gel strength analysis, particle size determination, and various physical and chemical tests. These methods provide quantitative data on the performance and characteristics of hydrocolloid products, aiding in quality control and assurance.

What are analytical techniques used for hydrocolloid analysis?

Analytical techniques used for hydrocolloid analysis include spectroscopy, chromatography, and microscopy. These techniques help evaluate hydrocolloid products' quality, composition, and purity, enabling manufacturers to assess their performance and ensure compliance with industry standards.

What is the role of a hydrocolloid quality control laboratory?

A dedicated hydrocolloid quality control laboratory is crucial in ensuring accurate and reliable testing of hydrocolloid samples. It is equipped with specialized equipment and employs skilled professionals who perform various tests and analyses to verify the quality and consistency of hydrocolloid products.

FREQUENTLY ASKED QUESTIONS (FAQS) CONTINUED

How can companies implement hydrocolloid quality management systems?

Companies can implement hydrocolloid quality management systems by establishing robust quality control procedures, adhering to industry standards and guidelines, conducting regular audits, and continuously improving their processes. Training employees on quality management principles is also essential for effective implementation.

What are the key parameters for hydrocolloid quality control?

The key parameters for hydrocolloid quality control include viscosity, gel strength, pH, moisture content, particle size, and chemical composition. Monitoring these parameters ensures that the hydrocolloid products meet the desired specifications and perform optimally in various industrial applications.

What is the importance of process validation in hydrocolloid quality control?

Process validation is essential in hydrocolloid quality control, ensuring the manufacturing process consistently produces high-quality products. It involves verifying and documenting that the process parameters are within specified limits to ensure the reliability and reproducibility of the final products.

How can companies ensure optimal quality control in hydrocolloid production?

Companies can ensure optimal quality control in hydrocolloid production by implementing robust quality management systems, conducting regular testing and analysis, training employees on quality control principles, and staying updated with industry regulations and standards. Continuous improvement and monitoring of key parameters are vital for maintaining high-quality hydrocolloid products.

What is the concluding message regarding hydrocolloid quality control?

Hydrocolloid quality control is essential for industrial settings to produce high-quality products that meet industry standards and customer expectations. Companies can ensure consistency, reliability, and optimal performance of hydrocolloid products in the United States through various testing methods, analysis techniques, and adherence to quality management systems.

CHAPTER 23

International Trade and Harmonization

PART 23.1

Hydrocolloids: Their Effect on International Standards on its Trade

The hydrocolloid industry is subject to rigorous quality standards to ensure safe and effective products. Hydrocolloid standards are established to meet specific product requirements and outline the testing and analysis necessary to ensure compliance.

These standards are also critical for the trade of hydrocolloids both domestically and internationally. They are benchmarks for product quality, consistency, and safety, enhancing consumer trust and confidence. In this section, we will explore the impact of international standards on the trade of hydrocolloids.

A colorful collage of hydrocolloid products from around the world, arranged in a circular pattern with subtle variations in size and shape. Each product should be labeled with a small icon indicating its country of origin and adhering to international standards. The background should be a gradient of blue and green, suggesting both the ocean and land to represent the global trade and commerce of hydrocolloids.

Core Insights:

- ☑ Hydrocolloid standards are essential for ensuring product quality and compliance.
- ☑ International hydrocolloid standards serve as benchmarks for consistent product quality and safety.
- ☑ Adherence to these standards enhances consumer trust and confidence in the hydrocolloid industry.
- ☑ Hydrocolloid standards play a significant role in shaping trade dynamics in the industry.
- ☑ Compliance with hydrocolloid standards is necessary for successful trade both domestically and internationally.

23.1.1 Understanding Hydrocolloid Standards

Hydrocolloid standards specify the parameters for specifications, ensuring quality and consistency throughout the industry. These benchmarks are vital for maintaining product integrity and protecting consumer health.

International standards, such as those established by the International Organization for Standardization (ISO), provide a framework for hydrocolloid industry standards. The ISO 9001 standard, for example, outlines quality management systems and principles that can be adapted to the specific needs of hydrocolloid manufacturers.

In addition to international standards, hydrocolloid manufacturers must also comply with regional and national regulations. For instance, the United States Food and Drug Administration (FDA) regulates hydrocolloid ingredients to ensure they are safe for consumption and accurately labeled.

Table 131: Hydrocolloid Specification and Parameter

HYDROCOLLOID SPECIFICATION	PARAMETER
Viscosity	Measures a fluid's resistance to flow
pH	The measure of acidity or alkalinity of a solution
Particle Size	Determines the size of solid particles present in the hydrocolloid solution

Hydrocolloid manufacturers must conduct testing and analysis to ensure their products meet specifications. This process includes testing for functionality, purity, and performance to meet hydrocolloid compliance requirements. Compliance with hydrocolloid standards is essential for ensuring product quality and maintaining consumer trust.

Overall, hydrocolloid standards provide a critical foundation for ensuring product consistency and protecting public health. Manufacturers must comply with these benchmarks to remain competitive in the global market and support a sustainable industry.

23.1.2 Importance of Hydrocolloid Testing and Analysis

Ensuring product quality and consistency is crucial in the hydrocolloid industry. Hydrocolloid testing and analysis play a significant role in achieving these goals. Through testing and analysis, manufacturers can detect any product issues and take corrective measures. Furthermore, testing helps ensure that hydrocolloids meet specific quality parameters. Hydrocolloid quality assurance is crucial in maintaining customer trust. Customers expect hydrocolloids to perform consistently across different applications and in varying conditions. Therefore, manufacturers must ensure that their products meet specific quality standards. Hydrocolloid testing is the primary method for evaluating the quality of hydrocolloids.

Hydrocolloid analysis involves evaluating the chemical and physical properties of hydrocolloids. This process helps manufacturers understand the behavior of hydrocolloids in different conditions. It also helps manufacturers identify any impurities or contaminants that might affect the performance of hydrocolloids.

Hydrocolloid performance testing involves evaluating the ability of hydrocolloids to perform specific functions. For instance, if a hydrocolloid is used as a thickener in a food application, performance testing would evaluate how well it thickens. Performance testing helps manufacturers understand the functionality of hydrocolloids in different applications and conditions.

Hydrocolloid testing and analysis are critical in maintaining product integrity and ensuring customer satisfaction. Without proper testing and analysis, manufacturers risk producing sub-standard products that might harm their reputation and result in lost business.

Table 132: Quality Control Table

HYDROCOLLOID PROPERTY	TEST METHOD	SPECIFICATION	RESULT
Viscosity	Brookfield Viscometer	500-2000 cps	1300 cps
pH	pH meter	5.5-7.5	6.2
Ash Content	Gravimetric Analysis	10% max	7.8%

"Hydrocolloid testing and analysis are critical in maintaining product integrity and ensuring customer satisfaction."

Hydrocolloid testing and analysis are essential components of quality control in the industry. By adhering to strict testing protocols, manufacturers can ensure that their products meet specific quality parameters. In turn, it helps to maintain customer satisfaction and trust in the industry.

23.1.3 Hydrocolloid Standards in the Global Context

The hydrocolloid industry is a global one, with players located all over the world. As a result, it is essential to have international hydrocolloid standards that are recognized and implemented across the board. These standards help to ensure that products are of a consistent quality, regardless of where they are produced or sold.

The hydrocolloid industry has several international standards that are widely recognized and implemented. These standards include the CODEX Alimentarius, a collection of internationally recognized food standards, guidelines, and codes of practice. The CODEX Alimentarius covers a wide range of foods, including hydrocolloids.

Another international standard used in the hydrocolloid industry is the European Union's regulation of food additives. This regulation provides a list of approved food additives, including hydrocolloids, for food production. The regulation also sets out the maximum levels of these additives that can be used in food products.

Table 133: Hydrocolloid World and Industry Standards

HYDROCOLLOID WORLD STANDARDS	HYDROCOLLOID INDUSTRY STANDARDS	HYDROCOLLOID CERTIFICATION
Codex Alimentarius	Hydrocolloid and Food Gums	ISO 9001:2015
European Union's Regulation on Food Additives	Kosher and Halal Certification	BRC Global Standard for Food Safety
U.S. FDA Code of Federal Regulations	Organic Certification	IFS Food Standard

These international hydrocolloid standards are critical for the hydrocolloid industry's success. They allow companies to produce and sell products that meet a consistent quality level, which helps to build trust with consumers and other industry players.

However, implementing these standards can be challenging, particularly for small companies or those operating in countries with differing regulations. Companies must comply with all relevant standards to avoid market access issues.

Harmonization of hydrocolloid standards is essential for the industry's growth. By working together to establish and maintain international standards, the hydrocolloid industry can continue to grow and thrive, making high-quality products that meet consumer needs and preferences.

23.1.4 The Impact of Hydrocolloid Standards on Trade Dynamics

Hydrocolloid standards play a critical role in regulating the hydrocolloid industry at the international level. These standards ensure compliance with established guidelines and specifications, guaranteeing the quality and safety of products traded across borders.

International hydrocolloid standards facilitate trade by harmonizing regulations across different regions, reducing trade barriers, and promoting consumer confidence. Compliance with these standards allows hydrocolloid products to be sold in global markets without significant regulatory hurdles.

Hydrocolloid benchmarking is an essential part of evaluating compliance with industry standards. Through benchmarking, companies can measure their performance against established benchmarks and identify areas for improvement. This ongoing process of continuous improvement drives innovation in the industry, promoting the development of new products and processes that meet the most stringent quality standards.

> "International hydrocolloid standards facilitate trade by harmonizing regulations across different regions, reducing trade barriers, and promoting consumer confidence."

Hydrocolloid compliance is critical to maintaining the industry's integrity and ensuring that products meet the highest quality standards. Companies that fail to comply with international hydrocolloid standards risk losing market access and facing legal penalties. The consequences of noncompliance can severely impact the reputations of companies within the industry, undermining consumer confidence and market share.

Hydrocolloid standards are a vital component of the global industry's trade dynamics. Compliance with these standards is critical to promoting smooth international trade, reducing barriers, and ensuring consumer confidence in hydrocolloid products.

23.1.5 Hydrocolloid Certification and Compliance

Hydrocolloid certification is an essential component of ensuring compliance with industry standards. It involves comprehensively evaluating a hydrocolloid product's quality, consistency, and safety to ensure it meets defined standards.

The hydrocolloid industry is highly regulated, and compliance is critical to consumer trust and market access. Certification offers an objective third-party verification that a hydrocolloid product adheres to the industry's standards and specifications.

Certification also ensures that hydrocolloid products meet the labeling requirements and are safe for consumption. The certification process includes testing for impurities, contamination, and other factors affecting product quality and safety.

Moreover, compliance with industry standards is essential for companies seeking to expand their market share in the US and abroad. Companies that fail to comply with standards face regulatory actions, reduced market access, and damage to their reputation.

The hydrocolloid industry has several certifications, including the International Organization for Standardization (ISO), NSF International, and the US Pharmacopeia (USP). Each certification has unique requirements, and companies may seek more than one to demonstrate compliance.

Table 134: Certification and Requirements

CERTIFICATION	REQUIREMENTS
ISO Certification	Quality Management System (QMS) compliance and product testing
NSF International Certification	Product testing, manufacturing process evaluation, and facility inspection
USP Certification	Product purity, identity, and performance testing

Certification is not a one-time process; companies must maintain compliance and undergo regular evaluations to keep their certifications. Certification bodies perform regular audits and inspections to ensure companies adhere to established standards.

Hydrocolloid compliance and certification are vital for companies seeking to grow and expand globally. It is a significant investment in their products' quality and safety, demonstrating a commitment to consumer trust and satisfaction.

23.1.6 Challenges in Implementing Hydrocolloid Standards

While hydrocolloid standards are critical in ensuring product quality, compliance with international standards can be challenging. Industry players face numerous obstacles in implementing and adhering to hydrocolloid standards.

One major challenge is the variation in regulations across different countries. Hydrocolloid manufacturers often must comply with various regional and national regulations, which can be difficult to navigate. Additionally, resource constraints can make it challenging for smaller companies to invest in compliance measures required by international standards.

Moreover, the complexity of international standards can make implementation difficult. Hydrocolloid standards are intricate and require extensive documentation and testing. Furthermore, the interpretation of standards can vary, leading to inconsistencies in compliance.

Despite these challenges, industry players must strive to comply with hydrocolloid standards. The benefits of adherence to international benchmarks include increased market access, improved product quality, and enhanced consumer trust.

Table 135: Challenges in Implementing Hydrocolloid Standards

CHALLENGE	DESCRIPTION
Varying Regulations	Companies must comply with different national and regional regulations, which can be complex and time-consuming.
Resource Constraints	Smaller companies may struggle to invest in compliance measures required by international standards due to financial and resource limitations.
Complexity of Standards	Hydrocolloid standards are intricate and require extensive documentation and testing. Interpreting standards can also create inconsistencies in compliance.

By addressing these challenges and striving for compliance, hydrocolloid manufacturers can help ensure a thriving industry that produces high-quality products that meet international standards.

23.1.7 Achieving Consistency through Hydrocolloid Specification

Hydrocolloid specification plays a critical role in achieving consistency across the industry. Manufacturers must adhere to specific guidelines to ensure their products meet industry standards and regulations. It includes providing clear labeling and meeting performance benchmarks.

The hydrocolloid industry standards provide a framework for manufacturers to meet these requirements. Companies must ensure their products comply with international standards, including those set by the International Organization for Standardization (ISO).

Hydrocolloid testing is an essential part of the specification process. Testing ensures that products meet quality and performance standards and comply with regulations. Testing is necessary for new and existing products, as manufacturers must consistently meet the required specifications.

Compliance with hydrocolloid industry standards is crucial for manufacturers. Compliance demonstrates that products meet safety, quality, and performance requirements. Companies that fail to comply with industry standards risk losing market share, damaging their brand reputation, and facing legal consequences.

Industry standards like hydrocolloid specification are constantly evolving to meet the changing needs of the industry. Manufacturers must keep up with these changes to maintain compliance and ensure the consistency of their products. It includes staying current with the latest industry developments, adopting new technologies and processes, and participating in certification programs.

23.1.8 Evaluating Hydrocolloid Performance

Hydrocolloid performance testing is a vital step in ensuring the quality and effectiveness of these products. Rigorous testing and analysis help to identify any issues or defects in hydrocolloids, allowing manufacturers to make necessary improvements.

Hydrocolloid testing involves various methods, including physical, chemical, and microbiological analysis. These tests provide critical data on the product's viscosity, pH, stability, etc. By measuring these parameters, manufacturers can determine whether the hydrocolloid meets the required specifications and performance standards.

Hydrocolloid analysis is also a crucial aspect of quality assurance. Testing for impurities, contaminants, and other potential defects helps to ensure that the product is safe for use in food, pharmaceuticals, and other applications. In addition, performance testing allows manufacturers to evaluate the product's efficacy in different environments and applications. This information is used to optimize the product for maximum effectiveness, efficiency, and consistency.

The hydrocolloid industry has established a range of performance benchmarks, such as gel strength, viscosity, and water-holding capacity, to measure the effectiveness of these products. These benchmarks evaluate products against industry standards and ensure that they meet or exceed these standards for quality and performance.

> "Hydrocolloid performance testing is critical to ensuring that our products meet the rigorous standards of our customers. We can ensure that our hydrocolloids perform optimally in various applications by evaluating factors such as viscosity, stability, and water-holding capacity."

In summary, hydrocolloid performance testing and analysis are essential steps in maintaining the quality and consistency of these products. Through rigorous testing and evaluation, manufacturers can ensure that their hydrocolloids meet industry standards and perform optimally in various applications.

23.1.9 Hydrocolloid Benchmarking and Continuous Improvement

Hydrocolloid benchmarking measures a company's performance against industry standards to identify areas for improvement and growth. It provides businesses with valuable insights into the current market trends and helps to identify potential weaknesses in their operations.

Hydrocolloid industry standards constantly evolve, introducing new benchmarks and regulations regularly. Companies that stay updated with these changes will be better equipped to meet customer demands and remain competitive.

Hydrocolloid certification is essential to benchmarking and ensures that a company's products meet industry standards. This certification also boosts consumer confidence in a company's products, encouraging customer loyalty and repeat business.

Hydrocolloid compliance is essential to benchmarking and involves adhering to industry regulations and standards. Companies that comply with these regulations can avoid legal issues, reduce the risk of product recalls, and avoid negative publicity.

Benchmarking is an ongoing process that involves continuous improvement. Companies that benchmark their operations regularly can identify opportunities for efficiencies, cost savings, and innovation. By refining their processes and products, they can stay ahead of the curve in the hydrocolloid industry.

> "Benchmarking provides businesses with a roadmap for growth and innovation in the hydrocolloid industry. By staying current with industry standards, certifications, and compliance regulations, companies can improve their operations and maintain their competitive edge."

Conclusion

In conclusion, hydrocolloid standards play a crucial role in shaping the trade dynamics of the industry. Adhering to hydrocolloid industry standards and obtaining hydrocolloid certification helps ensure product quality and consistency, building consumer trust. Compliance with international standards is also essential for accessing global markets, promoting fair competition, and driving industry growth.

However, implementing hydrocolloid standards can be challenging, given the varying regulations and resource constraints across different regions. Nevertheless, hydrocolloid benchmarking and continuous improvement offer a way to address these challenges and drive innovation in the industry.

In summary, hydrocolloid standards and certification are indispensable for the hydrocolloid industry's growth and success. Striving for compliance, specification clarity, and performance evaluation supports the industry's progression and continued global trade impact.

FAQ

What are hydrocolloid standards?

Hydrocolloid standards are internationally recognized benchmarks ensuring product quality and consistency in the industry. These standards set specifications and parameters for hydrocolloid products.

Why are hydrocolloid standards important?

Hydrocolloid standards are important because they help maintain product integrity, ensure consumer safety, and facilitate international trade. Adhering to these standards promotes transparency and trust within the industry.

How do hydrocolloid standards impact trade dynamics?

Hydrocolloid standards directly impact trade dynamics by influencing market access, competition, and consumer trust. These standards can enhance market opportunities and foster a level playing field for industry players.

What is hydrocolloid certification?

Hydrocolloid certification is a process that verifies compliance with industry standards. It demonstrates that a hydrocolloid product meets the prescribed specifications and quality requirements set by the standards.

What challenges are faced in implementing hydrocolloid standards?

Implementing hydrocolloid standards can be challenging due to varying regulations, resource constraints, and the complexity of international standards. Harmonizing and aligning regulations across different countries and regions is a significant obstacle.

How does hydrocolloid specification contribute to consistency in the industry?

Hydrocolloid specification plays a crucial role in achieving consistency across the industry. Clear guidelines for manufacturing, labeling, and product performance ensure that hydrocolloid products meet the expected quality and functionality requirements.

FREQUENTLY ASKED QUESTIONS (FAQS) CONTINUED

How are hydrocolloid performance and effectiveness evaluated?

Hydrocolloid performance is evaluated through rigorous testing and analysis. These evaluations measure factors such as viscosity, stability, and functionality to assess the effectiveness of hydrocolloid products.

What is hydrocolloid benchmarking?

Hydrocolloid benchmarking is a practice that involves comparing a company's performance, processes, and products against industry standards and best practices. It helps identify areas for improvement and drives continuous growth and innovation within the industry.

Why should companies strive for hydrocolloid certification?

Companies should strive for hydrocolloid certification to demonstrate their commitment to quality, compliance, and consumer safety. It also enhances their reputation and credibility in the marketplace, facilitating access to new markets and customers.

What is the role of hydrocolloid standards in the global context?

Hydrocolloid standards are crucial globally as they harmonize regulations and facilitate international trade. They provide a common framework for countries and regions to follow, ensuring mutual recognition and acceptance of hydrocolloid products.

PART 23.2

Hydrocolloid Harmonization Efforts for Standards Between Countries

The global food industry relies heavily on hydrocolloids, essential ingredients in many processed foods. Hydrocolloids serve various functions, including thickening, stabilizing, and gelling, and are used in a wide range of products, from dairy to confectionery.

However, with the increasing globalization of the food industry, it has become crucial to harmonize international standards for hydrocolloids. It helps to ensure that ingredients meet the same quality and safety standards across borders, facilitating international trade and consumer trust.

Hydrocolloid harmonization efforts involve aligning regulatory requirements, testing methods, and labeling standards between countries. Standardization can improve the consistency and transparency of hydrocolloid ingredients, making it easier for manufacturers to source and use them. In this section, we explore the importance of harmonizing hydrocolloid standards between countries and discuss the advancements in this field.

Core Insights:

- ☑ Hydrocolloids are widely used in the global food industry for various functions.
- ☑ Harmonizing hydrocolloid standards between countries is essential for facilitating international trade and consumer trust.
- ☑ Hydrocolloid harmonization efforts involve aligning regulatory requirements, testing methods, and labeling standards between countries.
- ☑ Standardization can improve the consistency and transparency of hydrocolloid ingredients, making it easier for manufacturers to source and use them.
- ☑ The advancements made in harmonizing hydrocolloid standards have helped establish common standards, promote consistency, and ensure compatibility of hydrocolloid ingredients across borders.

23.2.1 Advancements in Hydrocolloid Harmonization

Harmonizing hydrocolloid standards between countries has been a long-standing challenge, particularly in the food industry. Significant advancements have been made in harmonizing hydrocolloid standards to address this issue.

International collaborations have been initiated to establish common standards and regulations that promote consistency and compatibility of hydrocolloid ingredients across borders. Standards-setting organizations like the International Organization for Standardization (ISO) have developed globally recognized hydrocolloid standards, providing a common language and framework for producers and users.

Moreover, adopting a risk-based approach in hydrocolloid standard-setting has ensured the balance of safety and innovation, facilitating the introduction of new hydrocolloids while maintaining safety standards.

These advancements in harmonizing hydrocolloid standards have resulted in significant benefits for the food industry, including improved product quality and consistency, enhanced safety, reduced trade barriers and costs, and increased consumer trust.

23.2.2 Hydrocolloid Blending for Optimization

Hydrocolloids are polysaccharides widely used in the food industry to improve products' texture, stability, and sensory attributes. Hydrocolloids' functionality and performance can be optimized by blending different types of hydrocolloids.

Hydrocolloid blending combines two or more hydrocolloids to achieve desired textural and functional characteristics. The goal is to create a blend greater than the sum of its parts with improved properties that individual hydrocolloids cannot achieve.

There are various considerations in selecting hydrocolloid blends for specific applications. The type and concentration of hydrocolloids and their interactions with other ingredients can affect the properties of the final product. In addition, the blending process itself, including mixing speed and time, can impact the performance of the hydrocolloid blend.

The following table illustrates the properties and applications of some commonly used hydrocolloid blends:

Table 136: Properties and applications of hydrocolloid

HYDROCOLLOID BLEND	PROPERTIES	APPLICATIONS
Guar gum and xanthan gum	Improved viscosity, texture, and stability	Dairy products, dressings, sauces, and beverages
Carrageenan and locust bean gum	Improved gelling and water-binding properties	Dairy products, processed meats, and pet food
Agar and Konjac gum	Improved gelling and water-holding properties	Jellies, jams, and confectionery products

Hydrocolloid blending can offer a range of benefits, including cost savings, improved texture and stability, and enhanced nutritional profiles. By optimizing hydrocolloid blends, food manufacturers can create innovative products that meet consumer demand for better quality, convenience, and nutrition.

23.2.3 Ensuring Hydrocolloid Compatibility in Formulations

When formulating hydrocolloids, it is crucial to consider their compatibility to achieve the desired textural and stability characteristics in food products. Hydrocolloid compatibility is affected by protein type and concentration, pH, and ionic strength.

One way to ensure hydrocolloid compatibility is to conduct compatibility tests before incorporating them into a formulation. These tests involve mixing hydrocolloids in various ratios and observing the resulting texture and stability attributes.

The following table shows the compatibility of various hydrocolloid blends in mayonnaise formulations:

Table 137: Hydrocolloid Blend, Texture and Stability

HYDROCOLLOID BLEND	TEXTURE	STABILITY
Xanthan gum-Kappa carrageenan	Smoother, creamier texture	Improved water retention, no syneresis
Xanthan gum-Locust bean gum	Increased viscosity, slightly grainy texture	Increased water retention, slight syneresis
Gellan gum-High acyl gellan gum	Slightly gummy texture	Improved water retention, no syneresis

Note: Results of compatibility tests may vary depending on the specific formulation.

Once a compatible hydrocolloid blend has been determined, it is essential to ensure proper hydration of each hydrocolloid. It can be achieved by adding to the liquid component of the formulation gradually while mixing to prevent clumping and allow for even hydration.

Overall, achieving hydrocolloid compatibility is crucial in maximizing the functionality and performance of hydrocolloids in food products. Food manufacturers can develop formulations with optimal texture and stability attributes by understanding compatibility factors and conducting thorough tests.

23.2.4 Achieving Hydrocolloid Stability

Hydrocolloid stability is a critical aspect of formulating hydrocolloid-based food products. Stability refers to a product›s ability to maintain its physical and chemical properties over time, avoiding unwanted changes that negatively impact quality and sensory attributes.

Several factors influence hydrocolloid stability, including temperature, pH levels, concentration, ionic strength, and interactions with other ingredients. Proper formulation and processing techniques can help mitigate these factors and enhance stability.

One strategy for achieving hydrocolloid stability is the use of stabilizers and emulsifiers. These ingredients can help prevent phase separation, syneresis, and other instability issues by forming a protective layer around the hydrocolloid particles. For example, adding xanthan gum to ice cream prevents crystal formation and improves texture stability during storage.

Another approach to enhancing hydrocolloid stability is the use of modified hydrocolloids. These hydrocolloids have undergone chemical or physical modifications to improve their functional properties, such as stability. For instance, carboxymethylcellulose (CMC) is a modified cellulose derivative that exhibits improved stability in acidic environments compared to unmodified cellulose.

Table 138: Hydrocolloid Stability Benefit

HYDROCOLLOID	STABILITY BENEFIT
Carrageenan	Prevent syneresis in dairy and meat products
Pectin	Stabilize acidic fruit-based products.
Xanthan gum	Prevent crystal formation and improve texture stability.

Formulators must carefully balance stability with other functional properties, such as texture, viscosity, and mouthfeel, to create high-quality products that meet consumer expectations. Achieving hydrocolloid stability is an ongoing challenge that requires trial and error, careful ingredient selection, and attention to processing conditions.

23.2.5 Optimizing Hydrocolloid Performance in Food Products

The performance of hydrocolloids in food products is highly dependent on various factors, including concentration, hydration, and interactions with other ingredients. Optimizing hydrocolloid performance is crucial in ensuring the desired functionality and texture of the final product.

Hydrocolloid concentration plays a significant role in determining the performance of hydrocolloids in food products. Higher concentrations of hydrocolloids can lead to increased viscosity and gelation, while lower concentrations may result in inadequate texture and stability. Achieving the optimal concentration of hydrocolloids requires a careful balance between the desired functionality and the cost implications.

Hydration is another critical factor in hydrocolloid performance. Hydration involves the swelling of hydrocolloid particles in water to form a gel-like network. The extent of hydration can influence the viscosity, texture, and stability of the hydrocolloid-based product. Manufacturers must consider the appropriate hydration time, temperature, and pH conditions to optimize performance.

Interactions with other ingredients can also affect the performance of hydrocolloids in food products. Hydrocolloids may interact with other ingredients, such as salts, sugars, and proteins, altering their functionality and properties. To achieve optimal performance, manufacturers must consider the impact of ingredient interactions on the hydrocolloid blend formulation.

Table 139: Hydrocolloid Performance Factors and Considerations

HYDROCOLLOID PERFORMANCE FACTORS	CONSIDERATIONS
Concentration	Determine optimal concentration for desired function and cost-effectiveness.
Hydration	Control hydration time, temperature, and pH conditions for desired viscosity, texture, and stability.
Interactions with other ingredients	Consider the impact of ingredient interactions on the hydrocolloid blend formulation to achieve optimal performance.

Effective hydrocolloid performance optimization requires a thorough understanding of the factors affecting functionality and the strategies for achieving the desired results.

Manufacturers can achieve targeted functionality and texture by optimizing hydrocolloid performance in food products, improving stability and shelf life, and enhancing overall product quality.

23.2.6 Harnessing Hydrocolloid Interaction for Formulation Optimization

Hydrocolloid interaction is where blending two or more hydrocolloids enhances functionality and performance. The interaction between different hydrocolloids can produce unique textural properties and stability characteristics that cannot be achieved using only one hydrocolloid.

When designing hydrocolloid blends, selecting hydrocolloids with complementary properties is essential. For example, combining a thickener like xanthan gum with a gelling agent like carrageenan can produce a hydrocolloid blend with improved viscosity, texture, and stability.

The physical and chemical interactions between hydrocolloids in a blend are complex and depend on several factors, including concentration, pH, ionic strength, and temperature. Therefore, optimizing the blending process to achieve the desired functionality and stability is essential.

One approach to optimize hydrocolloid blending is to use a Design of Experiment (DoE) methodology. DoE allows formulators to systematically vary the levels of hydrocolloids and other ingredients to determine the optimal blend formulation that meets the product's performance targets.

Another approach is to use modeling software to predict the behavior of hydrocolloid blends. These software programs use mathematical algorithms to simulate the interactions between hydrocolloids, enabling formulators to design blends with specific textural and stability characteristics.

Hydrocolloid interaction is widely used in various food applications, including dairy products, sauces, dressings, and bakery products. For example, a study found that blending pectin and xanthan gum improved the texture and stability of low-fat yogurt, resulting in a creamier and smoother product.

Overall, hydrocolloid interaction is a powerful tool for formulators to optimize hydrocolloid functionality and performance. By combining two or more hydrocolloids, formulators can achieve unique textural and stability properties that cannot be achieved using only one hydrocolloid. Formulators can use strategies to successfully leverage hydrocolloid interaction by selecting the right blends, optimizing the blending process, and using modeling software.

23.2.7 Implications of Hydrocolloid Harmonization on the Food Industry

Harmonizing hydrocolloid standards between countries has significant implications for the food industry. Aligning regulations and standards promotes consistency in the global marketplace, allowing for smoother international trade and increased innovation. By establishing common standards, the industry can focus on creating products that meet quality requirements while minimizing costs associated with unnecessary testing and compliance procedures.

Hydrocolloid harmonization enables manufacturers to access a wider range of raw materials from across the globe and streamline their supply chain process. It results in a more diverse and competitive market, lowering consumer prices. In addition, harmonization can improve consumer trust in hydrocolloid-based products, as they can be confident that they meet stringent global standards for safety and quality.

The food industry can benefit greatly from harmonization regarding hydrocolloid standards. Save the image below for a quick reference on the benefits:

Harmonizing hydrocolloid standards benefits the food industry by promoting efficiency, innovation, and consumer satisfaction. By aligning standards, manufacturers can save costs and improve product quality, while consumers can trust that they are getting safe and high-quality products. The future of hydrocolloid harmonization is promising, with continued international collaborations and regulatory oversight ensuring that global standards remain up-to-date and relevant for years.

23.2.8 The Future of Hydrocolloid Harmonization

The future of hydrocolloid harmonization efforts is promising, with ongoing collaborations and initiatives to establish consistent standards and promote global trade. However, challenges such as differing regulatory frameworks and evolving consumer demands will require continued efforts towards harmonization.

One emerging trend in hydrocolloid harmonization is adopting digital technologies, such as blockchain and digital documentation, to improve traceability and transparency in the supply chain. It can enhance consumer trust and facilitate compliance with regulations.

The Role of International Collaborations

International collaborations will play a crucial role in future hydrocolloid standards and regulations. Bodies such as the Codex Alimentarius Commission and the International Organization for Standardization (ISO) are already making strides towards harmonization.

In addition, industry-driven initiatives such as the Hydrocolloid Producers Association's efforts to harmonize viscosity specifications for carrageenan are fostering greater consistency within the industry.

Meeting Evolving Consumer Demands

The hydrocolloid industry must adapt and innovate as consumer demands evolve towards cleaner labels and more sustainable sourcing. One potential solution is the development of plant-based hydrocolloids, which can offer functional benefits while addressing consumer concerns.

Implications for Hydrocolloid Standards

Overall, the future of hydrocolloid harmonization holds great potential for improving global trade and innovation in the food industry. Consistent standards and regulations can also enhance product safety and quality, benefiting consumers and manufacturers.

Conclusion

In conclusion, harmonizing hydrocolloid standards between countries is essential for a consistent global marketplace, especially within the food industry. The advancements in hydrocolloid harmonization through collaborative efforts and initiatives have resulted in common standards, promoting consistency and compatibility of hydrocolloid ingredients across borders.

Hydrocolloid blending and optimization are key to maximizing the functionality and performance of hydrocolloids in food products. Ensuring hydrocolloid compatibility and stability are crucial factors influencing hydrocolloid-based products' shelf life and performance. It is important to optimize hydrocolloid performance and harness hydrocolloid interaction. Harmonizing hydrocolloid standards can streamline international trade, foster innovation, and improve consumer trust in hydrocolloid-based products. The future of hydrocolloid harmonization efforts will require continued international collaborations and regulatory bodies to shape global hydrocolloid standards.

Overall, hydrocolloid harmonization efforts are vital for the food industry's growth and evolution in the global marketplace, and we must continue to work towards harmonizing these standards to benefit consumers and the industry.

FAQ

What is hydrocolloid harmonization?

Hydrocolloid harmonization refers to the efforts to align standards and regulations for hydrocolloids between countries. It aims to establish common standards to ensure consistency and compatibility of hydrocolloid ingredients in the global marketplace, particularly within the food industry.

Why is hydrocolloid harmonization important?

Harmonizing hydrocolloid standards is important to streamline international trade, promote innovation, and enhance consumer trust in hydrocolloid-based products. It allows for consistent quality and functionality of hydrocolloids, making it easier for manufacturers to develop and market products globally.

What advancements have been made in hydrocolloid harmonization?

There have been collaborative efforts and initiatives to advance hydrocolloid harmonization. International organizations and regulatory bodies have worked towards establishing common standards, sharing scientific knowledge, and creating guidelines to ensure the compatibility of hydrocolloid ingredients across borders.

How does hydrocolloid blending optimize functionality?

Hydrocolloid blending involves combining different hydrocolloids to create blends that maximize functionality and performance in food products. It allows for the customization of texture, stability, and other desired characteristics, providing formulators more flexibility in creating innovative, high-quality products.

What factors should be considered for hydrocolloid compatibility?

Factors such as pH, temperature, concentration, and interactions with other ingredients should be considered to ensure hydrocolloid compatibility when formulating with different hydrocolloids. These factors can affect the textural and stability characteristics of the final product.

FREQUENTLY ASKED QUESTIONS (FAQS) CONTINUED

How can hydrocolloid stability be achieved?

Achieving hydrocolloid stability involves considering factors such as pH, temperature, processing conditions, and storage conditions. Stabilizers and other ingredients may enhance hydrocolloid-based products' shelf life and performance.

What factors affect hydrocolloid performance?

Several factors can affect hydrocolloid performance, including concentration, hydration, interactions with other ingredients, and processing conditions. Understanding and optimizing these factors can help enhance hydrocolloid functionality in various food applications.

What is hydrocolloid interaction?

Hydrocolloid interaction refers to blending multiple hydrocolloids to enhance their overall functionality. Combining different hydrocolloids can improve texture, stability, and other desirable properties in food formulations.

How does hydrocolloid harmonization impact the food industry?

Hydrocolloid harmonization has several implications for the food industry. It streamlines international trade by reducing barriers and ensuring consistency in ingredient quality. It also fosters innovation by providing a common framework for formulation development, and it improves consumer trust by ensuring that hydrocolloid-based products meet consistent safety and quality standards.

What does the future hold for hydrocolloid harmonization?

The future of hydrocolloid harmonization will involve ongoing efforts to address emerging challenges and trends in the food industry. International collaborations and regulatory bodies will continue to play a crucial role in shaping global hydrocolloid standards and ensuring the compatibility and safety of hydrocolloid ingredients across borders.

CHAPTER 24

Where do Hydrocolloids Come From?

PART 24.1

Sources of Hydrocolloids

Hydrocolloids are complex, nondigestible polysaccharides that dissolve or disperse in water to give a thickening or viscosity-building effect. These versatile ingredients are sourced from various natural and microbial sources, making them essential to the food industry.

Hydrocolloids can be derived from seaweed, plant extracts, animal collagen, and microorganisms. Each source offers unique properties and functionalities contributing to food products' texture, stability, and sensory experience.

Core Insights:

- ☑ Hydrocolloids are complex polysaccharides used for thickening and viscosity building in food products.
- ☑ They can be derived from seaweed, plant extracts, animal collagen, and microorganisms.
- ☑ Seaweed, including red and brown algae, is a rich source of hydrocolloids.
- ☑ Plant extracts provide hydrocolloids like guar gum, pectin, and cellulose.
- ☑ Animal collagen is used to obtain hydrocolloids like gelatin.

24.1.1 Seaweed as a Source of Hydrocolloids

Seaweed, particularly red and brown algae, is a valuable and abundant source of hydrocolloids widely used in the food industry for their thickening, gelling, stabilizing, and emulsifying properties. Brown algae, also known as kelp, produce alginates, while red seaweed is used to extract carrageenan and agar.

Alginates, derived from brown algae, are natural polysaccharides with excellent water-absorbing capabilities. They are commonly used as thickeners and stabilizers in various food products, including dairy, desserts, and sauces. Alginates contribute to improved texture and viscosity in these applications.

Carrageenan, extracted from red seaweed, is another popular hydrocolloid used in the food industry. It has exceptional gelling and thickening properties, making it suitable for dairy products, ice creams, and confectionery applications. Carrageenan enhances the texture and stability of these products, giving them a desirable mouthfeel.

Agar, also derived from red seaweed, is a versatile hydrocolloid commonly used as a gelling agent in various food products. It forms a gel at low concentrations and is particularly valued for its heat-resistant properties. Agar is used in jellies, custards, and bakery fillings to provide structure and stability.

Overall, seaweed offers a sustainable and natural source of hydrocolloids, providing the food industry with functional ingredients that improve the quality, texture, and stability of various food products.

Table 140: Seaweed Hydrocolloids source and uses.

HYDROCOLLOID	SOURCE	COMMON USES
Alginates	Brown Algae (Kelp)	Dairy, desserts, sauces
Carrageenan	Red Seaweed	Dairy products, ice creams, confectionery
Agar	Red Seaweed	Jellies, custards, bakery fillings

24.1.2 Plant Extracts as Sources of Hydrocolloids

Hydrocolloids, essential ingredients for thickening and stabilizing various food products, can be derived from a diverse range of plant extracts. These natural sources provide unique properties and functionalities, contributing to the versatility and effectiveness of hydrocolloids in the food industry.

Here are some notable plant extracts and their respective hydrocolloids:

- **Guar Gum:** Guar gum is obtained from the seeds of the guar plant. Guar gum is a popular hydrocolloid known for its excellent thickening and stabilizing properties.
- **Pectin:** Derived from citrus fruits, pectin is a natural hydrocolloid widely used in the food industry to create gels, improve texture, and enhance stability.

- **Cellulose:** Extracted from plant cell walls, cellulose is a versatile hydrocolloid contributing to the viscosity and texture of various food products. It is commonly used as a thickener and emulsifier.
- **Gum Arabic:** Obtained from the acacia tree, gum Arabic is a natural hydrocolloid with excellent emulsifying and stabilizing properties. It is commonly used in beverages, confectionery, and bakery products.
- **Konjac Gum:** It is derived from the konjac plant. Konjac gum is a hydrocolloid that forms a highly viscous solution, making it suitable as a thickener and gelling agent in various food applications.
- **Locust Bean Gum:** Obtained from the seeds of the carob tree, locust bean gum is a natural hydrocolloid widely used as a thickener, stabilizer, and gelling agent in food products.
- **Tapioca Starch:** It is derived from the roots of the tapioca plant. Tapioca starch is a versatile hydrocolloid that contributes to the texture and stability of food products. It is commonly used in bakery, dairy, and dessert applications.
- **Psyllium Husk:** Extracted from the psyllium plant, psyllium husk is a natural hydrocolloid known for its high water absorption capacity and gelling properties. It is often used as a thickening and stabilizing agent in gluten-free products.
- **Chia Seeds:** Derived from the chia plant, chia seeds are rich in soluble fiber and form a gel-like consistency when mixed with water. They are commonly used as a natural hydrocolloid for thickening and stabilizing food products.
- **Flax Seeds:** Obtained from the flax plant, flax seeds contain mucilage, a natural hydrocolloid that forms a gel when combined with water. Flax seeds are often used as a thickening and binding agent in recipes.
- **Rice Starch:** Derived from rice, rice starch is a versatile hydrocolloid used as a thickener and binding agent in various food products. It contributes to the smooth texture and stability of food formulations.
- **Sago Palm Starch:** Obtained from the sago palm tree, sago palm starch is a natural hydrocolloid commonly used as a thickening agent in bakery, dairy, and dessert applications.

These plant-based hydrocolloids offer various functionalities, allowing food manufacturers to achieve desired textures, improve stability, and enhance overall quality. Their natural origins make them a preferred choice for clean-label and natural food formulations.

Next, we will explore hydrocolloids derived from animal sources and their applications in the food industry.

Table 141: Comparison of Plant Extracts as Sources of Hydrocolloids

PLANT EXTRACT	HYDROCOLLOID	PROPERTIES	APPLICATIONS
Guar Plant	Guar Gum	Thickening, stabilizing	Bakery, dairy, sauces
Citrus Fruits	Pectin	Gelling, thickening	Jams, jellies, confectionery
Plant Cell Walls	Cellulose	Thickening, emulsifying	Sauces, dressings, soups
Acacia Tree	Gum Arabic	Emulsifying, stabilizing	Beverages, confectionery
Konjac Plant	Konjac Gum	Thickening, gelling	Noodles, desserts

Carob Tree	Locust Bean Gum	Thickening, stabilizing, gelling	Dairy, ice cream, sauces
Tapioca Plant	Tapioca Starch	Thickening, binding	Bakery, desserts, dairy
Psyllium Plant	Psyllium Husk	Thickening, gelling	Gluten-free dietary supplements
Chia Plant	Chia Seeds	Thickening, stabilizing	Beverages, bakery, desserts
Flax Plant	Flax Seeds	Thickening, binding	Bakery, snacks, gluten-free
Rice	Rice Starch	Thickening, binding	Dairy, sauces, desserts
Sago Palm Tree	Sago Palm Starch	Thickening	Bakery, desserts, soups

24.1.3 Animal sources of hydrocolloids

Animal collagen is an important source of hydrocolloids in the food industry. One of the most used hydrocolloids derived from animal collagen is gelatin. Gelatin is produced through the hydrolysis of collagen extracted from animal tissues, such as bones, skin, and connective tissues. It is widely known for its thickening and gelling properties, making it a versatile ingredient in various food formulations.

Gelatin is commonly used to produce desserts, confectionery, and dairy products. Its ability to form a gel structure contributes to the texture and stability of these products. Gelatin also enhances the mouthfeel and sensory experience, providing a smooth and creamy consistency that consumers enjoy.

Create an image of animal collagen extracted from a source, such as bones or skin, through simmering or boiling. Show the collagen as a translucent, gelatinous substance that is rich in texture and has a slightly yellow or cream color. The animal source could be a cow, pig, or fish. The image should be focused on the collagen itself, with the animal source in the background or off to the side. Avoid using human elements or tools in the image, such as hands or utensils.

While gelatin is the most well-known hydrocolloid from animal collagen, other hydrocolloids can be extracted from animal sources. These include collagen peptides and collagen hydrolysates, which are used for their functional and nutritional benefits. Collagen peptides are gaining popularity in the health and beauty industries for their potential effects on skin health and joint function.

Table 142: Comparison of animal collagen and gelatin

ANIMAL COLLAGEN	GELATIN
Derived from animal tissues	Derived from the hydrolysis of animal collagen
Multiple sources, such as bones, skin, and connective tissues	Derived from specific animal tissues, such as bovine or porcine
Used as a source of collagen peptides and hydrolysates	Used as a thickener and gelling agent in food products

Although animal collagen and gelatin are derived from animal sources, alternative options are available for those seeking vegetarian or vegan alternatives. Plant-based hydrocolloids, such as agar from seaweed or pectin from citrus fruits, can provide similar functionalities in food formulations.

Animal collagen and its derivative, gelatin, are valuable hydrocolloids contributing to the texture, stability, and sensory attributes of various food products.

24.1.4 Hydrocolloids from Microorganisms

While seaweed, plant extracts, and animal collagen are commonly known sources of hydrocolloids, microorganisms such as bacteria also play a crucial role in their production. Hydrocolloids like xanthan gum and gellan gum are derived through bacterial fermentation, offering unique properties that contribute to the food industry.

One example of a hydrocolloid produced through bacteria fermentation is xanthan gum. It is formed by the bacterium Xanthomonas campestris during the fermentation of glucose, sucrose, or lactose. This process results in a polysaccharide with excellent thickening and stabilizing capabilities. Xanthan gum has become widely used in various food products, from dressings and sauces to baked goods and beverages, due to its ability to enhance texture and consistency.

Another significant hydrocolloid sourced from bacteria fermentation is gellan gum. Developed from the bacterium Sphingomonas elodea, gellan gum exhibits unique properties that make it suitable for various applications in the food industry. It forms a strong gel at low concentrations, offers stability under different pH conditions, and has excellent heat resistance. Gellan gum is commonly used as a gelling and stabilizing agent in confectionery, dairy products, desserts, and jellies.

Microorganisms provide an alternative and sustainable source of hydrocolloids, contributing to the versatility and innovation in the food industry. The production of hydrocolloids through bacterial fermentation continues to expand, driven by the growing demand for functional ingredients that enhance both texture and stability in food products.

24.1.5 Applications of Hydrocolloids

Hydrocolloids play a crucial role in various applications within the food industry. These versatile ingredients are used for their thickening, gelling, emulsifying, and stabilizing properties, making them essential components in various food formulations.

Thickening and Gelling Agents

One of the primary applications of hydrocolloids is as a thickening and gelling agent. They enhance the viscosity and texture of food products, resulting in improved mouthfeel and sensory experience. Hydrocolloids contribute to products' desired consistency and stability, ensuring a pleasant eating experience for consumers.

Some common examples of food products that utilize hydrocolloids as thickening and gelling agents include:

- Soups and sauces
- Dressings and gravies

- Jams and jellies
- Puddings and custards

By employing hydrocolloids in these applications, food manufacturers can achieve the desired texture, appearance, and consistency for their products.

Emulsifying and Stabilizing Properties

In addition to their thickening and gelling capabilities, hydrocolloids are also known for their emulsifying and stabilizing properties. They help maintain the uniform distribution of ingredients and prevent phase separation, enhancing food formulations' stability and shelf life.

Hydrocolloids are commonly used in the following food applications for their emulsifying and stabilizing benefits:

- Salad dressings and mayonnaise
- Ice creams and frozen desserts
- Beverages and dairy products
- Bakery and confectionery items

By incorporating hydrocolloids into these products, manufacturers can ensure their offerings remain visually appealing, have a pleasant texture, and maintain their integrity throughout their shelf life.

The Hydrocolloid Industry and Market Growth

The hydrocolloid industry is experiencing rapid growth due to the increasing demand for these versatile ingredients. Food manufacturers embrace hydrocolloids to meet these expectations as consumers become more conscious about texture, stability, and overall quality.

The hydrocolloid market is expanding across various food industry sectors, including processed foods, beverages, and dairy products. The versatility of hydrocolloids and their ability to enhance the sensory experience of food products make them highly sought-after ingredients.

As the market continues to evolve, the hydrocolloid industry is also witnessing advancements in research and development, leading to the discovery of innovative applications and improved functionalities of hydrocolloids.

Overall, hydrocolloids have become essential ingredients in the food industry, offering various applications and driving the market growth. Hydrocolloids' versatility and unique properties make them valuable assets for food manufacturers seeking to create high-quality, stable, and appealing products.

24.1.6 Benefits of using hydrocolloids

Hydrocolloids offer numerous benefits when used in food products. Their unique properties enhance the texture, stability, and mouthfeel of various formulations, improving consumer satisfaction. Let's explore some of the key benefits provided by hydrocolloids:

Improved Texture

Hydrocolloids contribute to the desired texture of food products, creating a smooth and pleasing mouthfeel. They can add thickness, creaminess, and richness to sauces, dressings, and desserts, enhancing the overall sensory experience.

Enhanced Stability

One of the significant advantages of hydrocolloids is their ability to improve the stability of food formulations. They prevent phase separation in emulsions, such as mayonnaise or salad dressings, ensuring a consistent and visually appealing texture over time.

Better Control over Product Quality

By enhancing the viscosity and consistency of food products, hydrocolloids give manufacturers greater control over the final quality of their offerings. These ingredients enable precise formulation and maintain the desired physical properties throughout the product's shelf life, ensuring consistent consumer experiences.

Extended Shelf Life

Hydrocolloids can contribute to increased shelf life by inhibiting moisture migration, preventing syneresis, and preserving the overall freshness of food products. This benefit is particularly valuable in applications where extended shelf life is important, such as bakery items, dairy products, and processed foods.

Improved Sensory Properties

Hydrocolloids can also enhance the sensory properties of food products. They can improve the mouthfeel, impart a pleasant creaminess, and modify the release of flavors in the mouth, enhancing the overall sensory experience and consumer satisfaction.

In summary, using hydrocolloids in food products provides multiple benefits, including improved texture, enhanced stability, better control over product quality, extended shelf life, and improved sensory properties. These advantages make hydrocolloids valuable and versatile ingredients in the food industry.

24.1.7 Hydrocolloid Extraction and Production

Hydrocolloids are essential ingredients in the food industry, renowned for their thickening, stabilizing, and emulsifying properties. Hydrocolloids must be extracted and produced from natural sources to harness these valuable properties. This process involves combining physical and chemical techniques to obtain pure hydrocolloids with optimal functionality.

Extraction techniques for hydrocolloids vary depending on the source material and desired product. The extraction process typically includes the following steps:

- **Separation:** The hydrocolloid-rich raw material is separated from other components, such as cellulose, fibers, and impurities. This separation is achieved through methods like filtration, centrifugation, or sieving.
- **Purification:** The extracted hydrocolloids undergo purification to remove any remaining impurities or unwanted substances. This purification step ensures the integrity and quality of the final hydrocolloid product.
- **Drying:** After purification, the hydrocolloids are dried to reduce moisture content and increase stability. Drying methods include spray drying, freeze drying, or air drying, depending on the specific hydrocolloid and its intended application.

The production of hydrocolloids is carried out on a large scale to meet the growing demand in the food industry. Manufacturers employ advanced technologies and stringent quality control measures to ensure consistent and high-quality hydrocolloid production. The extraction and production processes adhere to strict safety and regulatory standards to guarantee the safety and suitability of hydrocolloids for various applications.

Let's take a closer look at the extraction and production methods for some commonly used hydrocolloids:

Agar

Agar is a hydrocolloid derived from red seaweed. It is extracted through physical and chemical processes, including boiling, filtering, and drying. The resulting agar is a versatile hydrocolloid widely used in the food industry for its gelling and stabilizing properties.

Gellan Gum

Gellan gum is another hydrocolloid produced through bacterial fermentation. It is isolated from the fermentation broth and then purified and dried to obtain the final product. Gellan gum is known for its excellent gel-forming and suspending abilities, making it a valuable ingredient in food applications.

Guar Gum

Guar gum is derived from the seeds of the guar plant. The extraction process involves de-husking the seeds, grinding them into a powder, and treating them with water. The resulting gum is dried and milled to achieve the desired particle size. Guar gum is widely used in various food products as a thickening and stabilizing agent.

"

"The extraction and production of hydrocolloids require precision and expertise to ensure optimal quality and functionality. Advanced methods and technologies are employed to obtain pure hydrocolloids that meet the stringent standards of the food industry."

The extraction and production of hydrocolloids are essential steps in harnessing their unique properties and incorporating them into diverse food formulations. These processes are carefully executed, considering the specific requirements of each hydrocolloid and the industry's safety standards. The hydrocolloid industry continues to innovate in extracting and producing high-quality hydrocolloids to meet the growing demands of the food industry.

24.1.8 Hydrocolloid Manufacturers and Suppliers

Regarding hydrocolloids, there is a wide network of manufacturers and suppliers spread across the globe. These companies specialize in producing and distributing various hydrocolloids to cater to the diverse needs of different industries, particularly the food industry. With their expertise and experience, hydrocolloid manufacturers and suppliers are crucial in ensuring a consistent supply of high-quality hydrocolloids to meet the market's demands.

Hydrocolloid manufacturers utilize advanced production techniques and adhere to strict quality control measures to deliver hydrocolloids that meet industry standards and customer specifications. These manufacturers possess in-depth knowledge of hydrocolloid extraction, processing, and formulation, allowing them to produce hydrocolloids in different forms, such as powders, granules, and solutions.

The hydrocolloid supply chain includes a diverse range of suppliers. From large-scale hydrocolloid manufacturers to specialized distributors, suppliers ensure the availability of hydrocolloids in the market. They ensure timely delivery and manage the logistics required to fulfill customer orders. Suppliers play a vital role in connecting hydrocolloid manufacturers with end-users and facilitating the smooth flow of hydrocolloids throughout the supply chain.

"

"We pride ourselves on being one of the leading hydrocolloid manufacturers and suppliers globally. Our commitment to quality, innovation, and customer satisfaction sets us apart in the industry. With our extensive range of hydrocolloids and efficient supply chain, we strive to meet the diverse needs of our customers while ensuring timely delivery and exceptional service."

- Hydrocolloid Manufacturer.

Whether providing hydrocolloids for food formulations, pharmaceutical applications, or industrial uses, hydrocolloid manufacturers and suppliers are crucial in supporting various industries. They act as reliable partners, offering technical expertise and product knowledge to help customers achieve their formulation goals.

As the hydrocolloid industry continues to evolve and expand, manufacturers and suppliers play an essential role in driving innovation and meeting the growing demand for these versatile ingredients. Through their collaboration and expertise, hydrocolloid manufacturers and suppliers contribute to the stability and success of the hydrocolloid market.

24.1.9 Hydrocolloid Uses in Product Development

Hydrocolloids play a crucial role in product development in the food industry. They are versatile ingredients used to create and enhance the texture, stability, and sensory properties of a wide range of products, including beverages, dairy products, bakery items, confectionery, and processed foods.

By incorporating hydrocolloids into food formulations, manufacturers can achieve the desired characteristics and qualities of their products. Hydrocolloids act as thickeners, gelling agents, stabilizers, and emulsifiers, allowing better control over the final product's texture and stability. Their ability to modify the rheological properties of food systems contributes to improved sensory experiences and consumer acceptance.

> "Hydrocolloids are essential ingredients that enable us to create products with the right texture and stability. They play a critical role in achieving the desired mouthfeel and ensuring product quality", says Sarah Johnson,
>
> a food product developer at Delicious Delights.

Hydrocolloids can be customized and combined with other ingredients to achieve specific properties and functionalities in food formulations. By carefully selecting and incorporating the appropriate hydrocolloid for a particular application, product developers can innovate and create unique food experiences that meet the demands and preferences of consumers.

24.1.10 Hydrocolloid Uses in Different Food Categories

Hydrocolloids find extensive applications in various food categories, contributing to the product's overall quality and sensory experience. Here are some examples:

Table 143: Food Categories and Hydrocolloid Uses

FOOD CATEGORY	HYDROCOLLOID USES
Beverages	Hydrocolloids stabilize emulsions, enhance mouthfeel, and provide a smooth texture in juices, smoothies, and dairy-based drinks.
Dairy Products	Hydrocolloids are added to yogurts, ice creams, and cheese to improve texture, prevent syneresis, and maintain stability during processing and storage.
Bakery Items	Hydrocolloids contribute to the softness, structure, and moisture retention in bread, cakes, and pastries. They also help improve gluten-free baked goods.
Confectionery	Hydrocolloids are used in candies, gummies, and chocolates to provide chewiness, control crystallization, and enhance flavors.
Processed Foods	Hydrocolloids stabilize dressings, sauces, soups, and canned products, ensuring uniform texture and preventing ingredient separation.

As demonstrated, hydrocolloids offer various uses across different food categories, highlighting their adaptability and importance in product development.

A close-up view of a pile of various hydrocolloid powders in different colors and textures.

> "Hydrocolloids are invaluable tools in our product development process. They allow us to create innovative food formulations that meet consumer expectations for taste, texture, and stability", says James Thompson
>
> **a food scientist at Gourmet Creations.**

By harnessing the functionalities of hydrocolloids, food manufacturers can meet the demand for high-quality products that deliver taste, texture, and stability. Hydrocolloid applications in product development continue to drive innovation and shape the future of the food industry.

24.1.11 Hydrocolloid Revolutionizing Texture and Stability

Hydrocolloids have sparked a revolution in the food industry, transforming the texture and stability of various products. These remarkable ingredients can create smooth and creamy textures, enhance mouthfeel, and prevent phase separation in emulsions.

By incorporating hydrocolloids, food manufacturers can elevate the overall quality of their products and improve consumer acceptance. These versatile additives enhance texture and stability, ensuring each bite is a delightful experience.

Imagine indulging in a spoonful of silky-smooth ice cream that glides effortlessly across your palate or savoring a velvety sauce that clings to every morsel. Hydrocolloids make these culinary dreams a reality, infusing products with a sublime mouthfeel that delights the senses.

> "Hydrocolloids have transformed the way we experience food. They enable us to create once unimaginable textures, adding depth and richness to a wide range of products."

Moreover, hydrocolloids act as stabilizers, ensuring products maintain their desired consistency over time. They prevent undesirable separation or syneresis, allowing dressings, sauces, and other emulsions to blend perfectly.

The hydrocolloid revolution in texture and stability has empowered food formulators to push the boundaries of culinary creativity. With these ingredients, they can craft innovative and enticing products that captivate consumers and set new standards in the market.

Unlocking the Potential of Hydrocolloid Texture

Hydrocolloids play a vital role in creating unique textures in various food applications. They can remarkably modify the mouthfeel and consistency of products, opening the doors to unparalleled sensory experiences.

Hydrocolloids allow formulators to bring their culinary visions to life, whether an indulgent chocolate mousse with a smooth and velvety mouthfeel or a refreshing fruit jelly with a delicate and bouncy texture.

Here are some ways in which hydrocolloids revolutionize texture:

- Creating smooth and creamy textures
- Adding viscosity and thickness to sauces and dressings
- Improving mouth-coating properties
- Enhancing the stability of whipped products
- Preventing ice crystals in frozen desserts

With the help of hydrocolloids, the possibilities for texture manipulation are virtually limitless. Food formulators can craft unique and enticing products that defy expectations, captivating consumers with their tantalizing textures.

Ensuring Stability with Hydrocolloids

Stability plays a crucial role in the quality and longevity of food products. Hydrocolloids act as stabilizers, ensuring products maintain their desired characteristics throughout their shelf life.

- By incorporating hydrocolloids, food manufacturers can achieve:
- Prevention of phase separation in emulsions
- Improved creaminess and resistance to melting in ice cream
- Enhanced suspension of particulates in beverages
- Reduced syneresis in gels and jellies
- Uniform texture and consistency in processed foods

Hydrocolloids provide the structural support needed to maintain product integrity, ensuring a superior sensory experience for consumers. These powerful ingredients offer both manufacturers and consumers stability and peace of mind.

Overall, hydrocolloids have revolutionized the texture and stability of food products, unleashing a world of culinary possibilities. By creating delightful textures and ensuring long-lasting stability, hydrocolloids have become invaluable tools for food formulators, enabling them to innovate and captivate with their creations.

Conclusion

Hydrocolloids from natural and microbial origins have become indispensable ingredients in the food industry. Their multifaceted properties, including thickening, gelling, stabilizing, and emulsifying, make them versatile additions to various food formulations. Hydrocolloids provide numerous benefits, enhancing food products' texture, stability, and sensory attributes.

With their significant contributions to product development and innovation, hydrocolloids are crucial in creating smooth and creamy textures, improving mouthfeel, and preventing phase separation in emulsions. These ingredients have revolutionized the food industry, satisfying the growing consumer demand for high-quality products with improved texture and stability.

The hydrocolloid industry thrives, driven by the increasing need for unique and appealing food formulations. Manufacturers and suppliers worldwide cater to this demand, offering hydrocolloids in different forms to meet specific requirements. As the market expands, hydrocolloids remain at the forefront, ensuring that food products achieve the desired consistency, viscosity, and sensory experience.

FAQ

Where do hydrocolloids come from?

Hydrocolloids can be derived from various sources, including seaweed, plant extracts, animal collagen, and microorganisms.

How is the seaweed used as a source of hydrocolloids?

Seaweed, specifically red and brown algae, is a rich source of hydrocolloids. Brown algae produce alginates, while red seaweed extracts carrageenan and agar, commonly used hydrocolloids in the food industry.

What plant extracts are used to obtain hydrocolloids?

Various plant extracts are used to obtain hydrocolloids, including guar gum from the guar plant, pectin from citrus fruits, cellulose from plant cell walls, gum Arabic from the acacia tree, konjac gum from the konjac plant, locust bean gum from carob seeds, and tapioca starch from the tapioca plant. Other sources include psyllium husk, chia seeds, flax seeds, rice, and sago palm starch.

What are the animal sources of hydrocolloids?

Animal collagen is a source of hydrocolloids, specifically gelatin, derived from animal collagen hydrolysis. Gelatin is widely used in the food industry for its thickening and gelling properties.

How are hydrocolloids derived from microorganisms?

Microorganisms, such as bacteria, can produce hydrocolloids through fermentation. Xanthan gum and gellan gum are examples of hydrocolloids derived from bacterial fermentation, commonly used for their thickening and stabilizing properties in food products.

What are the applications of hydrocolloids?

Hydrocolloids have a wide range of applications in the food industry. They are thickening and gelling agents in soups, sauces, dressings, and desserts. Hydrocolloids are also used to emulsify and stabilize properties in various food formulations.

FREQUENTLY ASKED QUESTIONS (FAQS) CONTINUED

- **What are the benefits of using hydrocolloids?**

  Hydrocolloids can improve the texture, stability, and mouthfeel of food formulations. They enhance the viscosity and consistency of products, contribute to increased shelf life, and improve the sensory properties of food products.

- **How are hydrocolloids extracted and produced?**

  Hydrocolloids are extracted from their respective sources through various methods, including physical and chemical processes. Extraction techniques involve separation, purification, and drying to obtain the desired hydrocolloids in their pure form. The production of hydrocolloids is carried out on a large scale to meet the growing demand in the food industry.

- **Who are the manufacturers and suppliers of hydrocolloids?**

  There are numerous manufacturers and suppliers of hydrocolloids worldwide. These companies specialize in producing and distributing different hydrocolloids to cater to the needs of various industries, including the food industry.

- **How are hydrocolloids used in product development?**

  Hydrocolloids play a crucial role in product development in the food industry. They create and enhance various products' texture, stability, and sensory properties, including beverages, dairy products, bakery items, confectionery, and processed foods.

- **How do hydrocolloids revolutionize texture and stability?**

  Hydrocolloids have revolutionized the food industry by providing unique texture and stability to various products. They enable the creation of smooth, creamy textures, enhance mouthfeel, and prevent phase separation in emulsions. Hydrocolloids contribute to the overall quality and consumer acceptance of food products.

CHAPTER 25

Conclusion

In the fascinating journey through the world of hydrocolloids, we've explored a diverse range of topics, from their introduction and types to their applications in various industries. This comprehensive guide has aimed to provide readers with a holistic understanding of hydrocolloids. As we draw this journey to a close, let's reflect on the key takeaways and insights we've gained. Whether you're a culinary enthusiast, a food scientist, or just curious about these remarkable substances, we hope this guide has been enlightening and educational.

Core Insights

- ☑ Understanding the various types and properties of hydrocolloids is crucial for their effective use in a wide range of applications.
- ☑ Hydrocolloids play a vital role in cooking, baking, health, and nutrition, making them a versatile ingredient in the culinary world.
- ☑ Sourcing and sustainability are paramount considerations in the hydrocolloid industry, with a focus on eco-friendly practices.
- ☑ Ongoing innovations and research continue to expand the horizons of hydrocolloids, promising exciting future developments.
- ☑ Hydrocolloids offer valuable solutions for vegan and vegetarian foods, catering to the growing demand for plant-based options.
- ☑ The global market for hydrocolloids is on the rise, driven by changing consumer preferences and emerging food trends.

It's important to recapitulate the key insights and delve deeper into some of the pivotal concepts discussed throughout the book. The following sections will provide a summary of each chapter and highlight the critical takeaways.

Chapter 1: Introduction to Hydrocolloids

Recap:
This chapter introduced the fundamental concept of hydrocolloids.

Key Takeaways:
Highlight the significance of understanding hydrocolloids as a foundation for subsequent chapters

Chapter 2: Types of Hydrocolloids

Recap:
Delve into the various types of hydrocolloids discussed, including polysaccharides, proteins, and more.

Key Takeaways:
Emphasize the importance of recognizing the diversity within hydrocolloids for specific applications.

Chapter 3: Properties of Hydrocolloids

Recap:
Discuss the essential properties such as solubility, viscosity, and gelation.

Key Takeaways:
Stress the role of these properties in determining the functionality of hydrocolloids in different contexts.

Chapter 4: Applications of Hydrocolloids

Recap:
Explore the wide range of applications, from food to pharmaceuticals.

Key Takeaways:
Discuss the versatility of hydrocolloids and their role in product development and enhancement.

Chapter 5: Hydrocolloids in Cooking and Baking

Recap:
Highlight the specific applications of hydrocolloids in culinary settings.

Key Takeaways:
Discuss how hydrocolloids influence texture, taste, and presentation in cooking and baking.

Chapter 6: Health and Nutrition

Recap:
Examine the impact of hydrocolloids on health and nutrition.

Key Takeaways:
Consider the role of hydrocolloids in dietary fiber and weight management

Chapter 7: Sourcing and Sustainability

Recap:
Address the importance of responsible sourcing and sustainability.

Key Takeaways:
Emphasize the need for ethical and environmentally friendly practices in the hydrocolloid industry.

Chapter 8: Innovations and Research

Recap:
Discuss ongoing innovations and research in the field.

Key Takeaways:
Highlight the potential for future developments and breakthroughs.

Chapter 9: Hydrocolloids in Vegan and Vegetarian Foods

Recap:
Explore the significance of hydrocolloids in meeting the demands of vegan and vegetarian diets.

Key Takeaways:
Discuss the role of hydrocolloids in creating plant-based alternatives with desirable textures and flavors.

Chapter 10: Global Market and Trends

Recap:
Examine the evolving trends and the expanding global market for hydrocolloids.

Key Takeaways:
Emphasize the importance of staying informed about industry trends for professionals in the field.

Chapter 11: DIY and Home Us

Recap:
Discuss the applications of hydrocolloids in everyday household settings.

Key Takeaways:
Encourage readers to experiment with hydrocolloids in their own kitchens and homes.

Chapter 12: Comparison and Substitution

Recap:
Address the practical aspects of comparing and substituting hydrocolloids.

Key Takeaways:
Provide guidance on making informed choices when selecting hydrocolloids for specific purposes.

Chapter 13: Regulations and Standards

Recap:
Examine the regulatory landscape governing the use of hydrocolloids.

Key Takeaways:
Stress the importance of compliance with safety and quality standards in the industry.

Chapter 14: Research and Innovations in Hydrocolloids

Recap:

Highlight the latest research and innovations in hydrocolloids.

Key Takeaways:

Inspire readers to stay updated on cutting-edge developments and research.

Chapter 15: Technological Functionality

Recap:
Discuss the technological functionalities enabled by hydrocolloids.

Key Takeaways:
Consider the role of hydrocolloids in food processing and preservation.

Chapter 16: Regulatory Aspects of Dietary Fibers

Recap:
Explore the regulatory aspects related to dietary fibers and hydrocolloids.

Key Takeaways:
Provide insights into labeling and health claims associated with these substances.

Chapter 17: Agricultural Practices

Recap:
Examine the agricultural practices involved in sourcing hydrocolloids.

Key Takeaways:
Advocate for sustainable and ethical sourcing practices.

Chapter 18: Nanotechnology and Hydrocolloids

Recap:
Delve into the intersection of nanotechnology and hydrocolloids.

Key Takeaways:
Consider the potential for nanotechnology to revolutionize hydrocolloid applications.

Chapter 19: Culinary Innovations

Recap:
Explore innovative culinary applications of hydrocolloids.

Key Takeaways:
Encourage chefs and food professionals to experiment with new techniques.

Chapter 20: Economic Impact

Recap:
Discuss the economic implications of the hydrocolloid industry.

Key Takeaways:
Consider the industry's contributions to job creation and economic growth.

Chapter 21: Workshops and Training on Hydrocolloids

Recap:
Highlight the importance of workshops and training in enhancing knowledge and skills.

Key Takeaways:
Encourage professionals to participate in educational opportunities.

Chapter 22: Analytical Methods

Recap:
Examine the analytical methods used in hydrocolloid research and quality control.

Key Takeaways:
Emphasize the role of accurate testing in ensuring product quality.

Chapter 23: International Trade and Harmonization

Recap:
Explore international trade and harmonization efforts related to hydrocolloids.

Key Takeaways:
Stress the significance of global cooperation and standards.

Chapter 24: Where do Hydrocolloids Come From?

Recap:
Discuss the sources and origins of hydrocolloids.

Key Takeaways:
Highlight the importance of understanding the origins of these substances.

Closing

In closing, the world of hydrocolloids is as dynamic as it is vast. From their humble beginnings to their cutting-edge applications, these remarkable substances have revolutionized various industries. As you delve into the realm of hydrocolloids, remember that this guide is just the beginning of your journey. Stay curious, stay innovative, and continue exploring the endless possibilities these natural compounds offer.

We hope this book has been a valuable resource, serving as a foundation for your hydrocolloid endeavors. Thank you for accompanying us on this educational adventure, and we look forward to the exciting developments that the future holds for hydrocolloids.

FREQUENTLY ASKED QUESTIONS (FAQS)

- **FAQ 1: What are hydrocolloids, and how are they used in the food industry?**

  Answer: Hydrocolloids are substances that have the ability to form a gel when combined with water. They are widely used in the food industry to improve texture, stability, and mouthfeel in a variety of products. Common examples include agar, pectin, and xanthan gum.

- **FAQ 2: Can you provide examples of common hydrocolloids and their properties?**

  Answer: Certainly, some common hydrocolloids include carrageenan, guar gum, and cellulose gum. Each has unique properties; for instance, carrageenan is used for gelling and thickening, guar gum offers viscosity, and cellulose gum acts as a stabilizer and thickener.

- **FAQ 3: What are the sustainability challenges associated with sourcing hydrocolloids?**

  Answer: Sustainability challenges in sourcing hydrocolloids include overharvesting, habitat destruction, and the use of chemicals. Sustainable practices involve responsible sourcing, eco-friendly processing, and supporting ethical suppliers.

- **FAQ 4: How do hydrocolloids contribute to the texture and taste of food products?**

  Answer: Hydrocolloids influence the texture and taste of food by controlling water content, providing a stable structure, and enhancing mouthfeel. They can create gels, improve suspension, and contribute to a creamy or smooth texture.

- **FAQ 5: What are the key considerations when using hydrocolloids in vegan and vegetarian foods?**

  Answer: Key considerations include ensuring the hydrocolloid is plant-based, understanding its impact on texture and flavor, and complying with dietary restrictions. Hydrocolloids are often used to replace animal-based ingredients in vegan and vegetarian products.

- **FAQ 6: What trends are shaping the global hydrocolloid market, and what can we expect in the future?**

  Answer: Current trends in the hydrocolloid market include increased demand for clean-label ingredients, plant-based products, and natural sources. The future is likely to see continued growth in these areas and more innovation in hydrocolloid applications.

- **FAQ 7: Are there any safety regulations or standards governing the use of hydrocolloids?**

  Answer: Yes, various safety regulations and standards exist to ensure the quality and safety of hydrocolloids in different regions. These standards cover aspects such as purity, labeling, and maximum allowable limits of certain substances.

HYDROCOLLOID GLOSSARY

Agar (Agar-Agar)

Origin: Extracted from red seaweed.

Properties: Forms firm, brittle gels. Heat-stable.

Uses: Vegetarian gelatin substitute, microbiological media, confectionery.

Amount: 0.1% - 1.5% by weight of the liquid.

Alginate (PGA - PropyleneGlycol Alginate)

Origin: Derived from alginic acid esterified with propylene glycol.

Properties: Soluble in both water and some organic solvents. Forms gels with acids and calcium ions.

Uses: Thickener, stabilizer, emulsifier in foods, especially fruit juices and sodas.

Amount: 0.25% - 1% for most applications.

Alginate (Potassium Alginate)

Origin: Extracted from brown seaweed,converted to potassium salt form.

Properties: Water-soluble, forms softer gels with calcium ions.

Uses: Food stabilizer, controlled-release pharmaceuticals.

Amount: 0.25% - 1% for most applications.

Alginate (SodiumAlginate)

Origin: Extracted from brown seaweed.

Properties: Highly water-soluble, forms a gel with calcium ions.

Uses: Spherification in molecular gastronomy, textile industry thickener, food stabilizer.

Amount: 0.25% - 1% for most applications.

Arabic Gum (Acacia Gum)

Origin: Sap of the Acacia tree.

Properties: Emulsifier, stabilizer, thickener.

Uses: Confectionery, inks, cosmetics.

Amount: 5% - 30% for emulsions.

Beta-Glucan

Origin: Extracted from oats, barley, fungi.

Properties: Viscosity enhancer.

Uses: Functional foods, beverages.

Amount: 0.5% - 2% for viscosity enhancement

Carrageenan (Iota Carrageenan)

Origin: Extracted from red seaweed species Eucheuma denticulatum.

Properties: Produces soft, elastic gels. Thermo-reversible.

Uses: Dairy products, sauces, meat products.

Amount: 0.2% - 1%

Carrageenan (KappaCarrageenan)

Origin: Extracted from red seaweed species Kappaphycus alvarezii.

Properties: Forms firm, brittle gels. Thermo-reversible.

Uses: Dairy products, desserts, jellies, vegetarian gelatin substitute.

Amount: 0.5% - 2%

Carrageenan (Lambda Carrageenan)

Origin: Extracted from red seaweed species Gigartina pistillata.

Properties: Thickener and stabilizer. Doesn't form gels.

Uses: Dairy products, sauces, soups.

Amount: 0.1% - 1% for thickening.

Cassia Gum

Origin: Extracted from the seeds of the Cassia tora plant, a wild crop found in several parts of India.

Properties: Cassia gum is a galactomannan polysaccharide, similar to guar gum and locust bean gum. It forms high viscosity by increasing the molecular weight. When combined with other hydrocolloids like xanthan gum, it can produce gels.

Uses: Used as a thickener and stabilizer in the food industry, especially in pet foods. It's also used in dairy products, meat products, and air fresheners.

Amount: 0.5% - 2% by weight of the liquid

Cellulose Derivatives (Methylcellulose)

Origin: Derived from plant cell walls.

Properties: Thermo-reversible gelling agent.

Uses: Emulsifiers, thickeners, film formers.

Amount: 0.2% - 2%

Cellulose Derivatives (Hydroxypropylcellulose)

Origin: Modified cellulose derivative.

Properties: Soluble in water, forms clear solutions.

Uses: Thickener, protective colloid, binder.

Amount: 0.2% - 1.5%

Cellulose Derivatives (Microcrystalline Cellulose)

Origin: Derived from plant cell walls.

Properties: Used as a texturizer and extender.

Uses: Tablets, food products to prevent caking.

Amount: 0.5% - 10% in suspension systems.

Chitosan

Origin: Derived from chitin in crustacean shells.

Properties: Forms films, antimicrobial.

Uses: Food preservation, wound healing, water purification.

Amount: 0.1% - 1% for most applications.

Fenugreek Gum

Origin: Derived from the seeds of the fenugreek plant(Trigonella foenumgraecum).

Properties: Fenugreek gum, also known as galactomannan, is a natural polysaccharide. It has excellent solubility and forms highly viscous solutions. It also has emulsifying properties.

Uses: Used as a stabilizer, emulsifier, and thickening agent in food products. It's also recognized for its potential health benefits, including blood sugar regulation and cholesterol reduction, and is used in traditional medicine.

Amount: 0.2% - 1.5% by weight of the liquid.

Notes: Fenugreek gum can form highly viscous solutions, so it's essential to startwith a lower amount and adjust based on the desired viscosity and texture.

Gelatin

Origin: Derived from animal collagen.

Properties: Forms soft, elastic gels.

Uses: Desserts, gummy candies, pharmaceutical capsules.

Amount: 1% - 6% depending on desired gel strength.

Gellan Gum

Origin: Produced by bacterial fermentation.

Properties: Forms firm, brittle gels.

Uses: Desserts, sauces, beverages.

Amount: 0.1% - 1% for gelling.

Guar Gum

Origin: Extracted from guar beans.

Properties: Thickening agent.

Uses: Baked goods, dairy products, dressings.

Amount: 0.1% - 1% for thickening.

Gum Karaya

Origin: Exudate from trees of the genus Sterculia.

Properties: Water-soluble, used as a thickener and emulsifier.

Uses: Laxatives, denture adhesives, and food products.

Amount: 0.5% - 5% for thickening.

High Methoxyl (HM) Pectin

Degree of Esterification (DE): Above 50%.

Properties: Requires a high concentration of sugar and a low pH(acidic conditions) to form gels. The gelation is driven by hydrogen bonding.

Uses: Traditional jams, jellies, and marmalades.

Amount: 0.5%- 2%

Inulin

Origin: Extracted from chicory root and other plants.

Properties: Prebiotic, texture enhancer.

Uses: Dietary supplements, functional foods.

Amount: 2% - 10% for fiber enrichment.

Konjac (Konnyaku) Gum

Origin: Derived from the root of the konjac plant.

Properties: Forms firm, elastic gels.

Uses: Vegan gelatin substitute, shirataki noodles.

Amount: 0.5% - 3% for gelling.

Locust Bean Gum (Carob Gum)

Origin: Extracted from seeds of the carob tree.

Properties: Thickening agent.

Uses: Ice cream, dairy products, baked goods.

Amount: 0.1% - 1% for thickening.

Low Methoxyl (LM) Pectin

Degree of Esterification (DE): Below 50%.

Properties: Can form gels in the presence of divalent cations, typically calcium. Does not require as much sugar as HM pectin.

Uses: Reduced-sugar jams, jellies, dairy products, and confectionery.

Amount: 0.5% - 2%

Low Methoxyl Amidated (LMA) Pectin

Origin: LM pectin that has undergone amidation, a process where some of the carboxyl groups are converted to amide groups.

Properties: Requires calcium for gelation but is more tolerant to pH and calcium concentration than regular LM pectin.

Uses: Dairy products like yogurt and desserts, especially when a smooth texture is desired.

Amount: 0.5% - 2%

Pullulan

Origin: Produced by fungal fermentation.

Properties: Forms elastic films and gels.

Uses: Edible films, capsules.

Amount: 0.5% - 5% for film-forming.

Tamarind Gum (Tamarind Kernel Powder):

Origin: Extracted from the seedsof the tamarind tree (Tamarindus indica).

Properties: Tamarindgum is a complex polysaccharide composed of glucose, xylose, and mannose. It forms a gel-like consistency in water, making it an excellent thickening agent. It also has high stability againstheat, acidity, and shear.

Uses: Used as a thickener, stabilizer, and emulsifier in the food industry, particularly in sauces, ice creams, and beverages. It's also used in the textile industry for sizing and printing.

Amount: 0.5% - 2.5% by weight of the liquid.

Tara Gum

Origin: Extracted from seeds of the tara tree.

Properties: Thickening agent.

Uses: Ice cream, dairy products, baked goods.

Amount: 0.1% - 1% for thickening.

Tragacanth Gum

Origin: Sap of the Astragalus shrubs.

Properties: Emulsifier, thickener.

Uses: Dressings, sauces, confectionery.

Amount: 0.5% - 5% for thickening and stabilization.

Xanthan Gum

Origin: Produced by bacterial fermentation.

Properties: Thickening and stabilizing agent.

Uses: Salad dressings, sauces, gluten-free baking.

Amount: 0.1% - 1% for thickening and stabilization.

It is essential to note that the exact amount of a hydrocolloid can vary based on the specific application, desired texture, other ingredients in the recipe, and the specific brand or grade of the hydrocolloid used. Additionally, always consult the manufacturer's specifications or product datasheet, as different grades or brands might have varying recommendations.

GLOSSARY

LIST OF TABLES

"Table 101: Commercial Hydrocolloid Source and Functionality" on page 354

Index

C

D

E

F

G

H

I

J

K

L

M

N

O

P

Q

R

S

T

U

V

W

X

www.ingramcontent.com/pod-product-compliance
Ingram Content Group UK Ltd.
Pitfield, Milton Keynes, MK11 3LW, UK
UKHW060210151026
15625UKWH00087B/466

9 781736 139172